Elsevier *Global Energy Policy and Economics* Series

Reshaping European Gas and Electricity Industries

Regulation, Markets and Business Strategies

Editors

Dominique Finon

and

Atle Midttun

Elsevier Internet Homepage - http://www.elsevier.com
Consult the Elsevier homepage for full catalogue information on all books, journals and electronic products and services.

Elsevier Titles of Related Interest

CLARK and BRADSHAW
Agile Energy Systems: global lessons from the California Energy Crisis
ISBN: 008-044448-2

ARENTSEN and KÜNNEKE
National Reforms in European Gas
ISBN: 008-043687-0

MIDTTUN
European Energy Industry Business Strategies
ISBN: 008-043631-5

WANG
China's Oil Industry and Market
ISBN: 008-043005-8

MIDTTUN
European Electricity Systems in Transition
ISBN: 008-042994-7

Related Journals:
A sample journal issue is available online by visiting the homepage of the journal (homepage details at the top of this page). Free specimen copy gladly sent on request. Elsevier Ltd, The Boulevard, Langford Lane, Kidlington, Oxford, OX5 1GB, UK

Energy
Energy Economics
Energy Policy
Resources and Energy Economics
Resources Policy
Utilities Policy
The Electricity Journal

To Contact the Publisher
Elsevier welcomes enquiries concerning publishing proposals: books, journal special issues, conference proceedings, etc. All formats and media can be considered. Should you have a publishing proposal you wish to discuss, please contact, without obligation, the publisher responsible for Elsevier's Composites and Ceramics programme:

Henri van Dorssen
Publisher
Elsevier Ltd
The Boulevard, Langford Lane
Kidlington, Oxford
OX5 1GB, UK
E.mail: h.dorssen@elsevier.com

General enquiries, including placing orders, should be directed to Elsevier's Regional Sales Offices – please access the Elsevier homepage for full contact details (homepage details at the top of this page).

Elsevier *Global Energy Policy and Economics* Series

Reshaping European Gas and Electricity Industries

Regulation, Markets and Business Strategies

Edited by

Dominique Finon

and

Atle Midttun

2004

ELSEVIER

AMSTERDAM – BOSTON – HEIDELBERG – LONDON – NEW YORK – OXFORD
PARIS – SAN DIEGO – SAN FRANCISCO – SINGAPORE – SYDNEY – TOKYO

ELSEVIER B.V.
Sara Burgerhartstraat 25
P.O. Box 211, 1000 AE
Amsterdam,
The Netherlands

ELSEVIER Inc.
525 B Street
Suite 1900, San Diego
CA 92101-4495, USA

ELSEVIER Ltd
The Boulevard
Langford Lane, Kidlington,
Oxford OX5 1GB, UK

ELSEVIER Ltd
84 Theobalds Road
London WC1X 8RR
UK

First edition 2004

Library of Congress Cataloging in Publication Data
A catalog record is available from the Library of Congress.

British Library Cataloguing in Publication Data
A catalogue record is available from the British Library.

ISBN: 0-08-044550-0

ISSN: 1571-4985

♾ The paper used in this publication meets the requirements of ANSI/NISO Z39.48-1992 (Permanence of Paper).
Printed in The Netherlands.

Contents

Preface

Following a first period of deregulation of European energy markets, this book takes stock of what we have achieved, both in terms of governance, markets and business models.

In spite of strong EU attempts at integration and harmonisation, the complexity of European energy industry and energy markets still reflects a heritage of deeply nationally embedded business styles and regulatory cultures. When studying the evolution of the European energy market project, it therefore seemed appropriate to put together a research team that reflected some of this institutional variety.

While limited to an "old" west European focus, we are still glad to have attracted both South-, North- and Central-European scholars to take part in an interesting and creative dialogue, that hopefully also colours this book.

The broad composition of the research team has also allowed us to conduct a dialogue across disciplinary boundaries, between political science, economics and management science, which is reflected in the three sections of the book: regulation, markets and business strategies.

While each member of the research team has drawn on national resources, the Norwegian research project the "Reshaping of Energy Industry" along with a publication grant from the Norwegian Research Council formed an important platform for our work. We are also grateful for the support from the French-Norwegian exchange programme (Aurora), which has supported travel costs for the French-Norwegian editorship, as well as the German-Norwegian exchange programme (DAAD), which has supported extensive collaboration with our German colleague.

Paris-Oslo October 2004

Dominique Finon — Atle Midttum

Acknowledgements

This book is the product of collective effort from a broad European research team where each contributor has drawn on national resources. The Norwegian research project the "reshaping of energy industry" and a publication grant from the Norwegian Research Council has been an important platform for our work. We are also grateful for the support from the French-Norwegian exchange programme (Aurora), which has supported travel costs for the French-Norwegian editorship, as well as the German-Norwegian exchange programme (DAAD), which has supported extensive collaboration with our German colleague.

Contributors

Maarten ARENTSEN is associate professor energy policy and managing director of the environmental research institute of the University of Twente. He holds a Master's degree in political science (specialization in research methodology and political modernization) from Nijmegen University and a PhD (Public Administration) from the University of Twente, the Netherlands. His research focus is on energy policy, energy market reform and (green) energy innovation, with a special focus on technological and institutional change. Maarten Arentsen has an extensive publication of books and articles in these fields.

Matthias FINGER is currently Chair and Professor of Management of Network Industries as well as Dean of the School of Continuing Education at the Swiss Federal Institute of Technology (EPFL). He holds a PhD in Political Science, and a PhD., Adult Education (both University of Geneva). His research focuses on the liberalisation of the main network industries' sectors – postal services, telecommunications, energy, public transport, water, and air transport – on the changes undergone by the historical operators in these sectors, and in issues of regulation and public service. He is particularly interested in the implications of the new information and communication technologies. He has written numerous articles and books on this subject and consults with public enterprises, as well as with public administrations and political authorities in Switzerland and internationally.

Christophe GENOUD is a PhD candidate at the Government Departement at the London School of Economics and Political Science (LSE). He holds a Bachelor and Master in Political Science from the University of Geneva and has worked as both researcher and consultant in the network industries sectors. His research interests include the regulation of network industries and the role of regulatory authorities in the regulatory process. He has several international publications in these fields.

Jean-Michel GLACHANT is Permanent Professor and Head of the Department of Economics at the University of Paris XI. He heads the "Network Economics" team GRJM at ADIS research centre (www.grjm.net). He has a PhD in economics from La Sorbonne University where he has been research center director or deputy director from 1991 to 2000. His main research orientations are network economics, institutional economics, energy industry competition and regulation, and the making of a single market in the European Union. He is or has been expert for the Frenchindustry ministry, the French energy regulator and the European Commision (DG Research or DG Tren). He is partner of the Cambridge – MIT energy research project CMI. He has published extensively in these fields.

Lutz MEZ is Senior Associate Professor. at the Department of Social and Political Sciences, Free University of Berlin, and Deputy director of the Environmental Policy Research Centre. His major research area is environmental and energy policy with particular reference to nuclear and electricity policy. He is author of numerous articles and chapters in internationally edited books on these topics.

Augusto Rupérez MICOLA is Doctoral Candidate in Decision Science at the Energy Markets Group, London Business School. He holds a BSc in Economics (Barcelona), a MSc in Energy Management (Norwegian School of Management, BI) and a MPhil in Decision Science (London Business School). He is currently interested in the econometric and game theoretical modelling of liberalised energy markets, with special emphasis on the European natural gas industry. Augusto Micola has several publications in this field.

Maria Rosaria Di NUCCI is an independent consultant and associate senior research fellow at the Environmental Policy Research Unit of the Free University Berlin. She holds a masters' degree in economics from the University of Rome and a PhD in the areas of technology and energy policy from the University of Sussex. Previously she held an executive position at the Environment Ministry of the Land Berlin and was lecturer in industrial economics at the Technical University of Berlin. Her current research interests are: the liberalisation of electricity and gas markets, flexible instruments in climate mitigation policy, financing mechanisms and incentives for renewable energy.

Terje OMLAND has a Master of Business and Administration degree from the Norwegian School of Management (BI). He worked for some four years as a researcher at the Centre for Energy and Environment at BI focussing on structural changes and strategic developments in the European energy sector. He has several publications in this field. He is currently working as an Analyst with Europower in Oslo.

Maria Isabel Rebelo Teixeira SOARES is Full Professor of Economics at the Faculty of Economics, University of Porto. She has a Doctorat-d'Etat-ès-Sciences Economiques from the Université Louis Pasteur, Strasbourg, and she was a Post-Doctoral Fellow at the VPI-Virginia Polytechnic Institute and State University. Her research interests are in utilities regulation and utilities business strategies (energy, telecommunications and water) and she has an extensive publication in these fields.

Steve THOMAS is a Senior Researcher at the Public Service International Research Unit, University of Greenwich since 2001. He has a BSc(honours) degree in Chemistry from Bristol University and he has been working in energy policy analysis since 1976.

His main research interests are reforms of energy industries, economics and policy towards nuclear power, and corporate policies of energy industry companies. He has an extensive publication in these fields.

Dominique FINON is Director of research at CNRS (The National Centre for Scientific Research in France), and at the same time Deputy Director of the programme of Energy also at CNRS. Between 1991 and 2002 he was head of the Institute of Economics and Politics, of Energy, a joint CNRS and Grenoble University in France. He is also the author of a number of books and journal articles on energy modelling, public policies (nuclear, renewables, energy efficiency, innovation) and electricity and gas markets liberalisation

Atle MIDTTUN is a Professor at the Norwegian School of Management and Co-Director of the Centre for Energy and Environment. He is also Director of the Centre for Corporate Citizenship. He is author of several books in related areas and boosts a significant number of articles to promote his research in the energy and environment field.

Introduction

DOMINIQUE FINON AND ATLE MIDTTUN

Liberalisation of the energy industries in Europe and the introduction of new and more homogenous regulatory frameworks have for the last decade been part of the vision of an integrated internal market economy in Europe. The liberalisation process has modified and altered the traditional regulatory systems, which have historically been built around monopolies and ministerial regulation of dominantly public-owned companies.

The role model for the new liberal order spelled out by EU's electricity and gas Directives was taken from the Anglo-Saxon neo-liberal regulatory approach with a transparent third party access to network, clear unbundling, vertical and horizontal de-integration and specialised regulatory authorities at arm's length from government, with a mandate to enforce competition in the network industries unfamiliar with competitive forms of trade.

However, the actual functioning of an integrated and liberal European energy market after the first waves of reforms is still in the making. The European internal market vision is up against the strong historical path-dependency, and the first EU gas and electricity Directives passed in 1996 and 1998 have been transposed into national law and national regulatory practice in several forms, depending on the variety of historical institutional environment and the previous organisation of the electricity and gas industries. Protected by regulatory monopolies, the energy systems have traditionally been used for redistributive public service objectives, and realisation of environmental and national industrial policies that are persistently hard to give up and to adapt to the new market environment.

Differences in regulation systems and 'national styles' influence both the development of the markets and the business strategies. This creates differences in regulatory efficiency, effectiveness of competition and entry barriers. It also maintains asymmetries in market rules and industrial organisations between countries, which hinder cross-border competition and market integration.

But asymmetries can also trigger processes towards institutional convergence and market integration. Once the initial vision of a liberalised market has been seriously launched, it can trigger commercial arbitrage and create a basis for super profits and for cross-national strategies legitimated by the European Union's liberal initiatives. Given a minimum of opportunities, therefore, commercial processes themselves may become important instruments for liberalisation and market integration. Once commercial processes have taken the market some steps further, this may also legitimate renewed regulatory action such as the gradually stronger harmonised rules of network access.

On the basis of these preliminary observations, this book explores the extent to which we are moving towards Europeanisation of energy systems in regulation, market scope and business strategies. Furthermore, the book explores to what extent and in what sense the energy systems are really becoming more liberalised and competitively exposed. Finally, the book also explores to what extent the electricity and gas companies are integrating across the sectoral boundaries.

The period in focus is from the first wave reforms of 1995 to 2003, the year of the second Directive on gas and electricity markets, which mandates a converging and deepening of the reforms.

Given the complexity of the energy systems and the issues involved, we choose to analyse these issues from several analytical perspectives, including a regulatory perspective, a market perspective and a business perspective. The book therefore contains three major sections. A final section discusses the interplay between regulatory development, market evolution and business strategic adaptation in a synthetic, integrative analysis.

Under the assumption that the vision of European deregulation and market integration is consistent and implementable, the question is primarily one of policy implementation. However, to the degree that this vision itself contains inherent challenges, the question of fulfilling it might also need further analytical clarification. The final chapter therefore also discusses possible problems and challenges inherent in the pan-European liberal model.

The Regulation Perspective

The section on regulation basically views the Europeanisation and liberalisation of energy markets from a public policy perspective, with a focus on efficiency and equity and regulatory instruments and regulatory action.

Reflecting the general focus of the book, the section focuses on how and to which degree the liberalisation process has modified or altered the traditional (pre-liberalisation) regulatory systems, which historically were built around monopolies and the domination of public-owned operators. Furthermore, the section explores to what extent the design and implementation of the post-liberalisation regulatory frameworks are pan-European in their character, and to what extent they remain strongly nationally tainted. These questions are addressed with respect to regulatory functions, institutions and processes in both the gas and electricity sectors.

The section also discusses the main conflicts and political issues around the new regulatory frameworks with a focus on the main 'regulation games' that are played within the different national regulatory frameworks both in terms of institutions (e.g., the pace of liberalisation, third party access (TPA) regulation, regulatory institutions) and substance (e.g., the definition of public service obligations (PSO), the implementation of public policies).

A central analytical idea in this section is the distinction between the economic and political regulation on the one hand and the distinction between the general and more specific regulation on the other. We are thus left with four core regulatory functions, which can be organised as shown in Table I.

These distinctions highlight the complexity in evaluating the liberalisation and integration/Europeanisation of energy markets

Table I. Core regulatory functions.

Economic (first order) regulation	Political (second order) regulation
Creation and maintenance of a competitive market *Competition regulation*	Enforcement of public service obligations and implementation of public policies *Public service regulation*
Supervision and regulation of natural monopoly elements and production processes *Sectoral regulation*	Management of public investments and public ownership *Public ownership regulation*

which may refer to both a political and economic sphere and to both general orientations and institutions (as the competition law) as well as to more specific organisations and procedures (as the sectoral regulation).

At the most general level, the section discusses the legal framework design including both the liberalisation of the regulatory framework and the regulatory institutions. At this level the focus is on the adoption of the liberalisation acts and the set up of the regulatory procedures and institutions, which is the exclusive responsibility of government and political parties.

At a more detailed level the section discusses the regulation of natural monopoly elements and sectoral regulation, which are usually in the hands of specialised agencies to enforce the market rules. The discussion covers activities ranging from TPA, pricing and control to grid management and supervision of code of conducts, including to some extent competition policy and antitrust regulation.

The section on regulation analyses and discusses how institutional path-dependency, anterior industrial organisation and influence of interest groups (incumbents in particular) generate variety of regulatory frameworks. Some countries have stayed at the minimal level mandated by the European Directive, whereas other countries developed more advanced regulatory positions. The analysis focuses on how national interests penetrate the competition-oriented regulation, public service-oriented regulation and national industry-oriented regulation. Countries tend to make their specific trade-offs between these policy instruments. However, the extended commercial interplay across European boundaries also exercises a pressure for European harmonisation and regulatory convergence.

As pointed out by Majone (1996), the expansion of the Commission's powers mainly takes place with reference to the core mandate of market expansion and thereby has a built-in market bias. The regulation section also shows, however, how other drivers are at work, in particular the endogenous adjustment of market rules to make competition, even in contexts where incumbents hold dominant positions.

The Market Perspective

The section on market views Europeanisation and liberalisation through the lenses of market rules and institutions, market structures and price formation. As a point of departure the section acknowledges that liberalisation and Europeanisation are in direct conflict with the previous market integration principles. In the gas sector the previous regime was characterised by loose co-ordination between former national supply monopolies, by co-operation in international purchases and by long-hauled transit from large remote producers. In the electricity sector co-ordination by various planning mechanisms used to be a central element in capacity build-up for the reliability of supply. Against this background the section analyses the current market development with the concepts of market attractiveness and market accessibility as the core analytical foci (Table II).

The concept of 'attractiveness' describes the relative potential for profitable operation in the national electricity markets, which would determine the companies and the traders' selection of foreign markets for entries. More specifically, the section defines the attractiveness of markets as their business potential in static terms (market size) and dynamic terms (market growth and price differentials).

The concept of market accessibility focuses on barriers to entry, including technical barriers (interconnection capacity, number of entry points), regulatory barriers (deterring rules of access), commercial and industrial barriers (in particular the positions of incumbents, the potential opportunities to acquire firms). On this point, the analysis undertaken in this section uses elements from the section on regulation to characterise those aspects of market accessibility that are shaped by the performances of regulatory frameworks. However, whereas the section on regulation is primarily focused on the regulatory approach as such, and its motivation, the section on market focuses on the effects upon the market.

The assumption is that on the one hand the market developments will depend on business potential in each national market, resulting from market size and market growth. On the other hand, the market developments depend on the opportunities for entry defined by the new regulatory regime. The section here discusses both the accessibility of each market either for contractual sales or for company acquisition, and conversely the possibilities offered to incumbents for preserving their dominant position and deterring entries. Besides the regulatory framework, the accessibility of the markets is partly conditioned by the incumbents' position – a high level of market power among existing generation and supply companies is generally associated with a lack of liquidity of wholesale markets, or an absence of marketplace, which clearly impedes new entrants.

The section also analyses future market integration potentials of electricity and gas markets. These potentials appear to lie in the results of the various European rule-making processes both at the European Commission level and at the level of national regulators and system operators, when it comes to suppression of interconnection

Table II. Characters of the competitive profile of national electricity and gas markets.

Attractiveness		Accessibility	
Size	Overall, segment	Technical accessibility	Interconnections, internal congestions
Growth	By segments • industrial • commercial • power generation, etc.	Regulatory accessibility	Degree of openness • Type of TPA • System operation unbundling • Transmission pricing • Regulatory institutions
Profitability	Price level	Commercial accessibility	Wholesale price index Exchanges: • bourses • hubs • pools Balancing market
		Industrial and capital market accessibility	Horizontal integration Vertical integration (production/import-supply) Regime of ownership

bottlenecks and stimulating market competitiveness. The creation of marketplaces in electricity and (with some delay) gas industries, the traders' arbitrage between national wholesale markets, as well as foreign companies' entries through acquisitions or through supply development in accessible markets, indicates a dynamic move towards a greater integration. However, the rapid market concentration through mergers and acquisitions suggests the need for a strong competition policy.

Our analysis in the section on market recognises inherent differences between the electricity and gas markets, due to the specific endowment and remote location of gas resources. This leads us to start with a two-level market perspective on the gas market, while the electricity market is discussed in a more traditional single-level analysis. The first level refers to the trade and distribution of gas domestically at the national level. The second level refers to the oligopolistic negotiated balancing of the interests of the major producers and suppliers with strong geo-political overtones. In this game, company positions have partly been defined to support national interests.

The Business Strategy Perspective

The section on business strategy basically views Europeanisation and liberalisation of energy markets from a business strategy perspective, with a focus on value maximisation and the firm's strategic positioning in the energy markets.

In the numerous combinations possible, the business strategy section distinguishes between four ideal types – the mono-sectoral functional specialists, the vertically integrated mono-sectoral, the vertically integrated pluri-sectoral and the multi-sectoral functional specialists (Figure I) – each representing sectoral and functional combinations.

The section discusses their distribution across the European energy markets and the theoretical motivation that this variety might have in a competitive market context. Taking the core economics repertoire as a point of departure, the section focuses on drivers such as the production cost and transaction cost as well as the strategic positioning, inherent in oligopoly theory. Then, from an organisation theory perspective, the section discusses drivers for strategic configuration inside the firm, including drivers such as firm capabilities and learning. Adding on the insights from a broad institutionalist literature, the section includes national institutional contexts and business styles as major shapers of strategic configuration. Finally, on

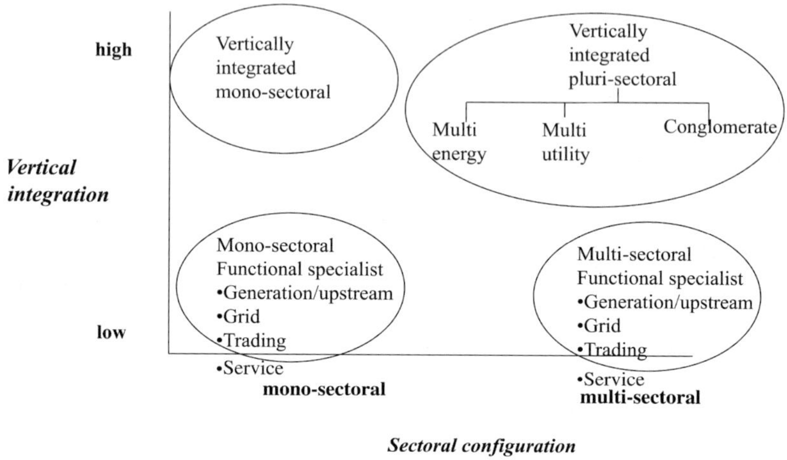

Fig. I. Four ideal type configurations.

the basis of a pragmatically motivated industry perspective, industry-specific factors are brought in as drivers of strategic configuration.

This juxtaposition of several theoretical approaches indicates the ambiguity of interpretation of the European business strategy development that characterises the section on business. Whereas a strict production cost and transaction cost analysis would lead us to expect homogenisation of business models across national boundaries, in response to Europeanisation and liberalisation, the national style, path dependency and capability perspectives suggest maintenance of configurational diversity even under a level competitive playing field. The business strategy analysis therefore, as far as the core questions of Europeanisation and liberalisation are concerned, views the development as shaped by the dynamic interactions with the regulatory and market context. The business strategy analysis, therefore, draws heavily upon the previous regulation and market analysis, but takes a distinctive business strategy outlook.

The Integrative Analysis

While Europeanisation, liberalisation and functional integration of gas and electricity markets may be discussed from specific regulatory, market-structural and business strategy perspectives, there are obvious interaction effects among the three domains/perspectives of fundamental importance.

Since the launch of the European energy market, deregulation policy in the late 1990s has evolved through a dynamic interplay between regulatory initiatives, business strategies and market structure. In a situation with limited European consensus on strong deregulation policy and extensive national subsidiarity in their implementation, the policy impact may be highly dependent on its ability to shape markets and trigger business strategies. The core question around Europeanisation, liberalisation and integration of European energy industry therefore obviously needs to be discussed in a broad and comprehensive way. Having discussed them partially, as related to deregulation, markets and business strategies, the final chapter moves on to an analysis across the three sections.

The first focus of the integrative analysis is on Europeanisation, contrasted against the continuation of national governance, as a counter position. The discussion here links the Europeanisation issue up to the national style versus globalisation debate, arguing for more nuanced and relevant interpretation of the two counter poles based on the evolutionary theory. This allows us, for instance, to explore various mixes of European and national styles, where variation in industrial strategies based on national/specific resources, competencies and traditions play together with a more Europeanised market selection mechanism. Retention at the institutional level may again have national peculiarities in form, but not necessarily in function.

The second focus of the integrative analysis is on Europeanisation and liberalisation. The focus here is on how market-based European integration may come about both in competitive and various non-competitive modes, notably through oligopolisation. Based on the joint evidence from the analyses of regulation, markets and business strategies, it is argued that non-competitive integration represents a possible outcome of the European energy deregulation exercise. This theme is pursued further in the final part of the chapter on the challenges of oligopolisation.

As a summing up of both the previous foci, the chapter presents the current European energy industry as an interplay between multiple political economies, where business strategies, market structure and institutional contexts co-evolve.

A brief discussion on the dynamic interplay among regulation, markets and business strategies is also included, indicating that the sequencing of politics, markets and business responses are central to the outcomes both as far as Europeanisation and liberalisation is concerned, and that the reform results have to be judged over considerable time.

Challenges

After a decade or more of deregulation (in pioneering markets) and half a decade for the rest of the EU area, European industry and policy makers are not only discovering the possibilities but also the challenges of the new market order which jointly question the economic and political regulations. Still focusing on the broader issue of the interplay between economic and political factors, the final chapter ends with a discussion of the following central challenges:

First, we note that the European energy industry has been exploiting its new freedom of international market configuration to scale up in concentration – in some cases – to a level where the competitiveness of the market itself could be affected. Facing such new oligoplistic configuration, the question raised here is if the European competition policy can follow suit.

Second, the chapter points out that the reforms have been conceived at a stage of overcapacity with a focus on market design and in some cases on industrial structures to promote static efficiency. Against this background we discuss the challenge of securing new investment as the dynamic capability of the market system is put on trial. The chapter also points out that the balancing of the power market in the long run is not only a question of a market equilibrium, but also of a social equilibrium, and these two equilibria do not always coincide.

Finally, the chapter focuses on the challenges arising as the energy sectors face extensive environmental demands as ambitious climate and renewables policies are targeting these industries. Major issues in focus here are not only the management of collective responsibilities across national boundaries, but also the management of the interface between the environmental policy field and the core energy markets. The chapter points out that, while the core markets are increasingly operated under a commercial/competitive mode, with increasing European standardisation, the environmental policies tend to have stronger national designs. Complex interaction effects between energy and environmental markets and policy instruments in both fields are also seen to constitute a challenge to the future management of the energy industry.

SECTION I
Regulation

Chapter 1
Regulation in Liberalised Energy Sectors: Introduction and Concepts

C. GENOUD, M. ARENTSEN AND M. FINGER

1.1. Introduction

Deregulation or liberalisation of network industries is a European, if not a global trend. Telecommunications, railways and airlines are increasingly being put under the pressure of competition. But, deregulation or liberalisation are inevitably followed by reregulation.[1] Such reregulation often takes the form of the introduction of competitive mechanisms and the setting up of new authorities in charge of supervising and regulating the sector. Many current discussions and recent publications on regulation in network industries indeed deal with institutional and pragmatic issues on how to best regulate these liberalised network industries (e.g., Newbery, 2000). Among the numerous issues addressed, many pertain to the scale and the scope of the regulatory frameworks, the nature of the regulatory institutions and the importance of public ownership in these newly competitive sectors. In this book, we exclusively focus on the regulation of the

[1]This is the reason why we will not use the concept of 'deregulation', which is obviously misleading. It is commonly accepted that 'deregulation' is a qualitative and/or quantitative reduction of government intervention in economic and/or social systems by means of cutting red tape and introducing competition into previously monopolistic or cartelistic sectors. This can lead to considering that regulatory activities are either reduced or abolished, which is clearly not the case, as we will see. We therefore prefer to use the concept of 'liberalisation', which does not suggest a quantitative reduction of regulatory pressure, but rather a qualitative transformation of its modes. For more information see Arentsen, M. and R. Künneke (1996). 'Economic organisation and liberalisation of the electricity industry.' *Energy Policy*, **24**(6): 541–552.

electricity and the gas sectors. The purpose of this first section is to briefly discuss the relevant regulatory issues, as they arise from the liberalisation of the network industries, as well as to present the concepts and theoretical instruments we use in this part and the next two empirical chapters on electricity and gas regulation in Europe.

1.2. Sectoral Specific or General Competition Regulation?

Antitrust and competition policies are already solidly institutionalised features of State intervention, even though they are still being significantly improved by governments of industrialised countries and more particularly by the members of the European Union (see Neven *et al.*, 1998). Indeed, the construction of a single European market not only requires efforts in building common European competition policies and institutions, but it also has important consequences on the Member States' policies regarding network industries. In this respect, the liberalisation of such network industries, and more particularly the electricity and gas sectors, is often understood as the logical consequence of the unification of markets in the European Union. Although, from a theoretical point of view, all competition and liberalisation policies lead to ever more competition in all sectors, from an institutional point of view, however, the question remains as to whether the regulation of the network industries can be assimilated or compared to the regulation of more traditional markets. More concretely, can or should third party access in electricity and gas be regulated by a general competition authority or is it more adequate to attribute this task to a sector-specific authority? Moreover, will general competition regulation be sufficient to regulate these sectors, once the transition phase (from monopoly to competition) will be over, provided that it ever will?

1.2.1. Independent or politico-administrative regulators?

The creation of regulatory (sector-specific) authorities is one of the most spectacular outcomes of the recent liberalisation policies. Labelled as 'national regulatory authorities' or 'regulators', these new institutions are now charged with regulating the recently liberalised sectors. But, although the 'independent regulator' model is the more frequently chosen option, many countries have set up political or administrative bodies, rather than more autonomous regulatory authorities. But in any case, the nature, the shape, the powers and the missions of these bodies and authorities do vary significantly, as we will see in the following chapters.

This relationship between the regulatory institutions and the political authorities has been the object of much attention by analysts.[2] They mainly ask why these institutions should be 'independent' from ministerial authorities and government or if they should be integrated into the administration? But this issue is in fact twofold: Indeed, this is not only the question of the relation between these 'regulators' and the political authorities, but it is also the question of the relationship between these 'regulators' and the regulated industries. Upstream, the issue revolves around the question of the degree of political or administrative independence of these regulatory institutions. Downstream, one faces the classical 'capture issue', i.e., how dependent or independent should the regulator be from the very objects of its activity? In both cases, the three following dimensions are involved, namely the regulators' legal status, their powers, and their resources.

Regulation is a multidimensional process and regulators are only one (key) element of a much larger system where multiple actors interact. The role of judicial and/or dispute settlement institutions, for instance, must also be carefully considered. In addition, the importance of consultative bodies (administrative or ministerial) and organised lobbies (consumer groups, unions) should be included into the wider analysis of the regulatory process.

1.2.2. National or supra-national regulation?

Liberalisation of the network industries in Europe is systemic in nature. Except in the case of the United Kingdom and some Nordic countries, the introduction of competition into the electricity and gas industries clearly is the result of regulatory pressures from the European Union. Such pressure in turn contributed to a retroactive feedback loop, whereby the dynamics unleashed by liberalisation at the national level requires in turn corrective answers from the European Commission and/or the EU policy makers. The harmonisation of network access conditions, the coordination of regulatory intervention on cross-border issues, along with the need for collaboration of national and EU competition authorities in mergers and acquisitions cases are just a few examples of direct consequences

[2]See Volume 25 Issue 1 of *West European Politics* and Volume 8 Issue 1 of the *Swiss Political Science Review*, Spring 2002.

of sectoral liberalisation. Although the creation of EU regulatory agencies is not (yet) on the agenda, many empirical developments indicate that the articulation of national and supra-national regulation will become one of the key issues to be addressed in the years to come (see Chapter 4). National and especially cross-border mergers and acquisitions in the electricity sector, for example, constitute real regulatory challenges for national regulators. Besides the question of the regulatory remedies (incentives or structure based) in such cases, one can doubt about the efficiency and the effectiveness of regulatory intervention on a purely national scale, when the tendency is clearly towards the 'globalisation' of problems and challenges, at least on the capital market as the chapters on market and business strategies show.

1.2.3. What future for public ownership?

Public ownership has been the dominant and traditional form of regulation. In the age of the liberalisation of network industries and the progressive globalisation of markets, public ownership is put into question. To recall, it has often been said that privatisation is just the logical consequence of liberalisation (e.g., Nestor and Mahboobi, 2000; Newbery, 2000: 176ss) since, it is argued, the merits of competition will only be fully reached with private operators. Not only has this hypothesis not yet been confirmed – in many countries public ownership is still important or even predominant, even though the market has long been liberalised (e.g., Norway) – but moreover, is this debate misleading, thus preventing one to fully understand the real issues underlying public ownership and regulation. Indeed, the question here is how public ownership is affected by the liberalisation process and whether the double role of the State as both owner and regulator is sustainable in the long run. Respectively, the question is also how public ownership affects and challenges the design of the regulatory framework? Does public ownership require specific arrangements and measures from a given regulatory framework? Indeed, as far as regulation is concerned, the question is not whether State-owned enterprises should or should not be privatised, but if privatisation matters when it comes to regulating these sectors (Genoud and Varone, 2002)?

1.2.4. Convergence or divergence?

Since its original formulation, Majone's hypothesis of the 'rise of the regulatory State in Europe' (Majone, 1997) has triggered lively debates but has led to few large-scale empirical studies and conclusive findings. With his seminal works, Majone has significantly contributed to the revival of sectoral studies on the emergence of a new form of state intervention, such as 'regulation by network' (Dehousse, 1997) and 'agencification' (Pollit and Talbot, 2003). He also helped to revive studies on European political and economic integration by focusing on network industries' sectors, which had been largely forgotten by political scientists and political economists. As a result and in addition to the impact of progressive economic integration on the political construction of Europe, several concurrent or complementary factors are now being identified. This is, for example, the case of the 'negative' integration by means of a gradual 'constitutionalisation' of competition in Europe, versus the 'positive' integration by means of implementing European public policies (e.g., energy, transports) (Scharpf, 1999). Another example is the possible emergence of a 'public service regime' within the European legal framework (Rodrigues, 1999). Behind these interrogations lies of course the question as to whether or not we observe converging or diverging tendencies in the design of liberalisation and regulation frameworks. This issue will be the object of the last chapter of this part (Chapter 4).

1.2.5. Research questions

The purpose of this part on regulation is to address five main questions, by investigating in particular the empirical developments in the electricity and the gas sectors of selected European countries on a comparative basis:

- How has the liberalisation process influenced, modified or altered the traditional (i.e., pre-liberalisation) regulatory systems, which, historically, were built around monopolies and the domination of State-owned operators? Here, we will determine in what way the regulation frameworks of these industries have evolved over time.
- Is it possible to observe general patterns in the design and implementation of the post-liberalisation regulatory frameworks in the electricity and gas sectors? Here, we will focus on the study of regulatory functions, regulatory institutions and regulatory processes in both sectors. In other words, we will investigate if 'regulatory styles' can be observed within and across these two sectors and, if this is the case, which factors may determine and explain these styles.
- What were and what are the main conflicts and political issues around these new regulatory frameworks? What are the main 'regulation games' that are played within the different national

regulatory frameworks both in terms of institutions (e.g., the pace of liberalisation, TPA regulation, regulatory institutions) and substance (e.g., the definition of PSO, the implementation of public policies)?
- Is it possible to observe dynamics of convergence and/or divergence within and across these two sectors, and what are their driving forces of these dynamics?
- Finally, based on the answers to the previous questions, how does or can regulation influence both the development of the markets and the business strategies?

The first three questions will be addressed in the next two empirical chapters on electricity and gas regulation. The remaining two questions will be treated in the final chapter of this part. Yet, in order to be able to describe, analyse, compare and assess the empirical developments when it comes to electricity and gas regulation, a comprehensive and integrated research framework is needed. We will therefore present and elaborate in the following paragraphs, our analytical assumptions, hypotheses, and descriptive tools on regulation design in post-liberalised contexts.

1.3. Regulation as a Public Policy Issue

Regulation is necessary in liberalised electricity and gas sectors for at least three reasons. First, because competition has to be introduced and created. Second, because natural monopoly elements are likely to remain in both sectors and thus require regulation, for example, when it comes to the so-called third party access (TPA).[3] Finally, because of public services and public policies issues. If the first justification is probably a transitory one, it is quite evident that whatever the evolution of market dynamics – even though technological innovation might further reduce the scope of the natural monopoly – the other two reasons will remain valid in the long run (Helm, 2001). Regulation is therefore a public responsibility. More precisely, it is the task of public authorities (governments) to design and implement regulatory frameworks following liberalisation. In this sense, regulation is an institutional public policy combining two complementary dimensions. As such it is concerned with institutions (i.e., regulators, ministries, courts, etc.) and instruments (i.e., unbundling, TPA, etc.).

[3]Third Party Access is the process by which an operator uses the infrastructure of another operator to provide its services.

Contrary to commonly held ideas, we consider that regulation is neither a substitute, nor a second best solution to competition (Newbery, 2000: 134ss), nor a temporary phenomenon (Bergman *et al.*, 1998: 8–9). Rather, it is in our view, an essential part of the regulation process of network industries and markets.

1.4. The Functions of Regulation

As a matter of fact, the regulation of the network industries is a process combining two potentially competing objectives (Eberlein, 1999), i.e., a first-order economic regulation and a second-order socio-political regulation.[4] First-order or economic regulation is concerned with market creation and market efficiency. As such, it covers the regulation of market imperfection and the promotion of competition. Also the efficient allocation of resources, as well as pricing and its theoretical justifications mainly fall into the realm of economic theory of regulation (e.g., Kahn, 1990; Lévèque, 1998; Newberry, 2000). The second order or sociopolitical regulation is concerned with public service obligations and public policy implementation. This type of regulation is not limited to market correction measures, as it more generally covers all the political actions that public authorities undertake in the name of general interest. In the specific context of electricity and gas sectors, environmental protection, energy savings measures, public service obligations (PSO[5]) in terms of access, quality and prices will typically fall into this category of regulatory intervention.

Regulation is also a complex system, where procedures are designed to deal with specific issues or problems. These issues or problems can be grouped into six distinctive regulatory functions, of which some are more technical and economical, while others are more political in nature.

1. *The sectoral–legal framework design.* This function covers both the liberalisation process and the regulatory framework design. It is exclusively the responsibility of governments, who adopt the liberalisation acts and set up the regulatory procedures and institutions.

[4]This does not imply that one function is more important than the other.

[5]PSO 'refers to specific requirements that are imposed by public authorities on the provider of the service in order to ensure that certain public interest objectives are met' (Green paper on Services of General Interest, COM 2003 270final, p. 7).

2. *The enforcement of public services obligations.* This function covers two distinctive issues, namely the definition and the financing of PSOs and their procurement. The first task is a governmental one, while the second is the operators' duty, even though placed under the scrutiny of the government or the regulatory institutions.
3. *The competition and antitrust regulation.*
4. *The regulation of natural monopoly elements or sectoral regulation.* This function covers activities ranging from TPA, pricing and control to grid management supervision through price control and code of conducts setting and enforcement.
5. *The implementation of public policies.* Generally speaking, this function groups together all the government activities related to the definition and implementation of public policies concerned or affected by electricity and gas, such as environmental and industrial policies.
6. *The evaluation of the liberalisation and regulation frameworks.* Often neglected, evaluation is a fundamental function of state activity. In this context, it is conceptualised as the evaluation of the regulatory framework, its coherence and its performance.

Though very important, functions 1 (legal framework design) and 6 (evaluation) differ in their nature from the other regulatory functions. They do not constitute the core activities of any given regulatory system, but rather their context. In other words, they refer to the regulatory design process, rather than to the regulatory activity in itself. We are thus left with four core regulatory functions, which can be organised as shown in Table 1.1 (Genoud and Varone, 2002).

The *competition* and *sectoral regulatory* functions refer respectively to functions 3 and 4, while the *public service regulation* is the result of the aggregation of functions 2 and 5. A fourth function, i.e., public ownership regulation function has been added here. Although it cannot be considered a core regulatory function of liberalised sectors, public ownership has represented – and in many respects continues to represent – the regulatory function *par excellence,* especially in monopolistic contexts and pre-liberalised sectors. Since liberalisation and privatisation policies have not been implemented in a straightforward manner – even though this would have been a 'logical combination', at least in the eyes of some (e.g., Lane, 1997: 3) – public ownership must be taken into account and thus be considered as a full-fledged regulatory function in any liberalised context (Genoud, 2001).

Table 1.1. Core regulatory functions.

Economic (first order) regulation	Political (second order) regulation
Creation and maintenance of a competitive market	Enforcement of public service obligations and implementation of public policies
Competition regulation	*Public service regulation*
Supervision and regulation of natural monopoly elements and production processes	Management of public investments and public ownership
Sectoral regulation	*Public ownership regulation*

1.5. The Regulation Actors

The nature, the shape and the quality of a particular regulatory framework is largely determined by the way the above functions are distributed among the different actors in any given sector. The following are generally the actors concerned by such a regulation framework.

- The political authorities (executive and legislative bodies)
- The sector regulator
- The competition regulator
- The public and private operators
- The courts and non-judicial settlement institutions
- The customers, the users and the citizens

Figure 1.1 summarises the positions and functions of each actor in the overall regulation framework, along with its responsibilities.

1.6. The Regulation Framework

In order to compare and analyse the different regulatory frameworks of electricity and gas in Europe, we will focus our analysis in the next two chapters on three dimensions, namely,

The *comparison of regulatory frameworks*, namely how the different regulatory functions have been designed and distributed among the different actors of the regulatory process.

The *comparison of regulatory authorities*, namely which organisations and institutions are in charge of one or more of the regulatory functions.

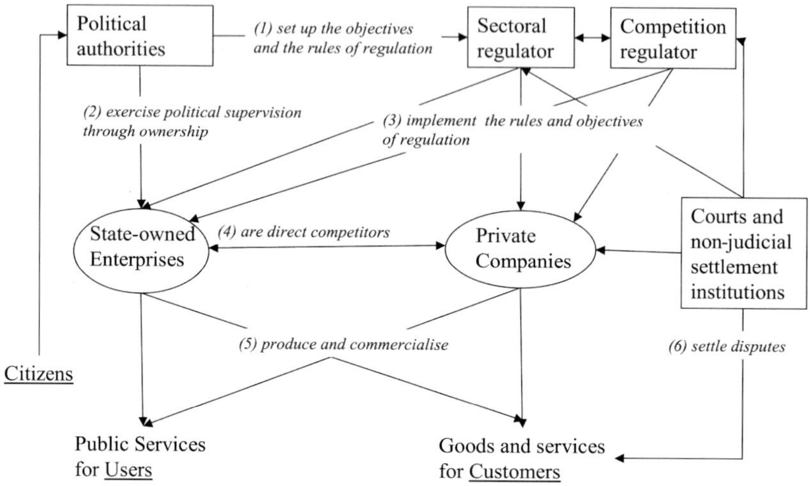

Fig. 1.1. Actors and process of a regulation framework.

The *comparison of regulatory games,* which have been and still are being played in each of the regulation frameworks.

On an empirical level we will consider the following dimensions, namely comprehensiveness, overlapping, competition, regulatory pressure, coordination, public ownership, regulatory authorities and regulation games.

- **The comprehensiveness of regulatory frameworks**: regulation frameworks can be more or less developed. Some countries might set up light regulatory institutions and consider one particular regulatory function as not being essential. For instance, PSOs can be more clearly defined and regulated in France, while in the United Kingdom, for example, the tradition of Public Service is less important. Therefore, the degree to which a regulatory framework is integrated and developed gives an indication of the regulatory style, as well as of the commitment to regulation. Some frameworks are based on the idea that all regulatory functions are equally important, while others focus more on specific regulatory functions such as TPA and competition policy, without, for example, paying much attention to PSOs. A highly comprehensive regulatory framework is therefore characterised by the fact that all the regulatory functions are taken on by at least one actor in the system, along with a strong institutionalisation of all the basic regulatory functions. Respectively, a non-comprehensive regulatory framework will be characterised by a selective focus on specific regulatory functions and/or weakly institutionalised or fragmented regulatory functions.

- The question of **overlapping competencies** between actors is another important aspect of institutional design. This is typically the case, for example, when it comes to setting the rules of cooperation between the sectoral regulator and the competition authority, especially on matters of mergers and acquisitions. Similar considerations apply to the relations between ministries and the sectoral regulator on issues of PSO supervision and implementation. In some cases, the division of labor among the involved actors is even more sophisticated: for example, the sectoral regulator formulates recommendations on access charges or tariffs to the responsible ministry, which in turn makes the final decision. The question here is whether two or more actors share one or more regulatory functions, and how the roles and competencies are distributed among them.
- **Concurrent or competing regulatory functions**: in some cases, certain actors take on multiple regulatory functions. For example, it is common to give Ministries or competition regulators the responsibility to supervise and regulate both TPA and mergers and acquisitions. This is particularly the case when no sector-specific regulator has been set up. As a result, the different regulatory functions are being distributed among a smaller number of actors, thus increasing the probability of attributing one actor several regulatory functions. This may result in centralisation or in a monopolisation of regulatory powers.
- **Regulatory pressure** is very often used in typologies (see Dumez and Jeunemaître, 1999) in order to characterise regulatory styles. Although regulatory pressure is a very valuable and critical dimension, it is very difficult to define, to operationalise, and to measure. Some analysts (e.g., Bergman *et al.*, 1998) use the concept of regulatory pressure without clearly defining it and predict that, over time, regulatory pressure will decrease proportionately to the emergence of market mechanisms, as shown in Fig. 1.2. Such an assumption clearly needs to be tested empirically, and, in order to do so, one needs a more precise concept. In other words, the issue here is to try to evaluate the level of regulatory intervention in the sector, addressing both competition and sector-specific regulatory pressure and both economic and political regulatory intervention. Two issues must be considered here, namely concentration on the one hand and scope and scale on the other.

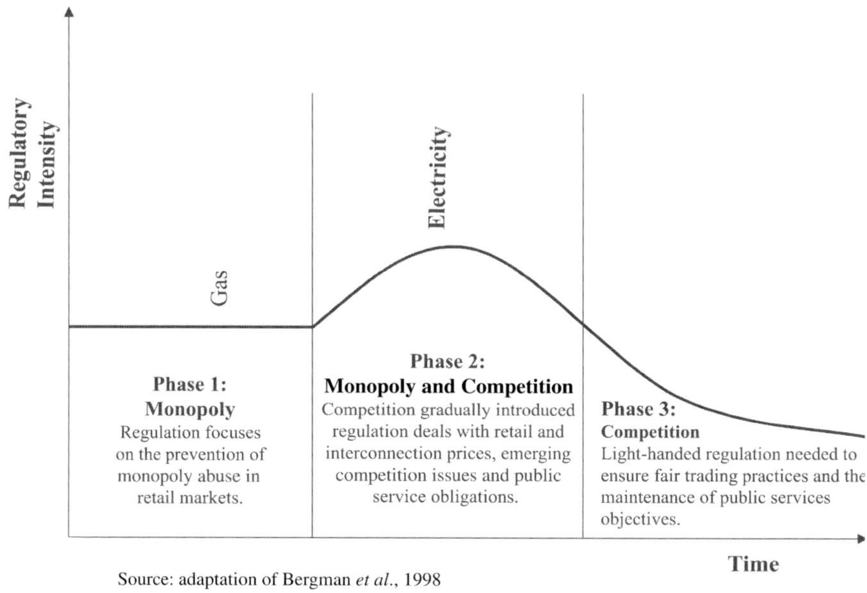

Fig. 1.2. Regulatory intensity.

A high regulatory pressure can be the consequence of a **concentration** of important regulatory functions and powers in the hands of a limited number of regulatory authorities in matters of sanctions, investigation and intervention. This raises the question of the powers, the resources and the nature of the regulatory institutions, as well as their number. But regulatory pressure can also be a function of political intervention within the regulatory process. In many cases, liberalisation has been seen as the best way to depoliticise in favour of a much more economic approach. However, empirically this 'withering away of political interference' has not been confirmed so far, as typically shown in the case of the United Kingdom (e.g., Young, 2001). As is indeed the case in UK, the intensity of the regulatory pressure can result from the continuation or even the increase of political intervention in the regulatory games.

The **scope** and the **scale** of regulation is also a good indicator of regulatory pressure. A 'light-handed regulation' would focus on limited and specific objects, like TPA or prices, letting market mechanisms address some of the technical issues as well as PSOs. Respectively, a 'heavy-handed regulation' would include prices, industrial structures, PSO and several others. This observation naturally points to the question of regulatory instruments and their impact on regulatory intensity. A traditional classification of regulatory instruments, inspired by the practice of competition regulation, distinguishes between incentive and structural remedies, the former being considered lighter instruments than the latter.

The regulatory process generally involves many actors, namely administrative bodies, ministries, regulatory and competition authorities, courts, etc. **Coordination** and planning among these actors are therefore also key factors. The larger the number of actors involved, the higher the possibility of fragmentation and regulatory incoherence and the weaker the regulatory pressure. Also, a high number of actors can either be a desired objective of the regulation design, or the result of structural and political factors, such as federalism and/or decentralisation. In other words, a large number of actors typically favours the setting up of fragmented regulatory frameworks, which in turn are characterised by slow and complicated regulatory processes, thus reducing regulatory pressure.

It has often been considered that **public ownership** is a paradigmatic illustration of a high regulatory intensity. It is indeed true that in pre-liberalisation times, regulation has mainly been conducted by means of the ownership of utilities. In liberalised environments, such practices are no longer possible. Public regulation and public ownership indeed have to be separated. This does not mean, however, that liberalisation systematically requires the privatisation of State-owned enterprises, but, as said above, the design of regulation frameworks has to take into account the specificity of the State's position as both the owner of operators and their regulator. A clear distinction between these two functions has to be made. In the case of public ownership (within a liberalised environment), the question is whether public ownership remains a (hidden) form of 'indirect' regulatory intervention. On a theoretical level, this question can be debated at length. It can be argued that the State will never be able to make this clear distinction and that, as long as public ownership exists, it will always constitute a perverse form of regulation. In other words, public ownership will always represent an excessive, albeit only indirect, regulatory pressure. It thus can be argued that, since public ownership requires specific regulatory framework arrangements, for example, in the form of arm's length regulators, it will favour higher regulatory pressure. All these hypotheses are highly debatable on a theoretical level. But the fact that public authorities are sometimes directly involved both in the management and the regulation of the very same operators has often led to higher regulatory pressure. On the other hand, and as the UK example shows, regulatory pressure has not decreased with the privatisation of State-owned enterprises.

- When comparing the **regulatory authorities**, one has to consider their autonomy in legal, political and financial terms, their resources (budget, staff), their instruments (hearings, setting codes of conduct, etc.), and their competencies (sanctions, decisions, etc).
- Finally, in comparing **regulation games** one has to consider the importance and the intensity of the main issues addressed in any given regulation framework. These games can be institutional in nature (pertaining to the institutional design of the regulatory framework) or oriented towards substantive issues as related to the existing public policies (environment, economic policy, etc.).

Structure of this part on regulation

In the next chapter, Genoud and Finger present and analyse the electricity regulation frameworks of 17 European countries, while in Chapter 3, Arentsen proceeds accordingly for the gas regulation framework of 13 European countries. Finally, in Chapter 4, Genoud, Finger and Arentsen compare, evaluate and discuss the analysis of these two sectors in the light of the issue of convergence.

References

Arentsen, M. and Künneke, R. (1996). Economic organization and liberalization of the electricity industry. *Energy Policy* 24(6): 541–552.

Bergman, L., Brunekreeft, G., Doyle, C., von der Fehr, N.-H., Newbery, D.M., Pollit, M. and Régibeau, P. (1998). *Europe's Network Industries: Conflicting Priorities, Monitoring European Deregulation 1: Telecommunications.* CEPR, London.

Dehousse, R. (1997). Regulation by networks in the European Community: The role of European agencies. *Journal of European Public Policy* 4(2): 246–261.

Dumez, H. and Jeunemaître, A. (1999). Les institutions de la regulation des marches: etudes de quelques modèles de reference. *Revue Internationale de Droit Economique* no1: 11–30.

Eberlein, B. (1999). L'Etat régulateur en Europe. *Revue Française de Science Politique* 49(2): 205–230.

Genoud, C. (2001). Privatization and regulation: The case of European electricity. *Cahier de l'Idheap* 196, p. 120.

Genoud, C. and Varone, F. (2001). Does privatization matter? Liberalization and regulation: The case of European electricity. *Public Management Review* 4(2): 231–256.

Helm, D. (2001). The assessment: European networks – competition, interconnection, and regulation. *Oxford Review of Economic Policy* 17(3): 297–312.

Kahn, A.-E. (1990 [1988]). *The Economics of Regulation, Principles and Institutions.* MIT Press, Cambridge.

Lane, J.E., Ed. (1997). *Public Sector Reform: Only Deregulation, Privatization and Marketization?: Public Sector Reform, Rationale, Trends and Problems.* Sage, London.

Lévèque, F. (1998). *Economie de la Réglementation.* La Découverte, Paris.

Majone, G. (1997). From positive to the regulatory state: Causes and consequences of changes in the mode of Governance. *Journal of Public Policy* 17(2): 139–167.

Pollit, C. and Talbot, C., Eds. (2003). *Unbundled Government: A Critical Analysis of Global Trend to Agencies, Quangos and Contractualisation*. Routledge, London.

Nestor, S. and Mahboobi, L., (2000). Privatisation of public utilities: The OECD experience. In OECD, *Privatization, Competition and Regulation*. OECD, Paris, pp. 13–49.

Neven, D., Papandropoulos, P. and Seabright, P. (1998). *Trawling for Minnows, European Competition Policy and Agreements Between Firms*. CEPR, London.

Newbery, D. (2000). *Privatization, Restructuring, and Regulation of Network Utilities*. MIT Press, Cambridge.

Rodrigues, S. (1999). *La nouvelle régulation des services publics en Europe. Energie, postes, télécommunications et transports*. Editions TEC&DOC, Paris, 694p.

Scharpf, F. (1999). *Governing Europe: Effective and Democratic?*. Oxford University Press, Oxford and New York.

Young, A. (2001). *The Politics of Regulation. Privatized Utilities in Britain*. Palgrave, London.

Chapter 2

Electricity Regulation in Europe

C. GENOUD AND M. FINGER

2.1. Introduction

In this chapter, we analyse and compare the electricity regulatory frameworks of seventeen European countries (fifteen European Union members plus Norway and Switzerland.[1]) In an introductory section we briefly present the European liberalisation framework and the two specific cases of Norway and England & Wales, which anticipated the EU reforms by several years. The inclusion of two non-EU members allows us to consider the influence of EU reforms on non-Member States. The next section is divided into three parts comparing and analysing regulatory frameworks, regulatory institutions and regulation games. Based on the construction of an electricity regulation systems typology, the third section of this chapter seeks to identify the factors that explain the national cases' position within the European context. We also discuss the possible evolutions of the near future. A final section will present and assess the critical factors that appear to drive the design of regulatory frameworks in the electricity sector.

2.2. Background of the Regulatory Challenge in Electricity

The establishment of a new industrial and regulatory model has been heavily determined by the characteristics of the electricity industry. It has resulted mainly from the European liberalisation process.

[1]A first liberalisation act was rejected in referendum in September 2002. A new law is in preparation (March 2004). Data for Switzerland presented in this chapter take into account the disposition of this new act.

2.2.1. The former industrial organisation and regulation

Before liberalisation the electricity supply chain was structured around four functions under a monopoly structure.

- *Production* involving transformation of primary energy into electricity;
- *Transmission* involving high voltage transport of electricity to the next level;
- *Distribution* completing transmission at the lower voltage to the final end user;
- *Supply* representing the retailing of electricity to end users (in a monopolistic situation this function is generally aggregated into distribution).

This type of industrial structure built around an integrated and coherent value chain is strongly subjected to vertical and horizontal integration. **Vertical integration** is linked to transaction characteristics such as asset specificity, externalities and measurement problems (Glachant and Finon, 2000). Asset specificity refers to the fact that production must be continuously linked to the end-user (site specificity) and that as a non-storable good, the electricity flow must be unbroken through the supply chain (time specificity). These two factors favour vertical integration. The natural monopoly situation in transmission and distribution reinforces vertical integration with production, as both cost functions can be interdependent. Finally, externality and measurement problems refer to the need for a 'system's operator', responsible for the maintenance of the stability of the network load. This network management function usually serves as a cooperation mechanism between the actors of the systems. A complete vertical integration of the system thus reduces transaction costs.

Horizontal integration in a monopolistic situation mainly depends upon economies of scale and scope, path dependency factors such as the historical heritage, the political structure and the legal order regulating the sectors in terms of ownership.

Table 2.1 presents the structure of the electricity sector in the countries studied.

The **structure of the ownership** is generally considered to be the third important dimension of an industrial structure analysis. The following table presents the current situation in five selected countries. Of course, England & Wales had, before their 1990 reforms, a completely public ownership (see Section 6) and an integrated industrial structure (Table 2.2).

Table 2.1. Electricity sector structure.

	Degree of horizontal integration			Vertical integration
	Production	Transmission	Distribution	
France	High	High	High	High
Germany	Moderate	Moderate	Low	Mixed
Norway	Low	High	Low	Low
Switzerland	Low	Moderate	Low	Low
England & Wales	Moderate	High	Low	Low
Austria	Moderate	Moderate	Moderate	High
Belgium	High	High	Moderate	Low
Denmark	Low	Moderate	Low	Moderate
Finland	Moderate/high	Moderate/high	Moderate	Moderate
Greece	High	High	High	High
Ireland	High	High	High	High
Italy	High	High	High	High
The Netherlands	Moderate	High	Low	Moderate
Portugal	High	High	Moderate	Low
Spain	Moderate	High	Moderate	Moderate
Sweden	Moderate	High	Moderate	Low

'Mixed' means that different utilities have very different degrees of integration.
'Moderate' means that the four main activities are not fully integrated within each utility
Source: IEA 1999.

Although each national electricity industry followed a specific historical path, general trends in electricity development and regulation can be observed (Stoffaes, 1994). During its first major development in the late 19th century, electricity production represented a complementary activity to the existing traditional industries (for instance, manufactures). When technological progress made it possible to extend the distribution of electricity outside the factory walls, municipalities very quickly showed an interest in having their citizen benefit from the wonders of this new energy. Technological standardisation and

Table 2.2. Ownership patterns in some selected electricity sectors in 1999–2000.

	Predominantly public	Mixed	Predominantly private
France	P, T, D		
Germany	D	P, T	
Norway	P, T, D		
Switzerland	D	P, T	
England & Wales			P, T, D

P, Production; T, Transmission; D, Distribution.

economic concentration in the sector reinforced the trend towards the regional integration of networks. It is generally during this period that the first national laws and antitrust regulations were passed in order to supervise this quickly developing industry. After the Second World War, the notion of electricity as a 'public service' was developed at the legal level in many countries. Price and investment controls through licensing or nationalisation were then widely implemented within the newly defined global energy policies frameworks.

Although the concept of 'public service' is mainly a French particularity, its principles and essence can be found, in very different forms and to very different extents, in almost all industrialised countries. In its broadest sense, the 'electricity public service' is generally understood as the recognition of electricity as an essential good for the well-being of society and economy, and as such should be regulated in the public interest in terms of price quality, security and access.

2.2.2. The European Liberalisation

The introduction of market elements into monopolistic network industries has progressively gained political attention during the past two decades. Antitrust and liberalisation reforms in the USA during the early 1980s and later in UK have initiated a wide discussion about the industrial organisation of these network industries. In Europe, with the exceptions of the Scandinavian countries and the UK, the liberalisation process has been mainly conducted directly or indirectly by the actions of the European Commission. Almost all liberalisation policies in Europe have been implemented under the strict scrutiny of sectoral directives defining the pace and the scope of the introduction of market mechanisms in network industries.

2.2.2.1. The European framework for electricity liberalisation

From the early stages of the integration process, energy has been an object of attention and action for the European authorities. The creation of a single energy market was and still is an economic and political objective. The real first steps towards the liberalisation of European energy markets were made in 1989, when the first 'Cardoso package' – named after the European Commissioner in charge of energy policy – prepared the adoption of two Directives (90/377/CEE and 90/547/CEE), respectively on price transparency for gas and electricity and on electricity transit on high tension grids. However, it was only in June 1996 that the European ministers passed the Directive 96/92 which gave momentum to the liberalisation of the sector in the Member States. The explicit goal of this Directive is the creation of a single energy market through the introduction of market mechanisms and the establishment of common rules for the generation, transmission, distribution and supply of electricity.

In generation, the building of new plants must henceforward be either subjected to an authorisation or a tendering procedure (Arts. 5 and 6). In either case, impartiality and openness are strictly requested. In transmission, a specific system operator must be designated for the operation, maintenance and development of the grid and its interconnection with others (Art. 7). Its activity should not bear prejudice to the grid users. Fair and transparent rules should not favour any particular producer. The same prerequisites are valid for distribution. However, public service obligations falling under specific categories (security of supply, regularity, quality and price of supply and environmental protection) can be imposed upon the distributor and the transmitter (Art. 3, 8, 10, 11).

For transparency and fair treatment reasons, integrated firms should at least introduce a separation of business and an unbundling of accounts (Art. 14). The core dispositions of the Directive are to be found in Articles 17, 18 and 19 on transmission and distribution networks access. Three access modes can be chosen from by Member States: the negotiated third party access (nTPA), the regulated third party access (rTPA) and the single buyer model. Within the nTPA, consumers and producers directly contract with one another and negotiate network access and prices with the system operator. Price publication and negotiations are subject to dispute settlement dispositions. The rTPA differs in the sense that the prices and access modalities are published by the operator and reviewed by the regulator. The single buyer model was introduced in order to satisfy France's willingness to protect its monopolistic operator 'Electricité de France'. At the last minute, however, France adopted the rTPA model.

Article 19 sets the minimal thresholds for the opening of the market and introduce, during the transition phase until complete opening, the distinction between eligible customers, who are allowed to choose their suppliers, and captive customers, who do not benefit from this option. Thirty-two percent of the market should be liberalised by 2006 and Member States should implement this Directive by 19 February 1999. The Directive should be reviewed and reformed in 2005. Since, the European Commission has declared its willingness to deepen and speed up the liberalisation of the sector. In a communication, adopted in May 2000, the Commission has presented a set of dispositions which could contribute to a quicker total opening of the European markets in 2005. Today, 65% of the European electricity market is already open. An increase of the eligibility thresholds and the reinforcement of the accounting separation of activities are expected. In order to facilitate electricity trade at the international levels, new access charging principles should be defined.

Finally, access modes regulate the natural monopoly elements of the sector. Various market mechanisms can be implemented in order to organise competition, such as energy pools, spot markets, term contracts, etc.

This Directive was the result of a long and difficult political process (see Schmidt, 1998; Héritier, 2001), characterised by tight political bargaining among the Commission, the Council of Ministers and the European Parliament. Not only were some countries opposed to this liberalisation process (e.g., France and Germany), but the electricity industry did not support, at the beginning, the opening of markets.

After a two-year process of bargaining, on 26 June 2003, a new Directive, 2003/54/EC, repealing the Directive 96/92/EC and a Regulation 1228/2003 strengthened and deepened the liberalisation process. According to this Directive, first, by 1 July 2004 all non-household customers should be eligible, while all the other customers will have to wait until 1 July 2007. Second, Member States will have to set up specific regulatory authorities in charge of regulating the TPA. This new disposition is particularly aimed at Germany. As a consequence, the German government announced the creation of a regulatory authority for July 2004.

Both the UK and Norway had reformed their electricity sectors before the EU adopted the Directive 96/02. Although, the United Kingdom is the forerunner in the implementation of liberalisation policies in network industries, electricity was among the last sectors to be restructured, liberalised or privatised under the Conservative government in the 1980s and early 90s. The Telecommunications industry was the first to be restructured (1984), followed by gas (1986), water (1989) and finally the railways (1993). Many of these reforms were undertaken before the European Union adopted its liberalisation directives. The distinguishing features of the British as opposed to the European reforms are the pace and scope of the changes. In the British case, these were planned and radical reforms, strongly oriented towards the divestiture of public ownership to the private sector and a shift from state intervention to market regulation. In the European case, reforms were and are more incremental and limited in scope. For example, privatisation policies have not been systematically implemented in European countries.

As noted by Midttun and Thomas (1997: 179), the British and Norwegian reforms illustrate two very different conceptions of a competitive market. Whereas in the former case, liberalisation has gone hand in hand with a full-scale privatisation programme, in the latter, reform set out to create the conditions for the introduction of competition without altering the structure of public ownership in the sector. For Magnus and Midttun (2000: 1), the Norwegian electricity market represents a type of 'anomaly' in terms of classical types of economic structures since it combines the capitalist model of market competition with one of public ownership traditionally associated with planned economies. In recent debates on the issue of liberalising the European electricity sector, Norway – and more generally the 'Nordic Model' including Sweden and Finland – has often been presented as a successful alternative to the strict and radical British model.

2.3. The Variety of Regulatory Frameworks and Institutions

In this section we compare the seventeen regulatory frameworks in regard to their objectives and objects. Our purpose here is to identify the goals and the focus of the liberalisation and regulation processes. We continue in the next section with the comparison and evaluation of the regulatory pressure and intensity, by studying the institutional arrangements between the main regulation actors. In a subsequent section, we will study the regulatory authorities, mainly the sectoral regulators in terms of mission, resources and powers. Finally, we seek to identify the main regulation games that have been or are being played in each regulatory framework.

2.3.1. Liberalisation goals

'This Directive establishes common rules for the generation, transmission and distribution of electricity. It lays down the rules relating to the organisation and functioning of the electricity sector, access to the market, the criteria and procedures applicable to calls for tender and the granting of authorisations and the operation of systems'. (Art. 1, Directive 96/92). Rather than the liberalisation of 15 national electricity markets, the explicit goal of this Directive was the achievement of a common European electricity market as a contribution to the constitution of the wider internal market. The existence of this common goal does not mean however that each Member State endorses it fully and exclusively. Indeed, Member States usually use their statutory autonomy during the transposition process of EU Directives into their respective legal orders to alter, sharpen or complete the general objective along the national political agenda.

Although this process is closely scrutinised by the Commission, important discrepancies and variations in terms of goals, scale, scope and objects can be observed. The electricity sector is no exception as Table 2.3 shows.

Apart from its rhetorical dimension, such comparison gives a first indication of the economic and legal tradition, the political priorities, and the political compromises from which these reforms resulted. Indeed, three 'models' of market and regulation design can be derived from this comparison.

The *market-oriented* model is characterised by the explicit reference to competition creation and efficiency as the primary objectives of the reforms. Emphasis is put on the promotion of competition for the benefit of customers in terms of price and quality. Although redistributive goals and public services objectives (PSO) are not forgotten, they constitute corrective mechanisms and as such are only secondary objectives. Regulation in this model is largely designed to create and then maintain competition in the sector in the interest of the consumer mainly through antitrust regulatory intervention. This model can be found in England & Wales, and to some extent in Germany and Luxembourg.

At the other extreme of the spectrum, the *public service-oriented* model emphasises the importance of electricity as a vital commodity that requires specific protections and safeguards in a competitive market. Generally, the concepts of 'competition' or 'market' are not used and if they are used, only on exceptional occasions. On the contrary, PSOs are very often defined and described in detail and referred to as the main goal of the regulatory reform designed for the citizen, not the consumer, as it is the case in France, Portugal and Spain.

In between these two ideal types, the *mixed model* is definitely the most popular one in European countries, although the differences between countries can be sometimes substantial. However, what characterises these countries is the explicit combination or tension between liberal measures like the creation of an efficient market and the willingness to implement large-scale redistributive and PSO objectives. Very often, both liberalisation and regulation design seek to create the condition for an efficient market taking into account the public interests. These interests can rank from ensuring the security of supply to implementing environmental measures.

2.3.2. Comprehensiveness of regulatory frameworks

Apart from these general objectives pertaining to market creation, regulatory reforms can vary as to their scale and scope. As we saw

Table 2.3. Goals of liberalisation and regulation of the European electricity sector.[1]

Country	Goals of liberalisation	Goals of regulation
England & Wales	Introduce competition in the electricity sector	Protect the interest of the consumers and promote competition (Art 13, Utility Act 2000)
Norway	'lay the basis for an efficient electricity market' (Energy Act 1990)	'control of network operations as a natural monopoly [. . .] ensure that electricity is transmitted at the right quality and price, and that the network is used and developed in a safe and rational way for society' (Energy Act 1990)
Germany	Delivery of secure, cheap and sustainable electricity and gas	Antitrust
France	–	Secure the supply on the national territory and the general interest
Switzerland	Create the condition of an electricity market	Secure the national supply of electricity and fair prices
Portugal	Provide the framework for the exercise of the activities of electric energy production, transmission and distribution	Contribute to economic and social development and to the well-being of the population
The Netherlands	Extend the possibilities for the generation, supply, import and export of electricity	Introduce new regulations relating to the production, transmission and supply of electricity, taking into consideration the importance of the reliable, sustainable and efficient supply of electricity
Italy	Liberalisation of electricity production and trading	Promotion of competition and efficiency while ensuring adequate service quality standards
Belgium	None explicit	None explicit
Spain	N/a	N/a
Luxembourg	None	Prevent market power abuse and predatory behaviour, Enforce PSOs
Ireland	None explicit	None explicit
Greece	N/a	N/a

(Continued)

Table 2.3. Continued.

Country	Goals of liberalisation	Goals of regulation
Sweden	Create competitive energy markets	Ensure competition within the electricity supply industry and create an ecologically sustainable and economically viable energy system
Finland	Ensure the preconditions for an efficiently functioning electricity market	Secure the sufficient supply of high-standard electricity at reasonable prices
Austria	Creation of an electricity market along EU dispositions	Provide cheap and quality electricity, increase the part of sustainable energy, ensure the provision of PSO (quality, security etc.)
Denmark	N/a	N/a

Sources: National legislations and official documents.
[1]In this table, we have only considered the goals that were to be found in the national legislation. We have either reproduced citations of the English translation or reformulated the objective, which were sometimes implicit. In some cases, we were unable to get an English translation of the legislation (e.g., Denmark). And in some other cases, the formulation of the legal text does not allow us to derive explicit goals (e.g., Ireland).
N/a, Not available.

above, market-oriented models, for instance, put a smaller emphasis on environmental issues or PSOs as do the public services or the mixed models. Therefore, when comparing regulatory frameworks it is interesting to observe how each country has used its statutory freedom in the transposition of an EU Directive to adapt or complete it along its own national agenda and priorities.

Table 2.4 compares countries on four issues. First, stands the institutional issue: are electricity and gas regulation integrated? In other words, is there a sectoral regulator for each sector or only one taking on both tasks? Second, the issue of the objects of regulation: what is regulated? What and how many aspects of the policy domain (electricity) are the objects of the regulatory process? Apart from the supervision of the TPA, regulation can extend to investment decisions of operators, PSOs (price, quality, universality) and environmental issues. Each of these four dimensions is ranked from highly regulated to no regulation. This evaluation is based on the appreciation of the precision of the objectives (general goals vs detailed and measurable outputs and outcomes), the type of regulatory instrument (hierarchic vs incentives based), and the nature of the control and monitoring instruments (voluntary vs compulsory and recommendation vs sanctions).

Table 2.4. Scale and scope of the regulatory framework.

Country	Integrated sectoral regulator (electricity and gas)	Scope of regulatory intervention		
		TPA[1]	Public service obligations	Environmental measures
England & Wales	Yes (OFGEM)	rTPA	Obligation to connect, standards of service performance, customers' complaint	Standard of performance on efficient use of electricity
Norway	No	rTPA	N/a	Gov. rights in hydro
Germany	No	nTPA[2]	Obligation to connect	Obligation to purchase renewable sources for grid operators
France	No	rTPA	Obligation to connect and supply, Social tariffs, Fund for PSO	No
Switzerland	No	rTPA	Obligation to connect	Public loans for stranded costs
Portugal	Yes	rTPA[2]	Obligation to connect and supply, Code of conducts and practice	Obligation to buy from renewable at governmental set prices
The Netherlands	Yes (Dte)	rTPA	Maximum prices for captive customers	Obligation to purchase green electricity, Green certificate
Italy	Yes (Autorità)	rTPA[2]	Obligation to connect, Social tariffs	Priority to renewables, Renewable production quotas, Levy on tarrifs for stranded costs
Belgium	Yes (CREG)	rTPA	Price caps for captive customers	Fund for stranded costs
Spain	Yes (CNE)	rTPA	Obligation to connect	Renewable support
Ireland	No	rTPA	N/a	N/a
Greece	No	rTPA	N/a	N/a
Sweden	Yes (OEGR)	rTPA	Obligation to connect	N/a
Finland	Yes (EMA)	rTPA	Obligation to connect	N/a
Austria	No	rTPA	Obligation to connect and supply	Priority dispatching from renewable, Stranded costs
Denmark	Yes (ESB)	rTPA	N/a	N/a

[1]This mainly includes grid management, prices and quality.
[2]These three countries also implement the single-buyer model on the local level.
Sources: Reports of the European Commission and national legislations.

The most striking feature of this comparison is definitely the almost universal diffusion of the rTPA model, although behind this quasi-unanimity, very different patterns and diverging implementations can be found in terms of regulatory process and tariffs supervision or approval, as we will see in Section 4.2. Germany is the only country that opted for the nTPA. As Eberlein (2001) shows, at least three reasons explained this German exception: the federal political structure of the country, the corporatist style of state–society relations, and the social market ideology that inspires many German institutions and policies. With this exception, the dominance of rTPA model all over Europe is striking. Various considerations could explain this. First, it is the logical model of market regulation, congruent with the economic doctrine of state intervention in case of market failure and natural monopoly elements. Secondly, it gives the state the opportunity to continue to have a word to say in the management and regulation of a sector where market imperatives and public interest have to be balanced. Finally, it is generally considered to be the most efficient model to promote competition in network industries.

The new Directive 2003/54/EC defines the universal service as 'the right to be supplied with electricity of a specified quality within the [Member States] territory at reasonable, easily and clearly comparable and transparent prices. To ensure the provision of universal service, Member States may appoint a supplier of last resort. Member States shall impose on distribution companies an obligation to connect customers to their grids under terms, conditions and tariffs set in accordance with the procedure laid down in Article 23' (Art. 3 al. 3). As Table 2.4 illustrates, access, translated into the 'obligation to connect' is the core element or the minimal standard of PSOs. Differences in terms of scope and scale do only partially depend upon the category of market and regulation design model identified in Section 3.1. It appears that the difference is not in terms of the nature of the PSO, but in terms of the perspective from which they are considered: from the citizen or the consumer point of view. If *market-oriented countries* focus the scale and scope of PSOs on this core element, quality of service considerations are not left aside and are generally dealt with by consumer protection measures. For instance, in England & Wales consumers are associated in the regulation of the sector through the institutionalisation of complaints or through the setting by the regulator of quality standards imposed on operators. On the other side of the spectrum, *public service-oriented countries* generally include social measures (social tariffs) and set up complex financial or regulatory mechanisms like funds to finance PSOs. This is the case of

France and to some extent, Portugal, where a distinction is made between the competitive electricity market and the public electricity system. In between, the mixed model countries combine customer-oriented dispositions (code of conducts) and public services measures (price caps for captive customers).

The situation is identical when it comes to environmental issues. A core element of environmental measures can be found around two dispositions, i.e., measures encouraging renewable energies (obligation to buy, price guarantees, standards and certification) and stranded cost recovery through levies, taxes or loans. If the second set of measures are transitional and aim at adapting the industry to the competitive environment, the first ones are designed to adjust energy policy instruments to the new context of competition. However, this adjustment process is rather limited and marginal considering the importance of energy policy issues, such as environmental protection, rational use of energy and security of supply. It is as if governments did consider that liberalisation and energy policies were not really linked together, both on the institutional and the substantial levels.

On the institutional level, a common model of regulation is slowly but clearly emerging, taking the form of an integrated electricity and gas regulatory framework, and resting upon the rTPA model. The picture is more uneven on the substantial level. Although, a 'universal' core definition of PSOs does exist, differences can still be important between countries focusing on consumer protection on the one hand and countries inspired by a more citizen oriented conception of public services on the other.

2.3.3. National regulatory institutions

Regulation is a multiactor process (see Fig. 1.1 of Introduction). In its wider sense, regulation is the meta-process including all the actors (governments, operators, regulatory authorities, customers, political parties, associations etc.) who interacts within the sector. In its strictest sense, it encompasses the relations among political authorities (government), regulatory authorities, appeal institutions and the industry, although in some cases, consumer and interests groups can also be included in the process. These actors form institutional arrangements determined as they are by statutory attributions (e.g., relations between the regulatory authority and the sponsoring ministry) and formalised in the law, by reputational or resource-based power distribution, and by informal relations. Table 2.5 summarises these arrangements in each country by

Table 2.5. Models of regulatory institutions in the electricity sector.

Model[1]	General features	Countries
Politico-administrative	• Concentration of regulatory functions in political hands • Existing independent regulatory institutions can act as consultative or advisory bodies • Competition regulation as a complement to sectoral regulation • Courts and judicial systems as appeal authority	Norway, Switzerland, Spain, Luxembourg, The Netherlands
Quasi-judicial	• Generally weak resources • Generally weak regulatory powers • Important role of ministries as coordinators and appeal authorities • Mainly ex post regulatory intervention • Weak or no ex ante regulatory intervention • Competition regulation is the sectoral regulation • Sectoral regulation is part of a judicial process	Germany
Independent regulatory authority	• Wide range of attributions to the independent authority • Combination of ex ante and ex post regulatory intervention • The regulatory authority generally takes decisions on its own • Coordination with competition authority • Ministries as appeal authority (1st order) • Courts and judicial systems as appeal authority (2nd order)	England & Wales, France, Italy, Portugal, Belgium, Ireland, Greece, Sweden, Finland, Austria, Denmark

[1]We consider the dominant regulatory actor, i.e., the actor responsible for the core regulatory functions (TPA and network access management).

considering three dimensions. The first dimension pertains to the relationship between the regulatory institution and the governmental authorities and identifies the legal and organisational nature of the regulatory institution (independent, administrative or judicial). The second dimension pertains to the type of decision-making process. Finally, we look at the nature of the intervention and its timing.

In the *politico-administrative model*, sectoral regulation is integrated into the governmental and political process. If an independent regulatory body exists, as is the case in Spain, this institution acts as an advisory body for the sponsoring ministry which monopolises most of the regulatory functions. In this model, the general competition regulation represents the second pillar of the sectoral regulation framework, if it is not the sole regulator, as in The Netherlands, where the sectoral regulation of electricity is just a division of the competition authority. The situation in The Netherlands is however unique, as the Ministry can give binding instructions to the competition authorities. This explains why this case is not included in the second model.

It is difficult to speak of a model since Germany is the sole example illustrating a *quasi-judicial regulatory authority*. However, in many ways, this framework can be considered as an alternative to the dominant independent regulatory authority model. At least, this is how the advocates of this model (i.e., the German government) saw it. It is generally considered that this model is more market oriented than the other models, as it relies on the dynamics of the markets on a self-regulatory and competition regulation basis. Operators and actors of the market act *'in the shadow of government intervention'* (Eberlein, 2001: 383), and regulatory intervention is only required to settle disputes or to prevent market power abuse. Regulation is therefore ex post, and competition regulation is an adequate substitute to costly and inefficient sectoral regulation.

The third and dominant *independent regulatory authority* (IRA) model can be considered as a combination of the first two others or as a mid-spectrum solution. It is however a genuine new model, whose rationale lie in the much debated and discussed issue of delegation (Thatcher and Stone Sweet, 2002). It is characterised by the nature of its relation with governmental authority (independence), its formal decisional attributions (e.g., TPA) and its ability to combine ex post and ex ante interventions. This model is definitely the most popular model in the electricity sectors. However, differences between countries can be important, as we will see in the next section.

2.3.3.1 Variety of competencies and powers of regulatory authorities

Regulatory authorities can differ in terms of nature as we saw in the previous section. In this section we compare them in terms of missions, competencies, powers, political autonomy and resources (Table 2.6).

The mission that a regulatory authority has to fulfill is often set in the law, although not always in a straightforward way. In some cases, it is very precisely specified, although most of the time it is described in very broad terms. For instance, one authority is there to promote competition and protect consumers, while another was set to regulate network access. But this wide diversity in the design of regulatory missions is often misleading. Almost all regulatory authorities share at least one regulatory function – the regulation of network access. Together with the supervision of tariffs, this task represents the core activity of the regulatory authority. On this foundation, other attributions can be delegated to the regulatory authority. In a first circle of regulatory competencies around this core, we usually find dispute settlement attributions. Then comes quasi-legislative powers, such as the setting of code of conducts or standards of performance. At the next level, only a limited number of regulatory authorities have the right to distribute licenses (Norway, Sweden, Finland, UK and Switzerland). Indeed this function is generally in the hands of the governmental authorities. Finally, all regulatory authorities act as policy advice bodies on regulatory or general issues related to energy matters for the government.

It is often considered that the more independent, the more powerful a regulatory authority. Although this affirmation is partially correct, empirical evidence shows that independent or not, regulatory authorities draw their power primarily from the formal dispositions of the law. It is possible to rank regulatory powers on a scale of five cumulative levels, from the less to the more powerful.

When the regulatory authority is deprived of one of these attributions, it is either the government, courts or the competition regulator that are given this task. The weakest regulatory authority is usually the one whose power is mainly an informational one. It acts as an information gatherer, investigator and advisor to the ministry. It is the case of the Spanish regulator. The second level is constituted of dispute settlement attributions: although it cannot act on its own initiative, the regulatory authority can intervene in the regulatory process on an ex post basis to settle dispute or act as an umpire. The third level is where the regulatory authority can make, on its own or on request, statutory and binding decisions that are considered as public intervention. The best illustration of this is the price review or

Table 2.6. Sectoral regulatory authorities.[1]

Sectoral regulator	Name	Missions	Competencies in electricity regulation	Powers	Political autonomy	Resources (2001)
England & Wales	Office of Gas and Electricity Markets (OFGEM)	Promote competition and consumer protection	Price control, Licensing, Network access enforcement, Quality standards setting, Dispute settlement	Inquiries, Financial penalties, Legal enforcement	Licensing decision by DTI	Board: 1 +(10) Staff: ~ 300 Budget: 35 m£ coming from licensee fees
Norway	Norwegian Water resources and Energy Administration (NVA)	Administering water and energy resources	Licensing, Network access regulation, Market monitoring	Legal enforcement (licenses), Recommendations	All decisions can be revised by Ministry (MPE)	N/a
Germany	Office of Cartels (Bundeskartellamt)	Enforce Cartel law	Dispute settlement and abuse of dominant power	Legal enforcement	Antitrust law	Board: –, Staff: 5–6 Budget: N/a
France	Commission de regulation de l'électricité (CRE)	Regulate network access	Access tariffs review, Dispute settlement, Advice to Minister, Grid management supervision	Legal enforcement, financial penalties	Access tariffs are set by the Ministry	Board: 6 Staff: 80 Budget: 60mFF from public budget

(Continued)

Table 2.6. Continued.

Sectoral regulator	Name	Missions	Competencies in electricity Regulation	Powers	Political autonomy	Resources (2001)
Switzerland	Office of Energy (OFEN)		Licensing, Benchmarking, Policy Advice, Codes setting	Inquiries, Sanctions	None. Disputes are dealt by a special independent commission	Board: – Staff: N/a Budget: N/a
Portugal	Energy Regulatory Authority (ERSE)	Regulation of electricity and gas sector	Transmission and distribution tariffs regulation	Secondary legislative power (Codes of conduct), Inquiries, Financial penalties, Legal enforcement	Decisions from ERSE can only be appealed before Courts	Board:3 Staff: 42 Budget: 4.5 m€ from 'use of system' tariffs
The Netherlands	Dienst Toezicht en Uitvoering Energie (DTE)	Regulation of network access	Policy advice (licenses), System tariffs approval, Codes setting, Grid management regulation	Inquiries, Sanctions	DTE can receive binding instructions from Ministry	Board:1, Staff: 33 Budget: 4 m€ from operators contributions and state budget

Italy	Autorità per l'energia electrica e il gas (AEEG)	Promotion of competition and efficiency and quality standards	Price regulation (price-cap), Enforcing quality standards, Grids management regulation, Policy advice (e.g., licensing), Dispute settlement	Inquiries, Hearings, Quality standards setting and enforcement, Financial sanctions	–	Board: 3 Staff: 63 Budget: 18 m€ from levy on operators
Belgium	Commission de regulation de l'électricité et du gaz (CREG)	–	TPA regulation, Grid management regulation, PSO regulation, Dispute settlement, Policy advice	Inquiries, Financial sanctions	CREG is responsible for the regulation of high tension grid, while regional regulation takes over regional grids.	Board: 4 Staff: 40 Budget: 9.4 m€ from levy on network users
Spain	National Energy Commission (CNE)	Regulation of Gas, Electricity and Petroleum	Advice to ministry; Dispute settlement, approve M&A	Monitoring, Expertise, Investigation	No decisional power. Propose decisions to Ministry	Board: 9 Staff: 120, Budget: 6.5 m$ coming from consumption tax

(Continued)

Table 2.6. Continued.

Sectoral regulator	Name	Missions	Competencies in electricity regulation	Powers	Political autonomy	Resources (2001)
Luxembourg	Luxembourg Regulation Institute (ILR)	Prevent market power abuse and predatory behaviour, Enforce PSOs	TPA tariffs and conditions proposition to ministry, Dispute settlement, Administration of compensation fund, Enforcement of PSOs	Inquiries, Sanctions	Most regulatory functions in the hands of the Ministry (licensing, tariffs etc.)	Board: N/a Staff: 1? Budget: coming from levy on regulated operators
Ireland	Commission for Electricity Regulation (CER)	Promote competition and protect the interests of consumers	Licensing Grids, Tariffs approval, Setting of performance standards, Dispute settlement, Policy Advice	Inquiries, Sanctions	CER can receive instruction on PSOs	Board: 1 (4?) Staff: 27, Budget: 5 m€ coming from levy on regulated operators
Greece	Regulatory Authority for Energy (RAE)	N/a	Tariffs proposition to ministry, Monitoring of law	Inquiries, Sanctions	RAE is mainly a consultative body to ministry	Board: 5, Staff: 10 Budget: 4.5m€ from energy taxes

Sweden	Office of Electricity and Gas Regulator (OEGR)	Supervision of natural monopoly elements	Licensing for network operations, Supervision of network tariffs, Data collection, Policy advice	Codes and guidelines setting, Recommandations	Licensing decisions can be appealed to ministry	Board: 1, Staff: 33 Budget: 6.5 m€ from State budget
Finland	Energy market Authority (EMA)	Monitoring of electricity markets	Licensing, Retail prices monitoring, Grid management monitoring	Investigation, Data collection, Sanctions (licence withdrawal, financial penalties)	Technical and economical regulations are fixed by the ministry	Board: N/a Staff: 15 Budget: 1.2 m€ from license fees and State budget
Austria	Electricity Control Ltd. (ECGmbH + Electricity Control Commission (ECC)	Monitor the liberalisation of electricity	Network tariffs approval, Dispute settlement, Technical standard setting	Investigation, Data collection, Binding decision, Financial sanctions	ECGmbH is a private law organisation working for the ECC, ECC is the judicial and decisional regulator	ECGmbH Board: 1, Staff: 43 Budget: 7 m€ from levy on network user ECC Board: 30 Staff: –
Denmark	Energy supervisory Board (ESB)	Monitor the liberalisation of energy markets	Tariffs supervision, Dispute settlement	ESB acts on an ex post basis	Licensing by Ministry	Board: 7, Staff: 30 Budget: 2.5 m€ from levy on consumption, Production and transmission

[1]When there is no formal sectoral regulator, we consider the actors which assume the sectoral regulatory functions (i.e., network access, technical and prices regulation).

Sources: European Commission reports, national legislations and regulatory authorities annual reports.

the price cap setting for the network access charge or final prices. The fourth level on the power scale is the ability for the regulatory authority to legislate on very specific issues, like the setting of code of conducts or performance standards setting. Finally the strongest regulatory authorities cumulate some or all the previous functions and the ability to decide and impose sanctions.

This classification is highly simplistic. It would indeed require more efforts and more space to proceed with a more refined analysis of regulatory powers. However, two conclusions can be drawn from this attempt. First, the design of regulatory authority does not follow a rational and cumulative scheme. For instance, a regulatory authority can be given sanctioning powers, but be deprived of almost all the other regulatory attributions (e.g., Greece). Second, this classification is a strictly static analysis, while power is a highly dynamic concept. Therefore, it is perfectly possible to imagine a very powerful static regulator, while being a weak actor within the regulatory process and vice versa.

The issue of independence of regulatory authorities is too complex to be dealt with here. Apart from the nomination of regulatory authority executives, which we do not deal with here, and the resource attribution, which we discuss in the next paragraph, we can make the following observations. The independence, or the 'zone of discretion' (Thatcher and Stone Sweet, 2002), of regulatory authorities can be reduced or circumvented by governmental supervision in at least three different ways. First, the sponsoring ministry can give binding instructions on specific matters, like for instance in Ireland and in The Netherlands, where the regulatory authority can be instructed respectively on PSOs and on competition issues. Second, the ministry can act as the appeal authority (Sweden). Third, the ministry can decide on its own to revise a decision from the authority (Norway). It is indeed very interesting to note that in many ways, political intervention is generally still very high in the regulatory process, although it takes different forms depending on the type of regulatory frameworks. Systems, which have a strong independent regulator, tend to be less concerned by direct political intervention, and vice versa.

The amount and the provenance of resources allocated to a regulatory authority is often an accurate indicator of both its power and independence. The higher the amount of resources, the more powerful is the regulator. In terms of political independence, the question is not the amount, but rather the origin of the resources. When the financial and/or staff resources come from the state budget and/or from the sponsoring ministry, the political dependence is higher than if the resources come from, for instance, a tax raised on consumption or on the industry itself.[2]

2.3.3.2. Intensity of regulatory powers

Regulatory pressure is not only determined by the quality and quantity of powers in the hands of the dominant regulatory authority. Rather, power is embedded in the structural arrangements implicating all the actors of the regulatory process. If we therefore want to evaluate the level of regulatory intervention (or regulatory pressure) upon the industry, we need to look at the relationships between the different actors of the regulatory process. Moreover, regulatory intervention addresses government intervention along with competition and sectoral regulatory pressure. In consequence of which the measurement of the regulatory pressure has to take into account the global regulatory intervention as a combination of these three forms of intervention. Thus, a high regulatory pressure is the consequence of important regulatory powers in the hands of the dominant regulatory authorities and of the others. To do so, we consider four main dimensions.

The first is related to the nature of the main regulatory authority in terms of power, attribution and resources as presented in the previous section. The second pertains to the structure of the institutional arrangements, the number and nature of the actors involved in the regulatory process. The third raises the question of the dominant actor, since, although there might be a sectoral regulatory authority, many regulatory functions can be concentrated in the hands of another institution (e.g., ministries), as we saw above. Finally, we consider the regulatory mode (i.e., ex ante or ex post regulatory intervention). Ex ante regulation is generally considered as a more intense regulatory intervention since it requires the activity of the regulatory authority in anticipation of the problem. This is typically the case with tariffs, which are generally approved by the responsible regulatory authority before they come into force. Ex post regulation generally covers dispute settlement activities and requires a less intense regulatory intervention.

Combined, these four dimensions allow us to evaluate the intensity of the regulatory pressure. The more powerful the dominant regulatory authority, the smaller the number of actors involved in the process and the more ex ante oriented regulatory mode, the higher

[2]The question is different then in terms of independence from the regulated industry.

the regulatory pressure. Respectively, the weaker the dominant regulatory authority, the higher the number of actors involved, and the more ex post oriented the regulatory intervention, the lower the regulatory pressure. To summarise, we refer to 'high', 'medium' and 'low' regulatory pressure.

It must be added that this evaluation is largely static and formal in nature, although in our appreciation we tried to include substantial elements and examples, which we cannot present and discuss within this chapter. Table 2.7 presents an evaluation for each regulatory framework.

We must first start with a methodological remark, which can also be extended to the discussion on regulatory authorities' powers. This analysis is a first exploratory investigation on regulatory pressure. In order to get sounder and more robust results, ranking methods based on quantitative evaluation should be elaborated. We, however, consider this first comparison as very insightful.

This being said, and as Table 2.7 shows, some regulatory systems put a heavier pressure upon the industry than others. Federal countries tend to put a lighter pressure upon the industry, due to the fragmentation or the weakness of their regulatory frameworks. Respectively, centralistic countries and/or countries where there is a strong incumbent appear to put a higher regulatory pressure. Countries which have liberalised their electricity sector for a longer period (UK, Norway) tend to put a higher pressure than newly liberalised countries. This raises the issue of the evolution of regulatory pressure over time. We develop this point in Chapter 4.

2.3.4. The interplay of regulatory issues

The opening of markets is a highly politicised issue. Political bargaining about the rationale, the objectives, the scale, the scope and the pace of liberalisation has widely affected the policy formulation process in each country. Liberalisation and regulation design are by definition the outcomes of the political policy process. Therefore, the study of the political issues or 'games' that have accompanied the design process and its aftermath provide valuable information on the characteristics of a specific regulatory framework. These political issues around national regulatory frameworks design can be classified into two categories, namely institutional policy issues (i.e., issues related to the design of regulatory institutions) and substantial policy issues (i.e., issues related to the design of regulatory instruments and public policies affected by liberalisation). In the next paragraphs, we present the main issues that have been the object of

Table 2.7. Intensity of regulatory powers.

Country	Actors involved in the regulatory process	Dominant regulatory actor	Nature of regulatory intervention	Intensity of regulatory pressure
England & Wales	• Department of Trade and Industry (DTI) • Competition Commission • OFGEM	Office of Gas and Electricity Markets (OFGEM)	Ex ante	High
Norway	• Office of Energy and Water (NVE) • Ministry of Oil and Energy	Office of Energy and Water (NVE)	Ex ante	High
Germany	• Ministry of the Economy • Ministry of Energy • Regional Economy and Energy ministries • Office of Cartels (Bundeskartellamt)	Bundeskartellamt	Ex post	Low
France	• Ministry of Economy and Finance • Ministry of Environment and Energy • Competition Commission • Commission de regulation de l'électricité (CRE)	Commission de régulation de l'électricité (CRE)	Ex ante	High
Switzerland	• Office of Energy (Department of Energy, Transports, Environment and Communications) • Dispute Settlement Commission • Cantonal Authorities • Local Authorities	Office of Energy	Ex post	Low
Portugal	• Director General of Industry • Ministry of Economy • Regulator of Energy Markets (ERSE) • Competition authority	Regulator of Energy Markets (ERSE)	Ex ante	Medium
The Netherlands	• Ministry of Economic Affairs	DTE	Ex ante	Medium

(*Continued*)

Table 2.7. Continued.

Country	Actors involved in the regulatory process	Dominant regulatory actor	Nature of regulatory intervention	Intensity of regulatory pressure
	• 'Dienst Toezicht en Uitvoering Energie' (DTE) inside the Competition Authority			
Italy	• Ministry of Industry • Authority for Electricity and Gas Energies (AEEG)	AEEG	Ex ante	High
Belgium	• Ministry of Economic Affairs • Commission de regulation de l'él. et du gaz (CREG) • Comité de contrôle de l'él. et du gaz • Regional Regulatory Commissions (3)	CREG	Ex ante	Medium
Spain	• Ministry of the Economy • National Energy Commission	Ministry of the Economy	Ex ante	Medium
Ireland	• Ministry of Public Enterprises • Commission for Electricity Regulation (CER)	CER	Ex ante	High
Greece	• Ministry of Development • Regulatory Authority for Energy (RAE)	RAE	Ex ante	Medium
Sweden	• Ministry of Industry, Employment and Communication • Office of the Electricity and Gas Regulator (OEGR) within the Swedish Energy Agency • Swedish Competition Authority • National Electrical Safety Board	OEGR	Ex post	Medium
Finland	• Ministry of Trade and Industry (MTI) • Energy Market Authority (EMA)	EMA	Ex post	Medium
Austria	• Ministry of Economy and Labour • Electricity Control Ltd • Electricity Control Commission • Regional authorities	ECGmbH + ECC	Ex ante	Medium
Denmark	• Ministry of the Environment and Energy • Energy Supervisory Board (ESB) • Competition authority	ESB	Ex post	Medium

the political debate and do not pretend to draw an extensive picture of the issues raised in each national context.

As far as the institutional policy issues are concerned, four questions have been widely addressed during the design process, namely the pace and scope of the liberalisation process, the attributions and powers of the sectoral regulator, the privatisation or incorporation of State-Owned Enterprises (SOEs), and the design or constant tuning of regulatory instruments. Although this last issue has been dealt with in almost every country, in some specific cases the discussion on the pace and scope of the liberalisation process has constituted the main or one of the dominant issues of the design process. In France and Switzerland for instance, the political debates were largely focused on the political desirability and feasibility of liberalisation. The actually limited scope of the French market opening (37% in 2003) is the logical result of the political consensus that arose between left-wing and major right-wing parties, not to weaken the national champion Electricité de France (EDF) and not to threaten the 'service public'. In Switzerland, the debates were focused on both the rationale for a market opening and its pace. Political fights between parties, unions and the industry were concentrated on the way to protect hydro-electricity in an open market. By opening the market too quickly (and quicker than the minimum requirement of the EU), one risks to weaken the hydro-power plants which require long-term investments. The result of this political debate is that the Swiss population rejected the liberalisation law in a public referendum in September 2002.

There appears to be three ways of linking liberalisation and privatisation. The first way is by considering privatisation as a pre-requisite for liberalisation. This is what was done in the UK, where the public monopolist was restructured, and then partially privatised while liberalisation took place. The second way is by considering the privatisation as an improvement or the logical consequence of the liberalisation process. This is to some extent what happened in Italy, Spain and Portugal. Finally, the third way is by considering the privatisation of public operators as a possible outcome of liberalisation but neither as a pre-requisite, nor as a logical consequence. This is what has happened or is happening in countries like Norway, Austria, Germany, and Switzerland. In these countries, privatisation is the last resort. The incorporation of public operator and their transformation into public joint companies have been considered as acceptable alternatives to the selling of public assets. In each case, the design of the regulatory framework was affected. One does not regulate a private sector, a mixed or an entirely public sector with the same instruments and within the same process.

2.4. The Variety of Regulatory Styles

The above sections have shown how the implementation of a single directive can produce quite different outcomes in terms of regulatory goals, institutions and systems on the national level. Is it nevertheless possible to draw a coherent typology of regulatory systems? In other words, is it possible to observe certain patterns in regulatory frameworks?

We focused our comparison on two critical dimensions of regulation. The first pertains to the regulatory pressure measured in terms of coherence, integration of regulatory mechanisms, scale and scope of the regulatory process and power of regulatory institutions: in other words, it is concerned with the degree of comprehensiveness of the regulatory framework. Indeed, national cases can clearly be placed on a continuum from weak regulatory and fragmented frameworks to strong and integrated regulatory ones. The second dimension pertains to the structure of the electricity industries. Electricity industries vary from cases where one (France) or few highly cartelised actors dominate (e.g., Germany), to cases where the industry counts a high number of operators and is heavily fragmented (e.g., Switzerland).

2.4.1 Comprehensiveness of the regulatory framework

Measuring the overall comprehensiveness of a regulatory framework helps to gather under a single concept a wide variety of dimensions and elements about liberalisation and regulation. It also contributes to assess the degree to which a regulatory framework covers the basic functions necessary to make competition a reality, simultaneously taking into account the policy and public services objectives. The 1996 Directive set the momentum for reforms and fixed the minimal requirements for liberalisation, while leaving a large room to manoeuvre for the Member States. Some designed full-fledged competition-oriented regulatory frameworks, while others chose to follow a more prudent and less ambitious path.

In terms of regulatory functions, the issue is not much whether all regulatory functions are indeed assumed in each regulatory framework – they all are in one form or another – but rather to what degree it has been developed, institutionalised and given credibility. *Sectoral (or monopoly and technical) regulation* is the critical issue, around which the regulatory process is organised. The almost universal diffusion of the rTPA model, although a good indicator of convergence, does not imply that regulatory frameworks give the same attention and care to this function. Again, some frameworks have designed, sophisticated instruments, institutions and mechanisms (UK, Norway, France, Portugal, Austria), while others have chosen simpler and lighter solutions (Germany, Switzerland). Similarly, *competition and antitrust regulation* are given more importance in specific frameworks for different reasons. Liberal systems (UK, The Netherlands, Ireland) have designed regulatory frameworks where sector specific and competition regulation are deeply intertwined, while continental and Nordic countries competition regulation is as separated as a secondary regulatory path. Germany is the only case where competition-regulation is a surrogate for sector-specific regulation. As far as *public services obligations* and *public policies* are concerned, differences can be substantial between liberal and public services-oriented systems, although in general terms the minimal requirement of universal services are to be found everywhere together with incremental adaptation of policy instruments.

In brief, although one cannot say that some regulatory framework suffers from a lack of comprehensiveness, it is possible to distinguish between more and less comprehensive ones. UK, France, Italy, Ireland, The Netherlands and Norway are countries where regulatory frameworks are more developed. On the other hand, Germany and Switzerland are characterised by light and sometimes superficial regulatory frameworks.

2.4.2. The combinations of industrial organisation and regulation modes

One does not regulate in the same way a market where the historical incumbent is still largely dominant and a fragmented market where entry is easy and the number of competitors high. Regulatory processes, instruments and institutions are generally designed according to the structure of the industry. Asymmetric regulation is more likely in quasi-monopolistic and oligopolistic markets than in competitive ones. In the electricity sector differences can be important.

The combination of these two dimensions (the industry structure and the regulation structure) structures the following typology in four regulatory systems as shown in Table 2.8.

Regulated markets are clearly the optimal situation of a market economy: competition is effective, due to the presence of a sufficient number of operators and the absence of abusive market power. Such a setting is generally the outcome of network industries where natural

Table 2.8. Combination of industrial organisation and regulation modes.

	Integrated industry	Fragmented industry
Integrated/Robust regulatory framework	*Regulated Oligopolies* • Strongly integrated or organised operators/ dominant incumbent • Strong and powerful regulators • Conflictual regulation France, Spain, Italy, Portugal, *Belgium, Luxembourg, Greece*	*Regulated Markets* • High number of operators/weakly integrated or organised operators • No abusive market power • Strong and powerful regulators • Coherent and stable regulatory framework England & Wales, The Netherlands, Norway, Sweden, Denmark, *Ireland, Finland, Austria*
Fragmented/Weak regulatory framework	*Self-Regulation* • Strongly integrated or organised operators/ dominant incumbent • Weakly developed or fragmented regulatory institutions • Incoherent regulatory framework • High risk of regulatory capture Germany	*Co-operative Regulation* • High number of operators/weakly integrated or organized operators • Weakly developed regulatory institutions • Incoherent and/or unstable regulatory framework • Regulatory capture possible Switzerland

Note: Countries in italic are debatable.

monopoly elements and residual market failures (Helm, 2002) remain, along with a strong and integrated regulatory frameworks characterised by a coherent and efficient distribution of regulatory functions among actors.

Regulated oligopolies differ from the previous model by the fact that although there is an integrated and robust regulatory framework, the structure of the industry is oligopolistic or quasi-monopolistic. In such situations, market concentration or the presence of a powerful incumbent creates a conflictual regulatory process, primarily focused on the confrontation of regulators and the regulated incumbent.

Self-regulation is the typical outcome of both the absence or the weakness of a regulatory framework and the presence of a dominant incumbent or group of strongly integrated operators. Regulatory capture is common in such contexts and regulatory functions are often concentrated in single hands, usually the industry itself.

Co-operative regulation emerges when both the regulatory framework and the industry are weak and fragmented. The regulatory institutions face numerous and unorganised operators, which result in unstable processes and regulatory outcomes due to the fact that negotiation and cooperation is necessary at many stages of the regulatory process.

Weak and fragmented regulatory frameworks appear to be typical of federalist countries, such as Switzerland, Belgium and Germany. The existence of multiple levels of regulation (federal, regional and local) complicates the overall regulatory process, multiplies the number of actors, and thus dilutes the responsibilities and increases transaction costs. Germany and Switzerland present here some common features. Both countries do not have strong federal regulatory authorities, except for a small division in the Cartel authority in Germany and a weak dispute settlement authority in Switzerland. Regulatory power is therefore largely divested to regional authorities. Belgium is quite different. There is a federal regulatory institution, parallel to three regional authorities. These institutional contexts result in the constitution of highly complex regulatory frameworks that largely weaken the coherence of the regulatory process. However, these three countries differ on the structure of their electricity industry, although in all cases concentration processes are occurring. Belgium appears to have the least fragmented industry structure, since Tractebel is dominating the sector. The German industry is more fragmented. However, contrary to Switzerland, the industry is highly organised. At each level of the electrical value chain, there are associations of operators ('Verbund'), which clearly act as cartels. Moreover, the emergence of giant companies such as RWE and E.ON clearly indicate the tendency towards both cartellisation of the sector and the development of dominant operators in the German system. Switzerland is obviously the paradigm of a country where the industry is highly fragmented (more than 1200 operators) and weakly organised. Although the sector is dominated by six producers and grid managers, they are not as well organised as in Germany, where many regulatory issues are directly dealt and managed by the operators themselves in old formal structures. In this sense, Germany would fit in the category of *the self-regulation* system. It is indeed the only case where the nTPA rules have been set up and implemented by the industry itself, through the constitution of a cartel agreement called 'Verbändevereinbarung'. Switzerland, and to some extent Belgium, are good examples of *co-operative regulation*, where both sides are weak and forced to cooperate. The case of Austria is interesting. In order to design a stronger regulatory framework, the federal government had to alter the federalist tradition and create a single federal regulatory authority. Austria therefore can no longer be categorised as a weak regulatory framework.

Logically, strong and integrated regulatory frameworks are found in more centralistic states, although important differences can be found between countries in this category. As shown above, the evaluation of regulatory powers is a difficult and delicate task. Indeed, highly independent regulators can have limited attributions and vice versa. Nevertheless, one common feature of all these cases is the existence of one integrated and coherent regulatory framework. Again, the regulatory styles differ mainly because of the industry structure and the level of effective competition that can be observed. Latin countries are characterised by the presence of one dominant incumbent (France) or of a limited number of actors that tend to develop concentration strategies, thus limiting the degree of competition (Spain, Italy, Portugal). As the chapter on markets shows, Spain and Portugal are moreover affected by the fact that they are mainly regional markets, which are furthermore poorly connected to the rest of the continental grid. Evidence shows, especially in the case of France, how regulation is a highly conflictual and asymmetric practice, mainly focused on the control and supervision of the incumbent, which is the characteristic of *regulated oligopoly* systems. On the other hand, Northern European countries (England & Wales, The Netherlands and Norway) perform better on the competition scale, at least in terms of market structure. The British and the Norwegian cases are particular, in the sense that liberalisation is already an 'old' phenomenon. Although both countries had very different starting points (integrated industry in Britain, fragmented industry in Norway), the industry has had the time to restructure under the active supervision of regulatory authorities acting in the perspective to create and maintain competition. The Dutch case is more puzzling, since regulation is mainly performed by the competition authority. All three cases would therefore fit into the *regulated markets* category. They present coherent regulatory frameworks in which a stable equilibrium appears to have been found between the structure and the dynamic of the industry and the institutional regulatory framework both in terms of rules and institutions. It is interesting to note that although Norway stands among the last countries – with Germany and Switzerland – not having set up an IRA, it is in the same category as England & Wales, which are very often considered as the models of an arm's length regulatory system.

2.5. The Drivers of the National Regulation Dynamics

Regulation is a process and as such its structures, goals, actors and games evolve over time. In this last section of this chapter, we would like to point out some observations and theoretical arguments about regulatory change after some empirical remarks.

Differences in dynamics

Political factors have a major influence on the regulation styles and regulatory frameworks and their paths of change. The political system variable (i.e., centralistic or federalist) is definitely the heaviest driver of the regulation design process. Federalist countries tend to have weaker regulatory frameworks and more decentralised and fragmented industries, while in centralistic countries regulatory frameworks and the industry prior to the liberalisation are more integrated. From this observation, two 'golden rules' can be put forward: (1) the more centralistic the political structure, the more robust the regulatory framework, and (2) the more integrated the industry before the liberalisation process, the higher the regulatory pressure. Obviously, these two factors are not totally independent from one another.

The history of the electricity industry has shown how the political structure has widely affected its development. Although electricity originated at the local level, large scale nationalisation policies were implemented in centralistic countries (UK, Italy, France, Belgium, etc.). In this respect, countries where both sectoral and competition regulation are highly developed are typical of situations where the pre-liberalisation stage was characterised by the existence of a unique or heavy dominant monopolist. In both the UK and France, the set up of a strong and coherent regulatory framework was required to make competition possible. On the other hand, countries where the industry was already quite fragmented before liberalisation (Germany, Switzerland), a high regulatory pressure was not perceived as necessary. This divergence refers to the general discussion on competition creation versus competition conservation. In contexts where competition can only be achieved through the weakening of the historical incumbent, regulatory intervention in the first stage is quasi-exclusively focused on this task. Largely asymmetrical, this process aims at creating competition through the unbundling of the incumbent and a constant scrutiny from the regulatory authority. In countries where the industry is more fragmented, such regulatory intervention is not necessary and is more immediately focused on the conservation of competition.

Existing industrial policies also play an important role in the design of regulatory frameworks. France and Germany are good illustrations. Electricité de France (EDF) has always benefited from a strong state support. Often considered to be 'a State within the State', EDF has been put at the centre of the French energy (nuclear) policies and to become the 'national champion' to defend and promote. It was backed in this mission by the French 'service public' doctrine at the root of the political pressure from unions, which explains the limited enthusiasm of the French government for a large-scale liberalisation of the sector. This also explains that critical regulatory decisions like grid tariffs are still a ministerial decision taken on the recommendation of the regulator (CRE). The weakness of the German regulatory framework can be explained by the tradition of cartellistic institutional arrangements in many industrial sectors (Eberlein, 2001). In the German electricity sector, contractual or informal arrangements between the operators at all levels under the patronage of professional organisations have always complemented legal dispositions, for instance on issue of tariffs, territorial demarcation, and concession agreements. This quasi-self regulation situation has always been tolerated, if not actively supported by the governmental authorities. It is therefore no surprise that the nTPA model was chosen in Germany. Similar conditions can be found in Switzerland, although not to the same extent, where federal intervention in the regulation of competition has been considered for a very long time as damaging for the economic well-being of the country. To recall, Switzerland was among the last countries to enact antitrust legislations in Europe.

Changes in the second-order regulation

Former public policies can also affect the regulatory framework design. For instance, when the British conservative government initiated the electricity reform, the future of the coal industry was not solved. A brutal liberalisation of the electricity sector in which coal still represented at that time an important fuel, would have hastened the fall of the coal industry. In order to soften the transition, the government forced the generating industries to sign three years contracts with British Coal. When the coal market collapsed in 1992, the government and the regulatory authorities forced the distributors to sign five years coal-backed contracts with the generators and authorised them to pass these extra costs entirely to the captive customers.

These examples suggest the importance of substantial public policies. Existing environmental and energy policies do considerably affect the design of both liberalisation and regulation policies, with respect to institutions, instruments and goals. The illustration of the importance of substantial policies on the design and implementation of institutional policies cannot be developed within these pages. However, some general comments and observations can be put forward here. Electricity represents in Europe a fraction of the total energy consumption. The challenge for policy makers in this situation is to adapt and/or (re)design existing policy instruments or objectives of energy policies according to the new competitive environment in the electricity industry. The question is: can energy policies seeking to rationalise electricity consumption or to promote 'green' electricity by design be implemented and monitored in the same way in monopolistic and in competitive environments? This question raises two issues.

First, how will the existing policies evolve in this new context? Second, what is an adequate energy policy in competitive environments? The first question pertains to the weight of existing mechanisms, policy arrangements and instruments on the transformation process of both policy design and regulatory design and introduces the issue of path dependency. The second question addresses the issue of the new policy instruments adapted to market contexts. A rapid empirical observation indicates interesting but paradoxical facts. On the one hand, it is evident that existing energy policies have affected the way regulation has been designed, as the UK example shows. But on the other, both existing policies and policy instruments have not been so much affected by the regulatory reform. It is as if policy tools designed in a monopolistic context can perform well in competitive environments. This observation is puzzling. In particular, because policy analysis literature has shown how the policy context and the policy problem can determine the policy instruments' choice and design (Varone, 1998). In this sense, the market would require instruments adapted to the specific policy problem, the behaviour of actors (consumers, producers, traders etc.) and the regulatory mechanisms. What we observe today are incremental and peripheral adaptations of existing instruments (taxes) and some innovations in terms of new policy instruments (green labels, quotas and green certificates markets, cap and trade on emissions). Some countries have however introduced taxes on consumption or on grid management for various reasons, such as stranded costs, encouragement of renewable energy sources, etc. Green labels and other instruments have been widely introduced, but such measures are often the initiative of operators themselves and the regulators.

Institutional policies generally have long-term impacts. The inertia of the political and administrative processes is only slowly altered by institutional changes, although we can reasonably expect that the setting up of new actors such as independent regulatory authorities can introduce a considerable amount of instability, novelty and a new power distribution in the policy and regulatory arenas (Genoud, 2002). It can be expected, for instance, that market-driven policy instruments might be designed and introduced by regulators, based on the assessment and evaluation of market failures.[3]

Regulatory learning effects

When UK implemented its first privatisation and liberalisation in the 1980s, large academic and political debates took place as to how to best regulate these newly privatised and liberalised sectors. Out of these discussions, came the famous British RPI-X model of price cap regulation,[4] designed in opposition to the American rate of return system. In the electricity sector, the ability of the system and the regulatory authority to learn from experience has been assessed as being a critical condition of efficient and coherent regulatory framework. Different substantial issues have been raised and more or less efficiently dealt with in different countries. One was the

[3]In Chapter 4, we discuss the hypothesis on the probability and possibilities of change in energy policies.

[4]For a retrospect analysis of the Littlechild report that introduced this formula, see Bartel, I. (2003). *The UK Model of Utility Regulation*, University of Bath, Bath, 234p.

competition mechanisms. In the Nordic systems and in the UK system, the pool system has quickly become a huge problem and a hot political issue. Another issue, whose importance is growing by the day, is how to find adequate and efficient regulatory instruments in order to guarantee security of supply in the long term. Norway, Sweden, Finland, and more recently the UK with the fall of British Nuclear have been confronted with this delicate issue, which is at the crossroads between a pure regulatory problem and an energy policy.

As far as the substantial issues are concerned, and as aforementioned, security of supply is a critical issue for the regulatory frameworks. It is also a concern for energy policies. In Sweden, the decommissioning of production plants due to falling prices and market pressure have now raised serious problems of power capacity. More generally, due to the pressure on surplus capacities, some countries are already faced with power supply problems and are more sensitive to unexpected events such as very cold winters, dry seasons or hot summers. Investment in capacity is both a regulatory (what instruments?) and an energy policy issue (what goals, programmes, and resources?). Together with other substantial issues like stranded costs (public loans, taxes) and environmental protection (taxes, priority to renewables), we can only observe limited and incremental changes in policy goals and policy instruments after the liberalisation, although problems are only starting to arise.

Market forces influence

Could countries move from one regulatory system to another? What drivers might contribute to making one country move from one position – for instance from the regulated oligopoly system – to another – for instance to the regulated market system? The answers we provide are largely exploratory.

After implementation two major forces drive the regulatory process. The first is a *market pressure*. Markets, as the chapters on markets and business strategies show, lead to concentration and market power. Through mergers and acquisitions or through cartel agreements, market players although advocating more competition, seek in fact oligopoly, monopoly, or at least market power. This market dynamic, if strong enough and/or not impeded, might affect the regulatory system and make it shift from a regulated market system to an oligopoly system, respectively from a co-operative regulation system to a self-regulation system. There is a second opposite dynamic, resulting from *regulation pressure*. One of the many goals of the regulation is to create and/or to maintain competition in a given sector. Through competition or sectoral regulation, anticompetitive behaviours are fought. The crucial question is now, which dynamic will prevail over time – market or regulation. It is too early to comment, but some substantial hypotheses can nevertheless already be formulated.

First, although we seem to observe a tendency towards the increase of regulatory pressure over time, market dynamics towards concentration and market power are already solidly engaged. Second and consequently, the probability that market pressure will prevail is definitely higher in weak regulatory systems (self and co-operative systems), and this for obvious reasons – regulatory intervention can be slow, inadequate and vain, and might most of the time be late. We might therefore very likely observe in the future shifts from the co-operative to the self-regulation systems. Third, we must take into account the pressure towards greater regulatory harmonisation and a better articulation between national and supra-national regulation. Indeed, the constitution of big market players on the European market (E.ON, RWE, EDF) constitutes a push to scale up regulation.[5] Therefore, the prevalence of one of the two dynamics is not just a national issue. It is indeed a European one.

Is it indeed possible to shift from a weak to a strong regulatory framework and vice versa? This issue deals with institutional design, where variables and factors are more static. This is not so much a question of which dynamic will prevail, than a question of institutional design dealing with politics and heavy political variables such as the state structure, the legal tradition, the governmental system, etc. It is therefore reasonable to think that changes at this level are less likely to happen, given that the inertia and the weight of political and structural variables are so important. They are more likely to happen as the result of a conflict between market and regulation pressures, with a considerable advantage for the former.

2.6. Conclusion

The comparison of seventeen regulatory frameworks in a single chapter is a difficult and challenging task. Descriptions can only be superficial and often simplistic, while analysis is limited to striking features and specificities. The evaluation and assessment of regulatory frameworks of network industries from an institutional and political

[5]We discuss this issue in Chapter 4.

perspective suffer from at least two weaknesses. First, coherent descriptive and analytical instruments are still missing. Economic approaches overshadowing institutional and policy analysis perspectives still largely dominate the study of network industries' regulatory processes. Second, solid empirical evidence is still lacking. Many researches are mainly preoccupied with the market design and competition regulation, but the overall picture integrating all the relevant actors of an industry is still absent. In this chapter and the next chapter on gas regulation, we contribute to the remedy of these failures. By focusing on institutions (regulators), issues (regulation games) and frameworks (regulation styles) we have set a coherent analytical platform along which countries' cases can be compared. We address the issue of regulatory framework change and of convergence in Chapter 4.

References

Eberlein, B. (2001). To regulate or not to regulate electricity: Explaining the German *Sonderweg* in the EU context. *Journal of Network Industries* 2001(2): 353–384.

Genoud, C. (2002). Independent regulatory agencies: Between institutional games and organisational dynamics. *Swiss Political Science Review* 8(1): 117–125.

Glachant, J.M. and Finon, D. (2000). Why do the European Union's electricity continue to differ? An institutional analysis. In C. Menard, Ed., *Institutions, Contracts and Organisation. Perspectives form New Institutional Economics*. Edward Elgar, Cheltenham, pp. 313–334.

Helm, D. (2001). The assessment: European network competition, interconnection, and regulation. *Oxford Review of Economic Policy* 17(3): 297–312.

Héritier, A. (2001). Market integration and social cohesion: the politics of public services in European regulation. *Journal of European Public Policy* 8(5): 825–852.

IEA (1999). *Electricity Market Reform. An IEA Handbook*. OECD, Paris.

IEA (2001a). *Competition in Electricity Markets*. OECD, Paris.

IEA (2001b). *Regulatory Institutions in Liberalised Electricity Markets*. OECD, Paris.

Schmidt, S.K. (1998). Commission activism: Subsuming telecommunications and electricity under European competition law. *Journal of European Public Policy* 5(1): 169–184.

Magnus, E. and Midttun, A. (2000). The Norwegian model. In E. Magnus and A. Midttun, Eds., *Electricity Market Reform in Norway*. Macmillan Press, London, pp. 1–24.

Midttun, A. and Thomas, S. (1997). Theoretical ambiguity and the weight of historical heritage: A comparative study of the British and Norwegian electricity liberalisation. *Energy Policy* 26(3): 179–197.

Stoffaes, C. (1994). *Entre Monopole et Concurrence. La regulation de l'énergie en perspective historique*. Editions P.A.U, Paris.

Thatcher, M. and Stone Sweet, A. (2002). Theory and practice of delegation to non-majoritarian institutions. *West European Politics* 25(1): 1–22.

Varone, F. (1998). *Le choix des instruments de politiques publiques*. Haupt, Bern.

Chapter 3

Politics and Regulation of Gas in Europe

MAARTEN J. ARENTSEN

3.1. Introduction

'Today, more than 60% of EU gas consumption crosses at least one border on its way to the final consumer. This is a logical consequence of the uneven distribution of gas resources in and around Europe and has meant that gas has always had to be carried through transmission systems over long distances from the source to the consumer and there is less freedom of choice regarding transportation route and distance for gas than there is in the case of electricity'.[1]

The extensive cross-border flow of gas in Europe suggests that establishing a single European gas market would be a relatively easy task compared to the establishment of a single electricity market where cross-border flows have been much more restricted (only 8%) in the pre-liberalisation era. However, it seems that the scale of gas supply across Europe acts counterproductive to the establishment of a single European gas market, due to technical, political and institutional restrictions. Since the early 1960s gas supplies developed within specific technical and institutional boundaries of high pressure pipelines directly connecting the gas field with the load centre on the base of long-term contracts ascertaining gas supply for twenty years and more. The settlement of the extensive number of long-term contracts in European gas supply (about 95% of current gas demand) is but one of the regulatory challenges of establishing an internal gas

[1]A Long-term Vision of a Fully Operational Single Market for Gas in Europe, a (Draft) Strategy Paper, prepared by the Joint Working Group of the European Gas Regulatory Forum for the 4th Meeting of the Madrid Forum on 2–3 July 2001, p. 8.

market in Europe. Natural gas appeals to national sentiment and vested national interest, strong ingredients for passionate European debates on gas market harmonisation.

Despite inert political decision making the EU did agree on the gas Directive (98/30/EC) in 1998, holding initial rules for gas market harmonisation. The Directive, only providing the basics of gas market liberalisation, is but the first step in the process to establish an internal market for gas. As this chapter will show, the establishment of a single European gas market is facing regulatory problems. Settling these problems will not be easy given that the European gas market '(...) has largely been compartmentalised into 15 separate national markets fragmented into even more regional markets'. Thus far, these markets are only technically connected through pipelines, but are still strongly separated institutionally and economically. The route leading to the single European gas market is still full of technical, economical and institutional barriers and pitfalls and their settlement will ask high quality regulatory skills of all the regulatory bodies involved.

This chapter analyses the process of regulatory change in European gas in a similar way as the previous chapter did for the regulatory change in electricity. The chapter only covers the first wave of change initiated by the first – 1998 – gas Directive. Section 3.2 briefly addresses some background issues of the reform processes – the past regulation and industrial organisation and the European liberalisation process. Sections 3.3 and 3.4 assess the state of the art in regulatory reform from a legal and empirical point of view. Section 3.5 explains national dynamics in regulatory reform and Section 3.6 analyses the dynamics at the EU level in terms of convergence and divergence. Finally Section 3.7 draws some conclusions.

3.2. The Background of the Regulatory Challenge

The establishment of the single European gas market was one of the final topics of the EU agenda on the completion of the internal market. Gas market harmonisation was debated in the slipstream of the debate on the harmonisation of European electricity markets. The gas debate was long and passionate and the resulting gas market Directive has been recognised as a weak compromise full of unsolved issues (Hancher, 2003). The Directive was, however, a first and necessary step in the process to harmonise the internal gas market, a step that could only be taken after many years of inert political debate.

The hesitation to harmonise the internal gas market finds a strong explanation in history (see for instance Stern, 1998), and therefore, it is relevant to briefly discuss some of the regulatory controversies of the past. Despite the short time span of only four decades of gas market development, natural gas concurred a very strong position in European energy supply (see also the chapter on gas markets). This strong position is not only reflected in the share of natural gas in national energy balances today, but more importantly, in the nationally focused industrial and institutional organisation underlying the rise and settlement of natural gas in Europe.

3.2.1. Past regulations and organisations in upstream and midstream gas industries

From an institutional perspective natural gas is a rather complex energy resource that gave way to complicated industrial and institutional organisations everywhere in Europe. A major reason for this complicated institutional framing is the commercial and national interests at stake for both gas producing and gas consuming countries. Gas resources are valuable resources for the few countries possessing them. In Europe only Norway, United Kingdom, The Netherlands and Denmark, are gifted with substantial gas reserves voluminous enough for combined domestic consumption and export or only export in case of Norway. It took tremendous investments in pipeline infrastructures to connect the few European gas field regions with European load centres. These investments gave way to specific transactional relations between suppliers and demanders of the gas, being long-term contracts to cover the risks of the asset-specific infrastructure investments.

Except for Germany, countries erected national organisations and put them in charge with the central co-ordination of domestic supply and demand. These 'national champions' developed a central co-ordinating position as national system operator of the high-pressure transmission pipelines of the country. In the gas producing countries they co-ordinated the production of gas for both domestic consumption and exports. In many countries the state developed a strong financial involvement in these co-ordinating companies, or became partial or full owner of them. In this way each European country ascertained its national gas interests. The national companies also became the major actors in the European co-ordination and exchange of supply and demand of natural gas. Within the national frameworks these companies made the export and import agreements and the investments in the infrastructures needed to bring the gas from the few production regions to the European load centres.

This mixed politicocommercial model became the standard model for exploitation of the European gas fields in Norway,

United Kingdom and The Netherlands. In each of these countries the state took a strong position in the production of natural gas. A position ranging from participation in production companies to state-dominated ownership. The public-private model in production was to balance the state's financial interests and the commercial interests of private gas producing companies. Countries settled the politico-commercial interests in complex arrangements between the state and private companies and developed national gas policies for depletion of gas fields and for securing gas supply.

Downstream the institutional model developed differently. Initially, natural gas was supplied and distributed by local and regional city gas networks. From there, the distribution network extended and developed, connecting more and more people and industries to the European gas fields. State-owned companies distributed the natural gas as a public utility to households and industry. Natural gas was considered a public utility and regulated as a monopoly. In the 1990s, in the overture to liberalisation, gas distribution became more and more integrated in the service deliverance of energy companies and the multiutility companies that developed all over Europe at that time. Gas consumption in Europe steadily increased, pushing natural gas as the leading energy carrier in the European energy balance. Europe developed new connections with gas fields outside the European continent (Russia, Algeria and oversee LNG) to satisfy increasing gas demand.

In this way the production and consumption of natural gas institutionalised within a national setting, irrespective of the cross-border gas supply from the few production locations to the many European load centres. The EU debate on gas market harmonisation made it clear that the strict national focus that had been dominating gas production and consumption for some four decades could not continue under liberalisation. Moreover, liberalisation would relieve national boundaries, which had been protected so effectively for so many years. This proved to be a rather unattractive perspective for both national governments and national gas industries. Gas market liberalisation therefore, for long faced strong inertia in decision making and strong resistance in all the Member States. Whereas the debate on electricity market harmonisation made progress, *'(B)y contrast, many of the established actors in European gas industries still regarded the introduction of liberalization as the equivalent of the end of civilization'* (Stern, 1998 p. 91).

3.2.2. European liberalisation process

In this strong national interest-focused environment it was hard to find any reasonable compromise on gas market harmonisation. Each country had its arguments to resist gas market harmonisation and liberalisation. Apparently, all countries put themselves in a position to loosen from harmonisation and therefore, nobody was willing to make any move in the harmonisation debate. National governments and national gas industries participated as strong national coalition in the EU debate, taking every opportunity to debate against harmonisation and liberalisation. For the EU Commission it was difficult to move the national interest positions and it took several EU Presidencies to solve the clash of interests in an acceptable formulation of an EU Directive on gas market harmonisation. The compromise was found in a rather restricted EU Directive, defensively arranging some of the basic requirements of liberalisation, leaving vital conditions for competition untouched for the moment. In a way the gas Directive copied the electricity Directive, but in a rather stringent way.

Opponents and advocates agreed that the Directive left many issues and problems unsolved and passed hot topics to national political decision-making. Even on the issue of regulation itself the Directive was restrictive, only requiring '(...) appropriate and efficient mechanisms for regulation, control and transparency so as to avoid any abuse of a dominant position, in particular to the detriment of consumers, and any predatory behaviour' (Directive article 22). The Member States were obliged to establish regulatory bodies for ascertaining adequate functioning of a competitive gas market. Furthermore, the Directive anticipated the problems and issues left untouched by the Directive, by conceiving regulation predominantly as a function of dispute and conflict resolution, without providing any further methodological guidance in terms of tariff structures and access conditions to pipe lines. Disputes on access and gas trade, in and between countries were left to be solved by national regulators and for the moment the EU did not manage to agree any pan-European rules for that.

After the enactment of the EU gas Directive in 1998, the debate continued at the national level. The Member States had to implement the Directive in national gas market regulation before the 10th of August 2000 and despite vested interests involved, all the Member States, except Germany, France and Luxembourg, managed to integrate the requirements of the Directive in national gas legislation before the required deadline.

From 10 August 2000, the legal gas order in the EU region changed along the lines prescribed by the gas Directive. Countries enacted rules for the opening of the national gas market, the access to the pipeline infrastructure, rules for unbundling of integrated companies and PSOs. In some countries, like The Netherlands, the EU Directive induced public gas legislation for the first time in history and in some other countries, like the United Kingdom, the EU Directive only codified the already established competitive market conditions. Moreover, the Directive required the establishment of a gas market regulator and these regulatory bodies where indeed erected in almost every EU Member State, in many cases as part of the sector regulator for electricity. Only Germany did not establish a sector regulator for gas, but relied on self-regulation of the gas industry backed up by the German competitive authority.

In this way, the EU managed to take the second step in the process of gas market harmonisation. It soon turned out that the legal changes initiated by the EU Directive were but a small step in the still long track to establish an open internal gas market. The next section will discuss the new legal regulatory order that came on the scene in August 2000. Our analysis will show that EU countries did change the legal order in line with the requirements of the Directive.

From a regulatory point of view, the gas Directive initiated a new demarcation in the European gas market. The lines of demarcation were twofold:

- Matured and emerging gas markets
- Central and peripheral gas markets

The first line of demarcation distinguished the European gas markets in a majority group of countries required to liberalise the gas market and a minority group derogating under this requirement. First, liberalising countries were required to open the national gas market in the way prescribed by the Directive. Free choice of gas supplier for long has been a passionate demand of European industry and was one of the core drivers in the debate on gas market harmonisation. Second, countries had to change the monopoly position of vertically integrated gas companies and the opening of the transmission pipelines for third parties. In line with the Directive, all the Member States initiated legal changes in this direction obliging the integrated companies to unbundle their transmission and supply activities. Furthermore, the Directive also provided legal ground for third party access (TPA) to gas storage facilities. Third, the Member States were required to establish a gas market regulator.

As emerging markets Greece and Portugal had negotiated a derogating position under the Directive recognising both gas markets as still immature emerging markets in Europe. Both countries therefore were exempted from the legal requirements of the Directive until 2008, giving both countries more time to develop their domestic gas market, before being exposed to the dynamics of competition. The second demarcation was between the centre and the periphery of the European gas market. Finland and Sweden nominated for a peripheral position, with no or minimal connection to the central European pipeline system. Finland furthermore was allowed to continue the monopoly position of the national vertically integrated company due to its strong import dependency of Russian gas. For that reason Finland was exempted from several requirements of the Directive. Sweden's peripheral position in the European gas market and its limited connection with European pipelines, were the reasons for the Swedes to consider requesting for derogation under the Directive.

In the mid-2000, the larger part of the EU region shared some common focal points in gas market regulation as required by the Directive. The legal order of national gas markets in the EU did change. Despite the degrees of freedom offered by the Directive, at face value, the variation in new national regulatory models was even less than initially expected. However, a closer look under the surface reflected national differences. As explained previously, the history of European gas holds a strong national bias in terms of politics, industrial organisation, ownership structures, regulation and technicalities. The transformation of the national heritage into harmonisation turned out to be more difficult than expected. As will be explained in more detail below, in this process of transformation the effectuation of the legal changes in national gas regulation turned out to be a minor issue.

3.3. Variety of Regulatory Frameworks

Except for the United Kingdom, for all other Member States the gas Directive acted as the reference point in the process of legal change. But as indicated above, the reach of the Directive was rather limited and held degrees of freedom allowing the Member States to make their own choices in regulatory reform. As this section will show the national regulatory change indeed varied.

3.3.1. Comprehensiveness of gas regulatory frameworks

The gas Directive initiated some harmonisation in the economic and political regulation of gas markets. In the realm of economic regulation, the Directive specified a set of preconditions for competition in terms of rules for access to pipeline infrastructures, the unbundling of transmission and trade activities and the opening of the market. In the proposal amending Directives 96/92/EC and 98/30/EC, the Commission more explicitly referred to competition in the gas market. It was said that 'It is generally acknowledged that third party access based on published and non-discriminatory tariffs, and a high level of unbundling, are not only conducive but necessary to ensure effective competition. To complete the internal market, it is not sufficient therefore to fully liberalise demand, but effective market structures must also exist (COM(2001) 125 final p. 3). In the realm of political regulation, the Member States addressed a set of public service obligations. Previous to liberalisation, gas market regulation had a strong public service (PSO) orientation and therefore, the Directive continued the PSO focus next to the introduction of competition. Countries indeed included PSOs in national gas market regulation but not all in the same way.

The first group of countries only regulated a minimal set of PSOs related to security of gas supply and protection of captive customers (Belgium, Italy and the United Kingdom). The second group of countries also included environmentally oriented PSOs or typical domestic PSOs (France, Denmark and The Netherlands). For instance in France the principle of equity of gas supply is a typical national PSO. In The Netherlands, the small field policy, integrating the efficient exploitation of the small gas fields at the Dutch part of the Continental Shelf with the exploitation of the large Groningen gas field, has been identified as typical Dutch PSO. The small field policy was very rewarding for the Dutch reserve position. The Commission's proposal amending the initial gas Directive, specified five PSOs – protection of vulnerable customers, protection of final customers' rights, social and economic cohesion, environmental protection and security of supply (COM(2001) 125 final, p. 9).

Regarding the access regime for third parties, the Member States were allowed to choose a negotiated or a regulated access regime for the downstream (transmission and distribution) pipeline systems and for access to gas storage facilities. Table 3.1 displays the access models adopted by the Member States in 2001.[2]

More than initially expected, the Member States opted for the model of regulated third party access (rTPA) for transport and distribution pipelines. Austria and Germany opted for the negotiated access regime, only prescribing minimal access conditions and indicative

[2] Unless differently indicated, country data in the following sections have been taken from EU Commission Staff Working Paper *First Benchmarking Report on the Implementation of the Internal Electricity and Gas Market*, Updated Version with Annexes, Brussels, March 2002.

Table 3.1. Access regimes transport, distribution and storage.[1]

Country[2]	TPA transport	TPA distribution	Access to storage
Austria	Negotiated	Negotiated	No
Belgium	Regulated	Regulated	Yes
Denmark	Regulated	Regulated	Yes
Finland	Negotiated	Negotiated	No storage facilities
France	Negotiated	Regulated	No
Germany	Negotiated	Negotiated	Yes
Ireland	Regulated	Regulated	No storage facilities
Italy	Regulated	Regulated	Yes
Luxembourg	Regulated	Regulated	No storage facilities
The Netherlands	Negotiated	Regulated	Yes
Spain	Regulated	Regulated	Yes
Sweden	Regulated	Regulated	No storage facilities
United Kingdom	Regulated	Regulated	Yes

[1] It should be noted that Table 3.1 figures reflect the access regimes of 2001 and not of 2000. Some countries have changed the access regime since the national inauguration of the gas market Directive.

[2] Data of Greece and Portugal, both derogating countries, are not included in the tables in this chapter.

tariffs leaving the actual access arrangement to be negotiated. The other countries adopted a system of rTPA, with some countries (France and The Netherlands) differentiating between the access regime for transport and distribution networks. The Dutch negotiated access regime for transportation is however close to the regulated system because the Dutch regulator prescribes binding guidelines for access conditions and tariffs.

The final column of Table 3.1 shows that Austria and France do not offer access to storage facilities. However, in only some countries access to storage is independent of standard flexibility services as part of the transportation contract, as required by the gas Directive. Only Denmark, Germany, Italy, Spain and the United Kingdom offer access to storage independent of flexibility service by means of auction.

A second set of regulatory requirements by the Directive relates to the market opening and the unbundling of transmission and trade activities. Here too, the Directive held degrees of freedom allowing countries to choose among different models. First in terms of market opening, the Directive required a 33% market opening in 2008, but countries were free to exceed this minimal threshold. With two exceptions (Denmark and France) all other countries went for a track leading to complete opening of the gas market before the initial deadline of 2008 as required by the Directive. Next to the United Kingdom, Germany was the second country fully opening the gas

Table 3.2. Current *legal* market opening and expected *legally* full market opening.

Country	*Legal* market opening in 2001	100% *legal* market opening in
Austria	49%	2002
Belgium	59%	2006
Denmark	30%	2008
Finland	Derogation	Derogation
France	20%	2008
Germany	100%	2000
Ireland	75%	2005
Italy	96%	2003
Luxembourg	51%	2007
The Netherlands	45%	2004
Spain	72%	2003
Sweden	47%	2006
United Kingdom	100%	1998

market. All other countries opted for a more gradual opening. However, contrary to the initial expectation, almost all the Member States went for full market opening instead of the minimal 33% as required by the Directive. In 2001 the legal market opening in almost all the countries already exceeded the initial final threshold of 33%. The update of the gas Directive is now heading for full market opening in 2004 at the latest. It should be noted that the data in Table 3.2 reflect the *legal* positions of the Member States on market opening. Data in Table 3.2 should not be read as *real* market opening.

Second, regarding unbundling of trade and transmission activities, most Member States initially followed the minimal requirements of the Directive by financially separating accounts, putting 'Chinese Walls' between transport and trade activities of the former integrated gas companies. Only a few countries (Denmark, Italy, Spain and the United Kingdom) legally separated transport and trade activities in independent organisations, with the United Kingdom in the most advanced position with separated ownership. Some countries, like The Netherlands, unbundled trade and transmission in different stages. Already in 1999, the then still vertically integrated company Gasunie started offering transportation services independent of volume sales. In 2000 the company unbundled financially and in 2001 also legally. At the beginning of 2002 the shareholders of Gasunie (Dutch state, Shell and Esso) were discussing the termination of Gasunie's trade activities and the take over of these activities by Shell and Esso and to bring Gasunie transmission including the pipeline infrastructure in

Table 3.3. Unbundiling regimes for transport and trade.

Country	Unbundling regime
Austria	Accounts
Belgium	Accounts
Denmark	Legal
Finland	Derogation
France	Accounts
Germany	Accounts
Ireland	Management
Italy	Legal
Luxembourg	Accounts
The Netherlands	Legal
Spain	Legal
Sweden	Accounts
United Kingdom	Ownership

full state ownership. At the beginning of 2004 the debate on this issue had not been decided yet (Table 3.3).

So the differences between the Member States regarding market opening and unbundling were rather significant. Only a few countries were reluctant to full market opening, whereas the others were not. Regarding unbundling most countries kept quite close to the minimal requirements of the Directive, whereas only a few countries went beyond. The example of Gasunie shows that the unbundling of the former vertically integrated companies was not consolidated yet at the end of 2001.

3.3.2. *National gas regulatory institutions*

The Member States were committed to establish appropriate and efficient mechanisms for regulation, control and transparency, but the Directive did not specify any additional conditions regarding the type, status and powers of the required regulatory bodies. Furthermore, the Directive required an independent competent authority for settlement of access disputes and negotiations in the scope of the Directive. In line with this requirement, all the Member States, except Germany, established sector regulatory bodies for the gas market. Many countries simply extended the regulatory mandate of the electricity market regulator and transformed them into energy market regulators. In all cases, national governments or the head of state appointed these regulatory bodies, sometimes with the national parliament in an advisory role. All Member States organised the regulatory function in an independent public organisation with a general mandate regarding its regulatory function. Only Austria and The Netherlands

Table 3.4. Characters of gas market regulators in Europe in 2001.

Country	Regulator(s)	Capacity	Annual budget	Funding
Austria	ECGmbH ECC	43	€7,000,000	Levy network users
Belgium	CREG 3 regional regulators	40	€9,420,000	Transport and supply licences
Denmark	ESB	30	€2,460,000	Levies transmission/ distribution of gas
Finland	EMA	15	€1,200,000	85% levies, 15% state budget
France	CRE	65	€9,150,000	State budget
Germany	No regulator			
Ireland	Ministry public enterprises			
Italy	AEEG	63	€18,000,000	N/a
Luxembourg	IR	1	N/a	Levies
The Netherlands	DTE	33	€4,000,000	60% fees energy companies, 40% State budget
Spain	CNE	140	€16,800,000	Levies
Sweden	NEA-OEGR right to intervene	33	€3,400,000	N/a
United Kingdom	OFGEM	306	€103,858,000	Fees energy industry

deviate from this general model. In Austria the regulator is a private organisation with the state as the single shareholder. In The Netherlands, the energy market regulator DTE is part of the national competition authority holding a typically Dutch legal status (called 'publiekrechtelijk bestuursorgaan' in Dutch).

Table 3.4 gives an overview of the national regulators and their capacities. It should be noted that the data in Table 3.4 cover the regulation of both electricity and gas. The available data do not allow for any differentiation between gas and electricity market regulation.[3] It shows that the capacities and budgets of national regulators differ, although these figures of course account for differences in regulatory function and scale and scope of national energy markets.[4]

[3]Attempts to get specified data for gas regulation only failed.

[4]Unfortunately, available data do not allow for more detailed comparison of gas market regulatory models and functions.

In general, countries were not that eager to develop the sector regulator for gas. In many cases staff members of the electricity regulator were put in charge to initiate and develop the regulatory function for gas. The EU-benchmark on regulation of 2001 shows differences in the maturity of regulation of gas and electricity, with electricity clearly in front. In some countries, like Austria, the regulator still did not have any regulatory authority regarding gas in 2001 and in other countries, the gas market regulator only had a rather limited regulatory mandate for the gas market.

Table 3.5 gives an overview of the regulatory mandate of the national gas market regulators. The table shows the competencies of the national regulator with respect to four key issues – access conditions, dispute settlement, licences and sanctions.

Regarding the regulation of access conditions and tariffs, four different models came into operation within the EU territory.

1. In the first model, adopted by Italy and the United Kingdom, access to the national gas network is independently and autonomously regulated by the national gas market regulator. The mode of regulation is ex ante, meaning that regulatory approved access conditions and tariffs are guiding actual transactions. The ex ante mode of regulation in fact reflects the model of rTPA. The Netherlands developed a variation of this model in the form of regulatory approved binding guidelines for access conditions and tariffs.
2. In the second model, adopted by Belgium, France, Ireland, Luxembourg and Spain, the ministry decides about access conditions and tariffs, with the national regulator in a role of advisor/approver. Here too, like in the first model, access is regulated ex ante, before the actual transaction takes place. Due to the federal system in Belgium, the federal government decides for access to the transmission network and the regional governments for access to the local distribution networks.
3. In the third model, adopted by Denmark, Finland and Sweden, the system operators decide about the access conditions and tariffs, with the regulator in a position to intervene ex post. In this model the regulatory function of the regulator is activated by individual gas transactions.
4. In the fourth model, adopted by Austria and Germany, system operators decide about access conditions and tariffs, without interventions of gas market regulators, because these regulators have not been established. Specific bodies settle access disputes in both countries.

Table 3.5. Competencies of institutions involved in national gas market regulations in 2001.

Country	Access conditions	Dispute settlement	Licences	Sanctions
Austria	nTPA	Ministry, civil and high court	Ministry	Financial
Belgium	CREG, federal and regional governments	Chambre de disputes, civil court, CREG and regional regulators	Minister EA proposal of CREG	Financial
Denmark	ESB, right to intervene	ESB, energy complaints board (except access)	Ministry	?
Finland	Operators, EMA, right to intervene	EMA	EMA	Financial, withdraw of network licence
France	CRE, Ministry	CRE	Ministry	Financial
Germany	nTPA	Ministry federal and regional competition authority	Ministry federal and regional	?
Ireland	Ministry	Ministry	Ministry	?
Italy	AEEG	AEEG	Ministry on proposal AEEG	?
Luxembourg	Ministry on proposal IR	IR	Ministry on proposal IR	?
The Netherlands	Binding guidelines of DTE	DTE, Competition authority, geschillen commissies	Ministry on proposal DTE	Financial
Spain	Ministry on proposal CNE	CNE and regional authorities (access tariffs)	Ministry	No sanctions by CNE
Sweden	NEA	NEA	Ministry	?
United Kingdom	OFGEM approves methodology	OFGEM, Energywatch	OFGEM	Legal and financial

Dispute settlement has been organised in three different models.

1. In the first model, adopted by France, Italy, Luxembourg, Sweden and the United Kingdom, the gas market regulator is the principal dispute settling body.
2. In the second model, adopted by Belgium, Denmark, The Netherlands and Spain, the gas market regulator shares its dispute settling competencies with specific bodies or courts. In the case of Belgium and Spain, the regulator shares the responsibility with regional authorities.
3. In the third model, adopted by Austria, Germany and Ireland, the gas market regulator is not involved in dispute settlement. In these countries, the ministry or the competition authority is the first step in dispute settlement, sometimes with involvement of regional authorities, like in the case of federal Germany.

With regard to licensing of new infrastructures and licensed gas supply of captive customers, three different models have been adopted.

1. In the first model the gas regulator autonomously licenses infrastructure investment and gas supply (Finland and UK).
2. In the second model the ministry is the licensing authority with the gas regulator in an advisory position (Belgium, Italy, Luxembourg and The Netherlands).
3. In the third model the ministry is the licensing body without any involvement of the gas market regulator (Austria, Denmark, France, Germany, Ireland, Spain and Sweden).

In the realm of sanctions all national regulators have been facilitated with some kind of sanctions in most cases financial penalties and in some cases the authority to withdraw licences.

3.3.3. The regulatory variations

The observed regulatory variation stems from the degrees of freedom offered by the gas Directive. The range between the minimum and the maximum regulatory requirements of the Directive indeed was substantive and this variety spread over Europe in the new legal frameworks of the Member States. All the Member States came from a system of monopoly regulation and close national gas markets and this too contributed to the regulatory variation. According to Bergman cs (1998) regulation matures in congruence with the transition from monopoly to competition, assuming a clear mutual and reciprocal relationship between regulation on the one hand, and changing market structure and business strategies on the other. In this perspective, the development of the regulatory function in the gas market is a dynamic process in which regulation, market structure and market behaviour jointly determine next steps in the process. In this process the regulatory changes initiated by the gas Directive can be considered as a first, necessary, step only.

The newness of the regulatory changes makes a comparative evaluation difficult for several reasons. First, the legal changes are no goal in themselves, but they serve the introduction of competition in national gas markets in Europe and the longer-term establishment of a European-wide competition-based internal gas market. In this prospect, even the legal changes made thus far, are still rather restrictive. Second, it is unclear how a competition-based internal gas market should be structured and how such a market can function adequately. This complicates any assessment of the effectiveness of the legal changes thus far and the regulatory changes still needed to facilitate competition. Third, EU and national authorities are not the only ones influencing the regulatory reform process. In this highly political process many interests compete and try to influence regulatory outcomes. The variety in regulatory models shows that the EU Directive initiated only some minimal regulatory harmonisation between national gas markets.

In this stage of regulatory change it is therefore only possible (and fair) to evaluate the regulatory changes in the prospect of the longer-term goal of establishing a competitive internal gas market. This evaluative assessment will be made in two steps. First we will analyse whether or not the legal changes made thus far meet the basic legal requirements generally recognised as essential for the introduction of competition in gas markets (see also the chapter on gas markets). A standard for this kind of evaluation has been introduced in Chapter 2, under the label of comprehensiveness of regulation. Second, we will tentatively analyse how open and accessible gas markets actually have become after the legal reforms. Both steps allow to evaluate the state of the art of regulatory change thus far, and to classify country's regulatory reforms in legal and empirical terms in the prospect of a competition-based internal gas market.

Figure 3.1 schematically specifies the conditional relationship between regulation, openness and competition of gas markets.

The figure shows that comprehensiveness of regulation implies regulatory functions and competencies to ascertain openness of gas markets. The degree of openness in its turn is a basic condition for the

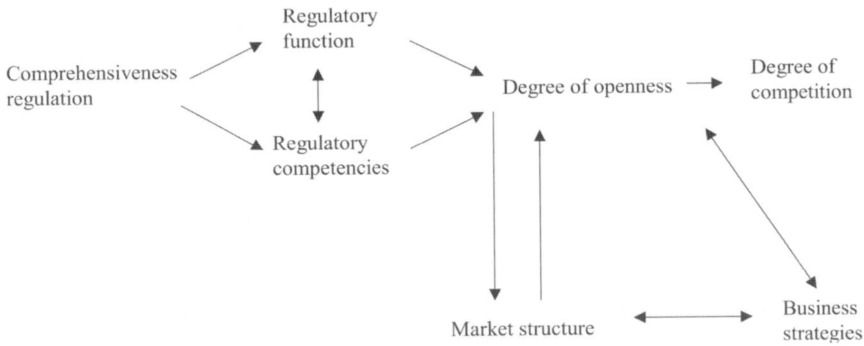

Fig. 3.1. Factors influencing the degree of competitiveness of gas markets.

degree of competition. But as the figure shows, competition is not only influenced by regulation but also by (changing) market structure and business strategies. The dynamics involved in the relationship between regulation, market structure and business strategies will be discussed in the final part of the book. Here we continue the two-step comparative evaluation of the regulatory reforms in the Member States. The next section analyses the regulatory changes in terms of comprehensiveness of regulation. Section 6 analyses the degree of openness of national gas markets. In this way it is possible to evaluate the regulatory changes both in legal and empirical terms.

3.4. Variety of Regulatory Styles: the Comprehensiveness of Regulations

With the help of the concept of comprehensiveness of regulation, introduced in Chapter 2, it is possible to assess whether or not the Member States' legal reforms meet the formal standards of a competition-based gas market. Following the discussion on comprehensiveness of regulation in Chapter 2, we conceive the concept here as a standard to assess the extent to which countries meet the necessary *legal* conditions for a competition-based gas market. For this purpose we conceive comprehensiveness of regulation as a two-dimensional concept: one dimension referring to necessary regulatory *functions* and one dimension referring to regulatory *competencies*. Competition requires certain regulatory functions and certain regulatory competencies and the way countries have organised these functions and competencies does influence the degree of openness and competition of gas markets. We first summarise the empirical findings on regulatory functions and competencies and then will aggregate the findings into a classification of regulatory models.

3.4.1. The establishment of necessary regulatory rules

The empirical overview shows that all countries in one way or another responded the requirements of the gas Directive with basic rules for redesigning national gas markets in line with the Directive. All countries introduced formal and basically procedure-oriented legislation allowing further specification of substantive issues on market structure and regulatory performance. But whereas the gas Directive pursued introduction of competition in the prospect of establishing the internal market for natural gas, the empirical overview indicates that the variation on almost all subjects covered by the Directive is still substantive in the first wave of reforms:

- TPA was regulated in four different ways, with still some countries opting for the negotiated model, lacking any ex ante or ex post intervention of sector regulatory authorities.
- Countries also differently unbundled transmission and trade activities; most countries opted for a financial or legal separation of trade and transmission activities of the former integrated companies. Only two countries (Italy and UK) established new independent transmission companies completely separated from the incumbent system operators.
- More countries than expected opened up their gas market beyond the minimal threshold of 33% in 2008, but the national trajectories to full market opening still vary significantly.
- Finally, except Austria and Germany, all Member States established a specific body for gas market regulation, in many cases by extending the regulatory mandate of the electricity regulator for the gas market. But again, as in the case of the other substantive issues covered by the Directive, the variety in the design, competencies and instruments of the regulatory bodies is still large.

Third party access of pipelines

Third party access to the pipelines is basic for the introduction of competition and our analyses showed that countries adopted different access regimes. Austria, Finland and Germany introduced a negotiated regime for access to transmission and transport pipelines. France and The Netherlands made a distinction between the access regime for transmission and distribution pipelines, introducing the negotiated model for the first and the regulated model for the latter. Furthermore,

Austria and France did not open their storage facilities for third parties as part of the flexibility services of transmission and transport. So regarding the access regime only the latter two countries display some legal deficit with respect to storage facilities. Therefore, with respect to access regime countries divide into two groups, one minor group of countries with the negotiated model and a larger group of countries with the regulated model.

Non-discriminatory access and unbundling

On this point, countries made legal provisions to unbundle supply and transmission activities of the former vertically integrated companies, but the legal provisions are quite weak in general. With only a few exceptions almost all the countries only required legal separation of accounts, which is a rather soft measure to prevent any monopoly abuse and cross subsidisation by system operators. The guarantee for this kind of abuses is the establishment of independent system operators. Only two countries established such an independent system operator – Italy and the United Kingdom. All other countries only made the second best legal provisions on this issue thus far. So here too countries divide into two groups, one small group with independent system operators and a large group of countries with no independent system operator.

Market access

Here too, all countries made legal provisions to open the gas market. It showed that almost all countries are ahead of the gradual schedule for market opening as required by the Directive. Only France and Denmark are still on the initial EU schedule. So on market access too countries divide into two groups, a small group not fully opening the gas market and a larger group of countries heading for full market opening in 2008 at the latest.

Dispute settlement

In all countries dispute settlement has been legally arranged but the models differ with respect to the position and involvement of the gas market regulator. In four countries the gas market regulator is not involved in dispute settlement. In the other countries the regulator is either the only or one of the bodies in charge with dispute settlement. So on this issue, countries deviate in three groups – a group with regular bodies in charge with dispute settlement, a group with the gas market regulator in charge with dispute settlement and a group with shared responsibilities in dispute settlement.

3.4.2. Coverage of necessary regulatory competencies, capacities and instruments

The second dimension of comprehensiveness of regulation refers to the competencies, capacities and instruments of the regulators. This dimension has two aspects – the type of competencies and capacities and the degree of autonomy of the regulator.

Competencies and capacities

On this point the empirical data show differences in the position of countries. The two dominant models here are the ex ante regulatory model, which in fact reflects the rTPA system and the ex post intervention model, reflecting the nTPA system. All have the competency to enforce competition rules and abuse of monopoly position and all regulators have been gifted with financial resources and sanctions. So on the aspect of competency and capacity countries divide in two groups – a group with the ex ante regulatory model and a group with the ex post intervention model.

Degree of political and economic autonomy of regulatory bodies

Only Italy, The Netherlands (to some extent) and the United Kingdom have legally independent and autonomous gas market regulators, with autonomous ex ante regulatory mandates. All other countries in one way or another share regulatory mandates with governmental bodies, in almost all cases the ministry for economic affairs or energy. Shared responsibility in economic regulation opens the regulatory process for political considerations, which might intervene with the regulation of fair trading conditions. For the same reason, the regulator should operate independently from the national TSOs. Ex post intervention types of regulation in combination with unbundling of accounts, are less favourable in this respect.

3.4.3. Comprehensiveness of national gas market regulation

In Table 3.6 and Fig. 3.2[5] countries have been classified in the two-dimensional space of comprehensiveness of regulation.

[5] Country positions have been calculated as the sum of the national scores on the characteristics of regulatory function and regulatory capacity, the two dimensions of comprehensiveness of regulation explained at the beginning of this section.

The upper right of the figure holds the countries with the most comprehensive regulatory model. Only two countries, Italy and the United Kingdom, meet this standard. Both countries did establish the regulatory conditions assumed to be necessary to ascertain competition in the national gas market. All other countries have less comprehensive regulatory models in this respect. The four countries in the bottom right of the table represent the group of countries with minimal regulatory models, both in regulatory function and in regulatory capacity. Germany is part of this group, because of its model of self-regulation by industry in combination with dispute settling regulatory facilities. Other countries such as Austria did not establish the regulatory mandate of the gas market regulator yet. All other countries listed in Figure 3.2 did develop regulatory functions and capacities, but in a mixed form. Sweden did develop its regulatory functions but hardly without backup in capacity. Ireland, Spain and Belgium did develop regulatory functions but without independent position of the national gas market regulator. The same holds for France and Luxembourg, but both countries differ in their

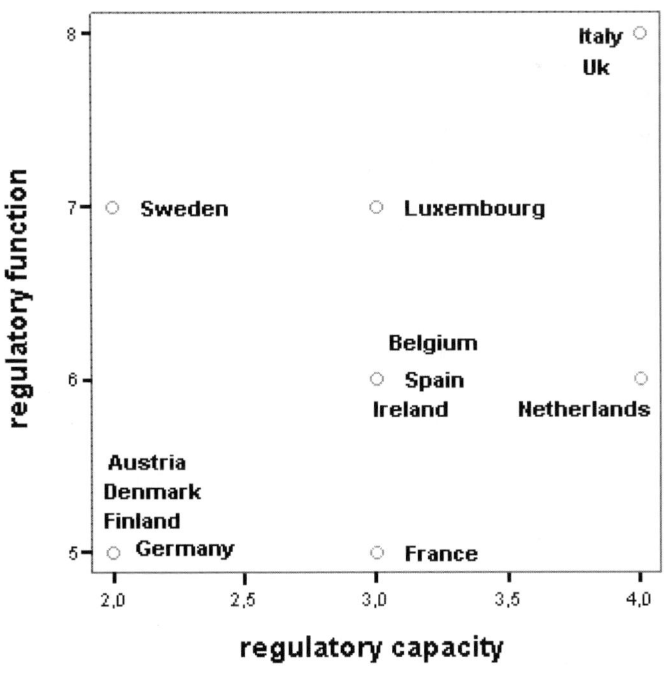

Fig. 3.2. Comprehensiveness of national gas market regulation.

development of the regulatory function. The Netherlands takes a deviating position, which is caused by the way the country has regulated pipeline access, which is a mixture of ex ante and ex post regulation. The Dutch regulator is however autonomous in decision-making, which is not the case in the five countries, listed in the middle of the table.

So in the perspective of the necessary regulatory conditions for competition, our conclusion is that legally three regulatory models currently are operational in the EU region:

- The *comprehensive regulatory model,* holding all necessary legal regulatory conditions for competition;
- The *emerging regulatory model,* still developing the necessary legal regulatory conditions for competition; and
- The *minimal regulatory model,* holding only minimal necessary legal regulatory conditions for competition.

The next section analyses the performance of these models to ascertain market access in practice.

3.4.4. Regulatory performance: access of pipelines and market

The previous section evaluated the regulatory reforms in legal terms. Regulatory change at the legal level is however, only a necessary condition but no guarantee for actual market opening and access to the gas system. The regulatory reforms at the legal level are not yet reflected in the actual change of the market. At both levels, the legal and the actual level, changes are incongruent. Only in the United Kingdom the change at both levels is congruent. In all other countries there are clear differences between the regulatory changes at the legal level and the actual openness of the gas market. Table 3.6 summarises the current national regulatory reform at the legal and the actual level.[6]

Vertically the table indicates the share of gas transport under TPA and horizontally the degree of comprehensiveness of national regulatory models. The table clearly shows the challenge countries are facing in developing their regulatory frameworks and the openness of the market. All countries have to move to the upper left cell of the table, now only holding the United Kingdom. The regulatory model

[6] In annexe we develop this aspect of development of TPA flows as index of competitiveness of the market and the perception of the players concerning the market accessibility.

Table 3.6. National regulatory performances in terms of competitive trade.

% of gas transport under TPA	Comprehensiveness of national regulatory model		
	Full	Emerging	Minimal
25–100	United Kingdom		
5 < 25	Italy	Ireland The Netherlands Spain	
<5		Belgium Luxembourg Sweden	Austria Denmark Germany France

of Italy is not yet reflected in the actual openness of the market. In Ireland, The Netherlands and Spain the actual state of the art of the regulatory reforms keeps track with the actual opening of the market, whereas in Belgium, Luxembourg and Sweden the opening of the market stays behind the regulatory reform. In Austria, Denmark, Germany and France the regulatory reform as well as the opening of the market, are still on a minimal level.

Table 3.6 furthermore shows that the reform process is ongoing in almost all countries. Only the position of the United Kingdom reflects the ambition of the EU gas reform politics of establishing a single, competition-based internal gas market. All other countries are still in the process of developing the necessary regulatory conditions for such a market. The next section develops some explanation for the different country positions in the regulatory reform process.

3.5. Drivers of National Regulatory Dynamics

Our empirical analysis shows that all countries, except UK, are still in the first two phases of regulatory reform as suggested by Bergman cs (see Chapter 1) in 2002–2003. The phases suggested by Bergman are helpful to describe the general pattern of regulatory change from monopoly to competition, if countries indeed are heading for competition in regulatory reform. But it might be questioned whether competition is the only orientation in the national regulatory reform process. National interest positions in gas have been and are strong drivers too in the reform process, next to the legal commitment to EU legislation. The national interest position in particular is reflected in the way countries are dealing with the conditions for access to the pipelines. Without adequate access conditions an open competitive-based gas market cannot emerge.[7] Our empirical analysis showed some differences in national access models. This section develops some explanation for these changes.

3.5.1. National regulatory dynamics

National regulatory positions in establishing adequate access conditions can be rather complex, because of the national interests at stake. Complexity stems from the fact that national governments are the legal authorities committed to EU legislation. They are legally obliged to integrate EU legislation in national gas market regulation. For that reason they cannot tolerate discriminatory access conditions within their territory and therefore they all initiated TPA regimes and they all established sector regulators to guarantee TPA. So being a member of the EU forced countries to open up the national pipeline infrastructure for third parties. But the regulatory models countries adopted differ and the question is why? Part of the answer may be found in national history.

Almost all national governments had strong interests in the vertically integrated companies, as shareholder or full owner of these companies. The vertically integrated companies unbundled trade and transmission activities and several Member States continued strong shareholder relationships with the transmission activities of the unbundled companies. The continued interest in national gas industry therefore, can intervene with the legal obligations to commit to EU legislation. This governmental interest in the national gas industry might be reflected in the national regulatory access model as a bias 'protecting' the interests of the national TSO. Such a biased regulatory model reflects the legal requirements of the EU Directive, but at the same time the model actually benefits gas industry. The industry bias of this regulatory model is reflected for instance by industry prescribing the access rules and conditions with the national gas market regulator in a position of authorising and legitimating industry's rules and conditions. In fact the negotiated TPA model does reflect this interest position.

National governments, however, also bear responsibility for the national security of gas supply meaning that they have to account for the diversity of interests of the national gas industry and the national gas consumers. Consumers, in particular large ones, want open and

[7] The logic underlying this argument is well explained in the chapter on gas market of this book and in the long-term vision of a fully operational single market for gas in Europe. See footnote 1 for reference.

unconditional access to transmission networks, they want to choose their own gas supplier and they want cost-effective and transparent transmission tariffs. This mixed national interest portfolio of national gas industry and national gas consumers therefore, can act as a third focal point in national gas market regulation. This type of regulation can be expected to be a referee-oriented type of regulation mediating between gas industry and gas consumer interests. In fact this situation is reflected in the model of ex post intervention.

So national governments have at least three different types of commitments in regulatory reform – a commitment to EU legislation, a commitment to national gas industry and a commitment to national gas consumers. Each of these commitments demands different types of regulatory orientations. Given these three different national interests, national regulation of the gas market can be:

- Competition oriented, in case the Member State's dominant focus is to commit to the harmonisation rules of the EU.
- National gas industry oriented, in case the Member State's dominant focus is to protect the national gas industry.
- Public interest oriented, in case the Member State's dominant focus is to balance the diversified interests of gas industry and gas consumers.

Furthermore, the Member State's specific interest portfolio also can serve as guidance for developing the regulatory capacity for the gas market. First, the national interest portfolio might decide the degree of political autonomy and political independence of the gas regulator. In case the regulatory focus is competition oriented, an independent and autonomous gas market regulator, operating at a distance of national political decision-making, would be the preferred model. Autonomy and independence of national political decision-making in combination with strong regulatory instruments is the best guarantee to design and to ascertain unconditional and non-discriminatory access conditions and tariffs. Under these circumstances, the national regulator is able to work independently from political interest mediation on the establishment of appropriate conditions for a competitive gas market.

In case the national interest portfolio is gas industry-oriented it might be expected that the distance to political decision-making is small to restrict the autonomy of the national gas market regulator for reasons of political intervention in the regulatory process. The governmental interests in the national gas industry might motivate this kind of political intervention. In this case therefore, the autonomy of the gas market regulator might be more restricted both in terms of institutional embedding and regulatory powers and instruments, than in the case of a competition-focused national regulation.

The same restriction on the political autonomy of the national regulator can be expected in case the national interest position is focused on both industry and consumer interests. So, as in the case of a gas industry-dominated interest portfolio, a national public interest focus too might restrict the political autonomy of the national gas market regulator.

The national interest portfolio might also decide the degree of market autonomy of the national gas market regulator. The degree of the gas regulator's market autonomy is basically decided by its dependency position vis-à-vis the national transmission company in determining and deciding access conditions and tariffs. This regulatory position vis-à-vis the national transmission companies can be:

- Independent from the national transmission company, meaning that the gas market regulator autonomously and independently prescribes access conditions and tariffs (ex ante regulation).
- Dependent of the national transmission company, meaning that the national regulator basically authorises the access conditions and tariffs of the national transmission companies (ex post intervention).
- Balanced with the national transmission company, meaning that the regulator and the transmission company compromise access conditions and tariffs, accounting for the interest of gas consumers (dispute settlement).

With the help of the distinction in regulatory orientation and the degree of political and market autonomy of the national regulators, it is possible to identify three different explanatory models for gas market regulation (see Table 3.7).

The first, competition-oriented regulatory model, can be perceived as the ideal regulatory model for further harmonisation of the internal gas market in Europe. This model is competition-oriented and operates independently from national, political and economic dynamics. Regulation is focused on the design and guarantee of the conditions for competition. The second and third regulatory model is focused on national, political and economic interests respectively and this might intervene with a competition focused regulation. The second, national gas industry-focused regulatory model, in fact favours the position of incumbent companies and therefore can be labelled as the *supply-focused regulatory model*, holding rather weak

Table 3.7. The position of the gas regulator in the three regulatory models.

	Political autonomy of gas regulator	Market autonomy of gas regulator
Competition-oriented model	Autonomous	Autonomous
National industry-oriented model	Restricted by national administration	Restricted by national gas industry
National- and public interest-oriented model	Restricted by national administration	Balanced vis-à-vis gas industry and consumers

incentives for competition. In the third model satisfaction of gas need and demand of consumers is the basic orientation and therefore, this regulatory model can be labelled as the *citizen/consumer-focused* or *demand-focused regulatory model.*

In sum, based on the national interest position it is possible to distinguish three different regulatory models:

- The competition-focused regulatory model
- The supply-focused regulatory model
- The demand-focused regulatory model

3.5.2. National styles of regulation

It might be expected that the *style of gas market regulation* differs in each of the three regulatory models discussed above. In the first, competition-focused model, the regulation style could be labelled as 'prescriptive and enforcing'. In this model the regulator acts independently and autonomously from economic and political dynamics to ascertain adequate conditions for competition. It prescribes the market and competition rules and enforces these rules in an effective way. The competition-oriented regulatory style reflects a strong position of the national gas market regulator.

In the second, supply-focused model, the style of regulation will be far less stringent and competition facilitating, because here the national gas industry is the dominant focus in gas market regulation. The regulatory style will be much more industry oriented, and therefore could be labelled as 'following and authorising'. The regulator follows the access conditions and tariffs of the transmission company and will not attempt to question these conditions and tariffs, but instead will only authorise them. For that reason the regulator will stay at distance of the parties negotiating access conditions and tariffs. The system operator controls actual access.

In the third, demand-focused model, the regulation style can be labelled as 'dispute settlement'. Here, the regulator basically mediates the interests of the transmission company and access demanding consumers. This type of regulation will be based on compromised access conditions and tariffs, reflecting the interests of the transmission company on the one hand and the customers on the other. Regulation will restrict to dispute settlement alternately deciding in favour of the interests of the transmission company or the interests of consumers.

3.5.3. National regulatory positions as mixed interest positions

Thus far our analyses suggest that national interest positions explain the national framing and style of gas market regulation, but the impact of the national interests on the framing of the national regulatory model is different. The commitment to the EU gas Directive requirements is legally obliged and our empirical analyses showed that all countries at least minimally committed to the requirements of the Directive. Therefore it might be assumed that the competition orientation serves as a kind of minimal threshold in regulatory change. But the strength of the national regulatory orientation on competition can be assumed to be mediated by the country's industrial and consumer interest orientation. Both the industry and consumer interest orientation may act as mediating guidance in national regulatory reform. The relationships between the three interest orientations in regulation are depicted in Fig. 3.3.

With the help of the logic displayed in Fig. 3.3 it is possible to explain current national regulatory positions. While making the regulatory changes, all countries were committed to the gas Directive and therefore were forced to change the regulatory framework of the national market in line with the minimal requirements of the Directive. Some countries indeed stayed at this minimal level, whereas other countries developed more advanced regulatory positions which all could be explained by the typical national interest portfolio.

With the help of the three different interest orientations it is possible to explain regulatory change of national gas markets as a two-sided change process:

1. On the one side, a process oriented on the establishment of competition but mediated by a tension to protect national industrial or national consumer interests.
2. On the other side, a process oriented on the protection of national industrial or consumer interests, but mediated by a commitment to the legal requirements of EU harmonisation rules.

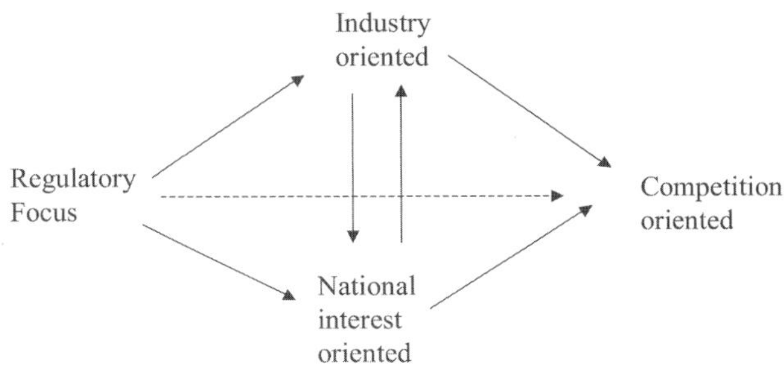

Fig. 3.3. National orientations in gas market regulation.

This two-sided regulatory change process is visible in the current gas reforms in Europe. Table 3.8 lists the different country positions in this respect.

Table 3.8 distinguishes horizontally three different *dominant* regulatory orientations and vertically three *mediating* regulatory orientations, with cells explaining the country positions as a mixture of dominant and mediating regulatory orientation. The table shows a distinction between two groups of countries – one group where competition is the dominant guiding orientation in regulatory reform and one group where national interest is the guiding orientation in regulatory reform. The UK is the only country where the dominant and mediating orientation is both competition oriented. In Italy, Spain, Sweden, Ireland and The Netherlands, the reform process is competition oriented, but mediated by either and industrial or a consumer interest bias. In Austria, Belgium, Germany, Luxembourg,

Table 3.8. Dominant and mediating regulatory orientation.

		Dominant regulatory orientation		
		Competition oriented	National gas industry oriented	National public interest oriented
Mediating regulatory orientation	Competition oriented	UK	Austria Belgium Germany Luxembourg	Denmark France
	National gas industry oriented	Italy Spain Sweden		
	National public interest oriented	Ireland The Netherlands		

Denmark and France regulatory reform is not competition oriented, but national interest oriented. In these countries regulatory change is still rather limited and in fact only reflects the minimal harmonisation threshold of the gas Directive.

Table 3.8 is not only helpful to explain current country positions in regulatory reform but it is also helpful to explain the next steps in the EU reform process. It might be expected that both groups of countries take different positions in the EU reform process. The next section develops some thoughts on this issue.

3.6. The EU Reform Process

3.6.1. The regulatory challenge

Major regulatory challenges at the EU level are the harmonisation of the still rich variety in the management of the European pipeline system and the occupation of transport capacity by long-term contracts. National transmission companies all introduced own access conditions and tariffs, which are considered as a most serious barrier for cross-border gas flows. Already in November 2000 the EU concluded that 'Access conditions including tariff structure as well as levels vary widely from company to company and from pipeline to pipeline', whereas 'Little explanation has been given by most companies of deviation of tariffs (methodology and quantitatively)' (TPA conditions of major European gas transmission operators. A preliminary overview and attempt at comparison, Draft discussion document. Prepared by DG Tren/A2 – European Commission, latest revision: 9 November 2000, p. 31). In the same document it was concluded that standardisation would be a minimal requirement to allow for any comparison between the applied access systems in Europe. System operators used different balancing regimes (hourly versus daily) and penalties. More importantly, system operators did not provide any information about capacity availability and were vague in their information on terms and conditions of access. Furthermore, cross-border transmission of natural gas resulted in cumulating costs known as 'pancaking'. 'Pancaking occurs when a transportation service involves or potentially involves more than one transportation system if (i) charges are added up that are not cost reflective and/or (ii) network users pay for part of the system they do not use, mainly because of swaps occurring 'on the way' between point of input and point of outtake' (A Long-term Vision of a Fully Operational Single Gas Market in Europe, a Strategy Paper, Prepared for the Joint Working Group of the European Gas Regulatory Forum of the 4th meeting of the Madrid Forum on 2–3 July 2001, p. 17).

Consequently, the cross-border transport of natural gas turned out to be extremely demanding and complicated. So at the European level too the limited reach of the gas Directive became visible and urged for the need to continue the debate on gas market harmonisation. One of the important mechanisms in the post-Directive harmonisation debate is the Madrid Forum, the meeting place of all actors involved in gas market harmonisation. The Madrid Forum is recognised as an important fuel in the reform process.

3.6.2. The Madrid Forum

The European Regulatory Forum of Madrid (Madrid Forum) was set up by the European Commission in 1999, to establish a platform of information exchange and discussion on the internal gas market. Twice a year the European Commission meets with national gas market regulators, representatives of the Member States, system operators, gas suppliers and gas traders and network users to discuss harmonisation issues. The forum has been established as an informal device to facilitate and support the harmonisation of gas markets in Europe. It acts as a marketplace of different perspectives and viewpoints on gas market harmonisation and regulatory issues and a meeting place of the variety of interests at stake. Since 1999 the Forum discussed a variety of themes and topics.[8] Harmonisation of access, terms and tariffs, European wide management of capacity and system management and effective and efficient cross-border trade of natural gas have been on the agenda of the Forum thus far. What could be the contribution of the Madrid Forum to further harmonisation of the internal gas market?

First, it should be noted that the Madrid Forum is a voluntary meeting place without any political or legal strength or power. It is a meeting place of viewpoints, perspectives and interests. All actor groups currently operating in the European gas market participate in the meetings and bring their points of view to the floor. So the Madrid Forum established a meeting place for the diversity of interests at stake. An important participant in the Madrid Forum is Gas Transmission Europe (GTE), the organisation of European system operators and transmission companies. Furthermore, the Council of European Energy Regulators (CEER), uniting the national gas market regulators

[8]See website of the Madrid Forum.

in Europe participates in the process. So all important actor categories are represented in the Madrid Forum and have the opportunity to bring their viewpoints to the floor and to exchange them with other participants.

Second, the Madrid Forum is an informal meeting place without any legal powers. The main focus is on knowledge production and information exchange and not on preparing or deciding new harmonisation rules. Given the complicated technicalities and the rich variation in the management of European gas pipeline systems, the exchange of information on system technicalities and management is important. Thus far, the Madrid Forum discussed several papers and reports providing information on the technicalities and management of the gas systems. This type of information is basic in order to derive at any common operational and workable rules on system access and management. The Madrid Forum initiates the production and availability of this kind of necessary information for regulatory harmonisation.

Third, the Madrid Forum allows all actors involved to express their specific viewpoints on issues of gas market harmonisation and liberalisation. For instance, CEER, addressed a paper to the Madrid Forum listing the important issues to be settled in the harmonisation process from the viewpoint of European gas market regulators.[9] The system operators united in GTE were invited to express their view on capacity and congestion management and access tariffs.[10] In this way the Madrid Forum facilitates the emergence of common elements in the discussion which can be processed into agreements on the further harmonisation of gas markets. The expressed viewpoints also highlight points of disagreement, which are put on the Forum agenda for further discussion. In Section 7 of the GTE Tariff Report, it is expressed that 'There is a consensus on the fact that in the short- to medium-term transparency, clarity of the rules and non-discrimination for each individual tariff systems are much more important for facilitating gas trade than any forced harmonisation between tariff systems' (p. 19). A quote, clearly demarcating GTE's position on harmonisation of the European tariff systems.

Fourth, the Madrid Forum allows developing a commonly shared perspective on a competition-based internal gas market in Europe.

[9] Practical steps for developing a competitive European gas Market. A paper by the Council of European Energy Regulators (CEER), October 2000.

[10] GTE Capacity and Congestion Report, 20 June 2001 and GTE Tariff Report, 20 June 2001.

Such a joint perspective has been developed by the Joint Working Group of the European Gas Regulatory Forum for the 4th Meeting of the Madrid Forum.[11] The paper discusses the basics of a full competitive internal gas market and the harmonisation requirements for its effectuation. This kind of strategic information can act as common reference in the actual debate on regulatory reform. It makes aware of the difficulties of establishing a single gas market and shows how current positions and activities hinder or facilitate such a longer-term perspective. Furthermore, a comparison between the ideal of a fully competitive internal gas market in Europe with the current gas market allows specifying the harmonisation agenda. In this way, the Madrid Forum facilitates the discourse among the actors and tries to move current interest positions.

Fifth, the conclusions formulated at the end of the meetings of the Madrid Forum resume the major results of the meetings and in this way the Forum develops its own collective memory. In this way, the points of agreement and disagreement become clear and allow to work on the agenda in next meetings. Between the meetings the actors can work on consensus building by bilaterally debating controversial issues, which hinder further steps in harmonisation.

Finally, the Madrid Forum might also be expected to be a forum of lobbying, exchange of interest and a place to develop coalitions in the European political harmonisation arena. The relevant question in this regard is the type of contribution to be expected from the Madrid Forum in terms of convergence or divergence in regulatory reform. The type of coalitions that will develop in Europe will give part of the answer and the next section develops some thoughts in this respect.

3.6.3. Coalitions in the European harmonisation debate

The regulatory contribution in the harmonisation process will be decided by the extent national regulatory positions will and can move from an industry or consumer-dominated orientation towards a competition-dominated orientation. Such a change in dominant regulatory orientation in its turn will be decided by the strength of the national coalitions supporting these national positions in the international arena. All the three regulatory models, previously discussed in Section 4, might be expected to be supported by different types of national coalitions.

---

[11] A Long-term Vision of a Fully Operational Single Gas Market in Europe, A Strategy Paper (draft), Prepared by the Joint Working Group of the European Gas Regulatory Forum for the 4th meeting of the Madrid Forum on 2–3 July 2001.

Table 3.9. Conditions for convergence/divergence in gas market regulation in Europe.

National regulatory orientation	International coalition	Outcome
Competition oriented	National regulators EU authorities, and international consumer interest organisations	Convergence of access and market regulation
Gas industry oriented	National regulators and International gas industry	Divergence of access (and market) regulation
National public interest oriented	National regulator, national gas industry and national consumer interest organisations	Divergence of market and access regulation

In the first, competition-oriented regulatory model, it might be expected that European-oriented advocates of gas market competition dominate the coalition. Large gas consumers might be expected to dominate this coalition. In the second, industry-focused regulatory model, it might be expected that the national gas industry in general and the transmission company in particular dominates the national coalition. In the third model there might be a nation-wide coalition of politicians, government, gas industry and consumers supporting the consumer-oriented regulatory model.

For the further harmonisation of the internal gas market these national coalitions are important, because each of them will strive for different international coalitions. Table 3.9 specifies the assumed relationship between the dominant national regulatory orientation, the international coalition the national interest coalition will seek and the type of outcome that might be expected in terms of convergence/ divergence of gas market regulation.

The table indicates that competition-oriented national regulatory models and an international coalition consisting of competition focused national regulators, international consumer interest organisations and EU agents will support regulatory convergence and gas market harmonisation. The second and third national regulatory models will search for international coalitions opposing convergence in regulation and further harmonisation of the internal gas market. Both types of regulatory models will maintain the existing regulatory divergence in access regimes and thus will hinder a further openness of markets. In the second case, national regulators will develop coalitions with the international gas industry advocating the continuation of diversified access conditions to pipelines. In the third case, the national regulator will develop coalitions with domestic gas industry and national consumer interest organisations in this way ascertaining continuation of the specific national regulatory model.

Only in the first case, dynamics towards convergence will be stronger than towards divergence.

3.7. Conclusion

Not only legal regulatory reform will decide the further harmonisation of the internal gas market. Economic and political forces will put pressure on the harmonisation process too. Consumers everywhere in Europe might induce economic pressure. European competition authorities also are important in this regard. In case countries cannot decide about harmonisation of access conditions, the competition authorities might tackle this problem from the perspective of competition instead of gas market harmonisation. Furthermore, increasing political pressure might be expected from the EU. Here the increasing import dependence of the EU might be an incentive for harmonisation. So the joint interest of EU countries in good access conditions to gas fields outside Europe might be another – longer-term – incentive for gas market harmonisation.

Annexe of Chapter 3

Perceptions of market actors of accessibility of national markets in 2001–2002

1. A first indicator of actual regulatory performance is the perception of market actors of the accessibility of national transmission networks. Table a[12] illustrates the rather wide discontent on this issue, irrespective of the regulatory regime. Only market actors in the United Kingdom favourably evaluated the access to the UK pipeline system. It should be noted however that access demand was rather restricted in the first year of liberalisation.

Market actors also were rather discontent on the actual access to storage facilities. In countries offering TPA to storage, the actual access conditions turned out to be rather poor, irrespective the regulatory regime (see Table b).

Comparison of Tables a and b shows that access to storage was even worse than access to the transportation pipelines. Here too, only the conditions in the United Kingdom were perceived rather favourable. In all other countries these conditions have been perceived with great

[12] The data displayed in Tables a and b are taken from DRI-WEFA, Report for the European Commission Directorate General for Trade and Energy to determine changes after opening of the gas market in August 2000, Volume 2 Country reports, July 2001. Data are based on restricted interviewing of market actors in several European countries.

Table a. Perception access to gas pipeline infrastructure.

Perception access conditions			
Major obstacle	Unfavourable	Rather favourable	Favourable, but not perfect
Belgium Germany France	Austria Italy Spain	The Netherlands	United Kingdom

Table b. Perception access of gas storage facilities.

Perception access conditions			
Major obstacle	Unfavourable	Rather favourable	Favourable, but not perfect
Austria Belgium France Germany The Netherlands Spain			United Kingdom

discontent thus far, no matter the type of regulatory model. This is not surprising, knowing that some countries did not open storage facilities for third parties and knowing that access to storage is still heavily debated in other countries, like The Netherlands.

A final EU-wide discontent was observed on the theme of national balancing regime (see Table c). On this issue too, the United Kingdom is the only positive exemption.

The tables show the problems market actors faced in the first year of liberalisation in getting access to the pipeline system which is necessary to switch supplier. The data collected by the EU furthermore show that the discontent was basically caused by the incumbent national gas companies. Despite the requirement of unbundling, the incumbent companies continued their rather powerful market position simply by means of controlling the access conditions for the transmission pipelines. Quotes of large gas consumers in different countries intending to switch gas supplier and gas traders trying to enter the market are informative in this respect. The box gives some examples of the observed barriers.[13]

[13]All quotes are taken from the country reports of the DRI-WEFA report.

Table c. Perception of national balancing regime.

Perception balancing regime			
Major obstacle	Unfavourable	Rather favourable	Favourable, but not perfect
Belgium Germany The Netherlands	Austria France Spain		United Kingdom

- 'Despite unbundling of activities by the main incumbent, access capacity by newcomers only exist in theory' (Austria).
- 'Although liberalisation of the market was to be expected, sooner or later, there was – and still is – too much legal uncertainty to conclude short-term contracts, given the danger of no alternative suppliers'(Belgium).
- 'The deregulation and the opening of the gas market in Denmark does not meet the requirements of the Gas Directive. Although a new legislation on the paper implements the Directive, the state-owned gas monopoly DONG A/S is still a monopoly, which controls the entire gas market. This is due to a regime based on negotiated access with very high fees and a strong protection of DONG against any attempt of competition exercised by the Ministry of Environment and Energy, who is the owner of DONG and also – at the same time – the regulatory body of the entire energy sector' (Denmark).
- 'GDF has published tariffs on voluntary basis available to all market participants. Access conditions are published for all injection points in France. However, out of all entry points into France (five in all), entry for new participants is limited, in practice today, to one injection point (Taisnières H)' (France).
- 'Access fee for the grid is too high. The biggest problem is pancaking. Each grid operator wants individual fees, what they ask for is unreasonably high. Often the price of transport in Germany is as high as the price at border level (up to 1.5 Euro cents/kWh)' (Germany).
- 'There is considerable regulatory uncertainty as access proposals have not been finalised, never mind implemented. Access is only available through negotiation, rather than on published terms as proposed by the regulator. There is little transparency in the market in relation to terms and conditions, a factor which is exacerbated as tariffs are currently only provisional and may be

subject of rebates. Gaining access to capacity is also extremely difficult due to the alleged degree of congestion (...)' (Italy).

- 'Gasunie should apply the same rules towards its own customers as it does to third parties. The current situation is an obvious example of a discriminator system in practice' (The Netherlands).

The quotes in the box show that the access problem was EU-wide and predominantly caused by the dominant position of the national transmission companies. Of course it is not surprising that the unbundled vertically integrated companies, controlling European gas supply for so many years, resisted change that affected their market position. The access conditions offered by the system operators were discouraging in terms of transparency of the cost components, the location of entry points maximising distance-based tariffs, the information demands regarding the specification of gas qualities and the penalties in case the balancing requirements were not met.

The quotes furthermore show the typical national manifestation of access problems. For instance, Dutch industrial gas consumers accused Gasunie of favouring customers buying both volume and transport over customers only buying transport capacity from Gasunie. Industry also accused Gasunie for choosing the entrance points of the Dutch at large distance of the attractive load centres in the Western part of the country, in this way maximising the distance related costs of the gas transports of third parties. In France GDF was exposed to the same type of criticism. In France too, the distance-related transmission tariffs were maximised by the chosen locations of entrance points in the French pipeline system. A general criticism heard all across Europe was that the transmission companies did not give any information on the availability of transmission capacity. Simply saying that there was no capacity available, was enough to refuse specific access demands. In many cases the transmission commitments of the long-term contracts were used as the ultimate argument to refuse access.[14]

2. A second indicator of regulatory performance is the volume of gas transported under the third party regime and to see whether this volume is higher in countries with a comprehensive regulatory framework.

[14] Long-term contracts indeed minimise available transmission capacity in Europe. See also the chapter on gas markets.

Table d. Regulatory model and gas transport under TPA.

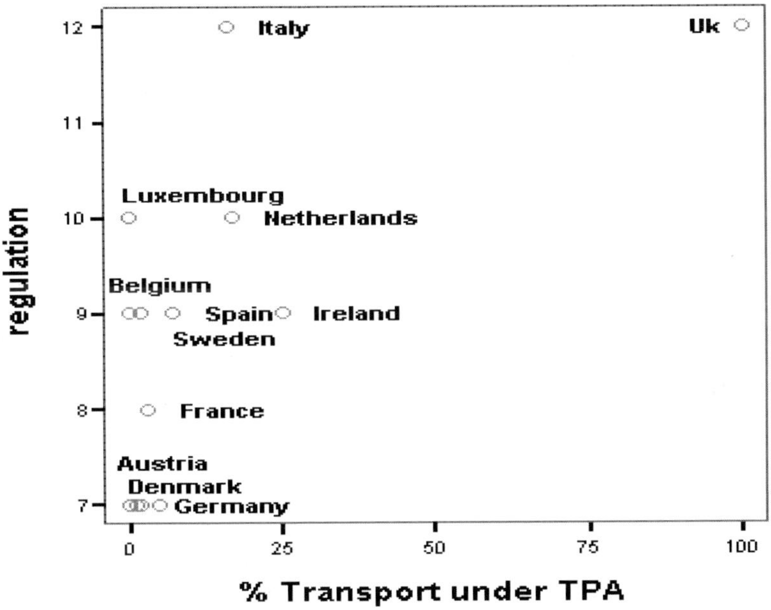

Table d shows the percentage of gas transport under TPA in different regulatory settings.[15] The United Kingdom strongly deviates from the general picture displayed in Table d. All gas in UK is transported under TPA, because of the country's independent system operator. In Italy, the second country with the most comprehensive regulatory model, transport under TPA is not that well-developed yet. Ireland and The Netherlands take middle positions with respect to TPA in transport. In the other countries gas transport under TPA is still restricted, whereas the degree of comprehensiveness of regulatory models differs.

A similar undefined relationship appears between comprehensiveness of national regulatory model and the switch of supplier by large industrial consumers and electricity producers. See Table e. Here too, the United Kingdom takes a derogating position, with switching of 90% in the large consumer segment.

[15] Country positions on regulation have been calculated as the sum of the scores on both dimensions of comprehensiveness of regulation. Data on transport under TPA are taken from Commission of the European Union, Commission Staff working paper, First benchmarking report on the implementation of the internal electricity and gas market, Brussels, March 2002, p. 21.

Table e. Comprehensiveness of national regulatory model and % switch of supplier by large consumers.*

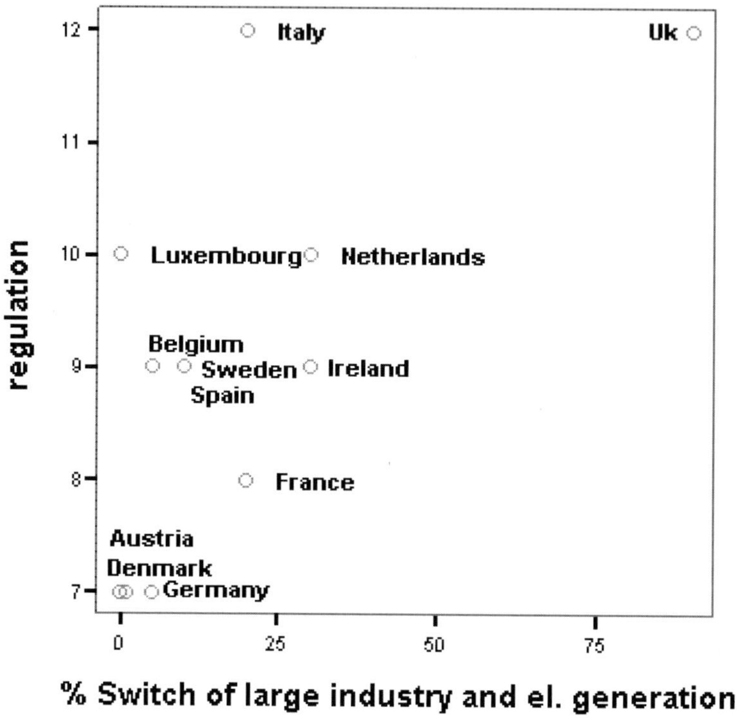

*Country positions are based on the same data and have been calculated in the same way as in table d. Data on % switch represent upper limits reported in the EU Benchmark report.

Next to the United Kingdom, switch of supplier develops clearly in Ireland and The Netherlands. France takes an in between position with some 20% of switch in the large consumer segment. Switching of supplier hardly appears in Austria, Denmark, Germany and Luxembourg.

Tables d and e hardly show correlation between the comprehensiveness of the national regulatory model and openness of market as indicated by gas transport under TPA and switch of supplier.[16]

[16] The significant observed correlation ($r = 0.63$) between comprehensiveness of regulatory model and gas transport under TPA was caused by the United Kingdom. The same holds for the correlation ($r = 0.65$) between regulatory model and switch of supplier. Without UK both correlations disappeared.

The lack of correlation clearly indicates the influence of other factors on the openness of gas markets, for instance lack of transport capacity or lack of choice of supplier. There are some indications that lack of capacity and market dominance of incumbents indeed are serious barriers in the gas markets in Europe.

References

Bergman, L., Brunekruft, G., Doyle, C., von der Fehr, N.-H., Newbery, D.M., Pollitt, M. and Régibeau, P. (1998). *Europe's Network Industries: Conflicting Priorities, Monitoring European Deregulation 1: Telecommunications*. London: CEPR.

CEER, Practical steps for developing a competitive European gas Market, Madrid, 2000.

Commission of the European Communities, Commission Working Paper Completing the Internal Energy Market, Brussels SEC (2001) 438.

Commission of the European Communities, Gas Directive gd/30/EC, 22/6/98.

Commission Staff Working Paper, *First Benchmarking Report on the Implementation of the Internal Electricity and Gas Market*, updated version with annexes, Brussels, March 2002.

DRI-WEFA, Report for the European Commission Directorate General for Trade and Energy to Determine Changes after Opening of the Gas Market in August 2000, Volume 2 Country reports, July 2001.

GTE, Capacity and Congestion Report, Brussels, 2001.

GTE, Tariff Report, Brussels, 2001.

Hancher, L. (2003). Harmonization of EU Gas Markets: the EU Gas Directive. In Arentsen, J. Maarten and Rolf W. Künneke, Eds., *National Reforms in European Gas*, Elsevier, London.

Joint Working Group of the European Gas Regulatory Forum, *A long-term vision of a fully operational single market for gas in Europe*, a (Draft) Strategy Paper, Prepared by the Joint Working Group of the European Gas Regulatory Forum for the 4th meeting of the Madrid Forum on 2–3 July 2001, p. 8.

Stern, J.P. (1998). *Competition and Liberalization in European Gas Markets. A Diversity of Models*, The Royal Institute of International Affairs, London.

Chapter 4
Energy Regulation: Convergence Through Multilevel Technocracy

CHRISTOPHE GENOUD, MATTHIAS FINGER AND MAARTEN ARENTSEN

4.1. Introduction

This chapter addresses the final question raised in the introductory chapter on regulation, the question on dynamics of convergence and/ or divergence within and across the electricity. With Bennett (1992: 219) we consider convergence as a process of 'becoming' rather than a condition of 'being' more alike. In this perspective, we are not so much interested in assessing similarities but in identifying elements where dynamics towards greater similarities or differences in the future can be expected in and between electricity and gas regulation. Since this study is among the first of its kind and regulatory change in the electricity and gas sector only recently started, the content of this chapter is largely exploratory in nature.

The chapter is structured as follows. Sections 2 and 3 summarise the convergence–divergence dynamics in electricity and gas regulation respectively. Section 4 explores convergence–divergence dynamics of regulation across electricity and gas and finally, Section 5 draws conclusions on regulatory harmonisation in European electricity and gas.

4.2. Converging and Diverging Dynamics in the Electricity Sector

At least two converging dynamics can be identified from the analysis of the 17 case studies presented in Chapter 2 on electricity regulation. The first pertains to the post-liberalisation regulatory frameworks, while the second refers to regulatory institutions. To recall, three models of electricity sector liberalisation are observable in the 17 studied countries, namely the market oriented, the public service oriented and the mixed models.

- The *market-oriented* model is characterised by its explicit reference to competition creation and efficiency as the primary objectives of the reforms. Emphasis is put on the promotion of competition for the benefit of customers in terms of price and quality. Although redistributive goals and public services objectives (PSO) are not forgotten, they constitute corrective mechanisms and as such are only secondary objectives. Regulation in this model is largely designed to create and then maintain competition in the sector in the interest of the consumer mainly through antitrust regulatory intervention. This model can be found in England & Wales, and to some extent in Germany and Luxembourg.
- At the other extreme of the spectrum, the *public service-oriented* model emphasises the importance of electricity as a vital commodity that requires specific protections and safeguards in a competitive market. Generally, the concepts of 'competition' or 'market' are not used, and if they are it is only on exceptional occasions. Respectively, PSOs are defined and described in detail and referred to as the main goal of the regulatory reform designed for the citizen, not the consumer, as it is the case in France and Portugal, Spain.
- In between these two ideal types, the *mixed model* is definitely the most popular in European countries, although the differences between countries can sometimes be substantial. However, what characterises these countries is the explicit combination or tension between liberal measures such as the creation of an efficient market and the willingness to implement large-scale redistributive and PSO objectives. Very often, both liberalisation and regulation design seek to create the condition for an efficient market taking into account the public interests. These interests can rank from ensuring the security of supply to implementing environmental measures.

As shown in Chapter 2, these models largely influence the shape and content of the post-liberalised regulatory frameworks on substantial issues such as public services obligations (PSOs) and environmental measure. This observation tends to indicate that political and institutional variables represent important factors explaining differences among countries. These factors can in fact slow down or even prevent the emergence of convergence dynamics on the long term. However, one element of these regulatory frameworks is widely shared and indicates the presence of a convergence driving force in spite of the importance of these models on the regulatory design. Indeed, the almost universal diffusion of the regulated Third Party Access (rTPA) model (except for Germany) is the most spectacular convergent feature, and other elements of the regulatory frameworks might be the object of further convergence, as we will discuss below, even if rules, scope and scale of the regulatory frameworks do still widely differ. The 2003 Directive indeed moved the minimal set of regulatory requirements to a higher level compared to the 1996 Electricity Directive. The acceptance of this new Directive has forced the countries still on the minimal level of regulatory change, to move ahead in establishing the necessary regulatory conditions for competition. The Directive requires the independent system operator and the introduction of the regulated model for third party access (TPA) to transmission.

Quite surprisingly, it is also on the institutional level that similarities and convergence can be observed in the electricity sector. The large domination of the 'independent regulatory authority' (IRA) model over the two others (respectively the 'politicoadministrative' and the 'quasi-judicial' models) can be considered as the illustration of a strong and coherent converging dynamic, at least for two reasons. First, because these IRAs were especially created for the purpose of regulating the newly liberalised sector. As such, their design is therefore less subject to institutional and organisational path dependency pressures as it could be the case if they were the outcome of the transformation of an existing entity. Second, because all of them, with the exception of UK and Sweden, have been created in a short and congruent time period – between 1999 and 2000 – this similarity in time and context (transposition of the EU Directive), makes it easier to detect converging mechanism in their design process.

Differences in regulatory authorities between and within the three models can be substantial in terms of power, missions, instruments, legal status and autonomy (see Chapter 2). But, contrary to the situation on the general regulatory frameworks level, where convergence is present but weak, convergence dynamics are here already very powerful. First, it is striking to note in many regulatory authorities activity reports that they plan to increase both their financial resources and staff in order to be fully operating. In other words, regulatory authorities consider that their importance and centrality within the regulatory process should and will increase in the future. An explanation of this phenomenon can be found in the second convergence dynamic – the tendency towards the integration of gas and electricity regulation within the same authority for effectiveness and efficiency reasons. Finally, the 2003 Directive requires the establishment of an independent regulator and no longer accepts the ex post regulatory model, as in the case of Germany.

As for other dimensions such as the missions, the power and the independence of the regulatory authorities, it is too early to observe any distinctive existing convergence and go beyond what has already been discussed in Chapter 2.

4.3. Converging and Diverging Dynamics in the Gas Sector

Gas market reform in the EU region undeniably has been initiated by EU regulation. Without the gas Directive, the current reforms would not have taken place. The Member States had to be forced by the EU legislation to initiate any change in gas market regulation. The gas Directive as such was already a weak compromise. Compared to the electricity Directive, the gas Directive allowed much more variety in regulatory reform and went less far in establishing the necessary conditions for liberalising gas markets and introducing competition. In short, the starting point for gas market reform and regulatory harmonisation was not that promising and clearly worse as in the case of electricity. In the perspective of this EU wide initial resistance, the regulatory reforms processed in the two years after the inauguration of the gas Directive, therefore, appear rather astonishing. In 2000 nobody could foresee the legal changes that have been initiated some two years later. Since August 2000 the minds of public authorities as well as large parts of the gas industry have clearly changed.

The gas Directive clearly provided a strong converging dynamic in gas market reform, but, given its weak and minimal content, it cannot provide for the overall explanation for the state of the art of gas market reform at the end of 2002. All countries could have gone for the minimal options offered by the gas Directive, but only some countries did, whereas the larger part of the countries did not. At least at the legal level, more countries than initially expected, processed the non-minimal options of the Directive. The negotiated TPA-model is now the minority model in the EU region, whereas everybody expected it to become the majority model. The enthusiasm for the 'Chinese Wall' unbundling model is declining, in favour of the legal and ownership unbundling, gradually pushing the necessity of an independent system operator. Almost all countries developed plans for the full opening of the gas market, whereas the Directive only required partial opening. So at the legal level – the surface – the converging dynamics have been rather strong in the last two years and at some points clearly overruled the diverging dynamics in the gas market reform.

The convergence and divergence are shown in the similarities and differences between country positions in the change process. Post-reform national regulation of the gas market can be classified in three models:

- Competition oriented, in case the Member State's dominant focus is to commit to the development of competition in view of efficiency and price decrease,
- National gas industry oriented, in case the Member State's dominant focus is to protect the national gas industry,
- Public interest oriented, in case the Member State's dominant focus is to balance the diversified interests of gas industry and gas consumers.

Countries can be positioned on a scale of comprehensiveness of regulation, indicating the degree of progress made in establishing the necessary regulatory conditions for competition and to assure non-discriminatory and transparent access to the national gas market. The analysis showed that, thus far, only two countries meet the 'ideal' of the full comprehensive regulatory model (Italy and UK) while all other countries are still in a process of establishing the necessary regulatory conditions for competition. But there is a clear division in this group of countries, between those heading for the full comprehensive regulatory model and those countries still on the minimal regulatory level. The countries heading for comprehensive regulation share a willingness to establish competition in the home market, but have not consolidated the required regulatory conditions yet, either in terms of regulatory function or in terms of regulatory capacity. For instance, Belgium, Ireland, France and Spain did develop the necessary regulatory functions, but without facilitating independence and autonomy of the gas market regulator. In these countries the government is the ultimate decision-making regulatory authority, which is generally conceived as a second best option for gas market regulation.

With respect to regulatory capacity, the professional, independent and autonomous gas market regulator is generally recognised as the ideal model, based on professionalism and expertise of the regulatory authority and not driven by political, industrial or national interest.

The openness for these kinds of interests is the core argument contra a government-dominated regulatory function and pro the independent gas market regulator. In Chapter 3, the industrial and national interests' mediation in gas market regulation has been identified as an important driver in the regulatory change process. Both types of interests not only prove helpful to explain the current country positions in regulatory change, but also hold diverging dynamics. The dominant orientation in regulatory reform provides the line of demarcation in the EU region, differentiating between countries focusing on competition, but without ignoring national and industrial interests and countries focusing on national and industrial interests, but without ignoring harmonisation requirements of the EU. The reforms in the first group of countries are already beyond the minimal levels of the gas Directive. In the latter group of countries, reforms are still on the minimal level of regulatory change and these countries probably only will move in the reform process by legal force of the next generation of the EU legislation.

Institutionalisation of convergence

The recent amendment of gas Directive[1] indeed moved the minimal set of regulatory requirements to a higher level compared to the gas Directive 98/30/EC. The acceptance of the amendment will force the countries still on the minimal level of regulatory change, to move ahead in establishing the necessary regulatory conditions for competition. The amendment requires the independent system operator for gas (and electricity) transmission and announces the introduction of the same model for gas (and electricity) distribution in 2005. It furthermore requires the introduction of the regulated (ex ante approval) model for TPA to transmission pipelines and extends TPA to storage and LNG-facilities. Finally, the amendment requires the establishment of an independent gas market regulator and no longer accepts the regulatory model based on competition authorities only, as in the case of Germany. The gas market regulator should be established and should operate as an independent regulatory body, prescribing or approving ex ante the access conditions and tariffs. Together with the full opening of the gas market in 2005, these amendments are clear improvements of the initial gas Directive and clearly move the regulatory reform further in

[1]Proposal for a directive of the European Parliament and of the Council amending Directive 96/92/EC and 98/30/EC concerning common rules for the internal market in electricity and gas, Com (2001) 125 final, Brussels, 2001.

the direction of competition. Moreover, as the gas Directive, the amendment too pushes convergence in gas market reform. The amendment clearly moves the Community-wide regulatory reforms one step further in the direction of the ideal set of regulatory requirements (necessary conditions) for competition. So at the legal level, EU legislation continues to catalyse regulatory convergence and the empirical analysis in Chapter 3 shows that community legislation indeed continues to be necessary for making progress in regulatory harmonisation.[2]

Next to EU legislation there are two more catalysts mechanisms pushing convergence in gas market rules – the informal debates and agreements of the Madrid Forum and benchmarking. As explained in Chapter 3, the Madrid Forum contributes to the harmonisation process in many ways and clearly contributes to convergence in gas market reform. The added value of the Madrid Forum in this respect basically lies in the technical or expert approach of controversial regulatory issues. For instance, the Madrid Forum discusses access problems in terms of methodologies. It considers possible options and compares current access practices in Europe, initiates research on specific issues, invites interest groups to express their views, suggests guidelines for good TPA practice and monitors and assesses the commitment of European system operators to these guidelines. Each meeting is concluded with clear findings regarding what has been accepted as regulatory standard or practice and what unsolved issues continue to be on the agenda. In this way, the Madrid Forum irreversibly pushes the regulatory reform in European gas and a comparison of the minutes of the first and the latest (sixth) meeting clearly shows the progress made in this respect. In this way, the Forum, for instance, has contributed to push the entry–exit system as the standard tariff system in gas transmission in Europe. The Forum also continues to repeat specific, rather controversial, issues as transparent information on capacity availability in the European pipeline system and congestion management. In this way, the expert instead of the political approach of the Forum clearly pushes convergence in European gas market reform and gas market regulation.

Finally, the benchmarking practice of the Commission also might release a converging dynamics in gas market regulation. The Commission increasingly monitors and compares regulatory practices

[2]At the end of 2002 the EU had not finally decided yet about the amendment of the gas (and electricity) Directive.

in Member States and publishes the state of the art in regulatory reform. In this way the 'weak spots' in the European reform process become visible and the Commission does not hesitate valuing practices deviating from favourable standards. The Commission has already published two benchmarking reports on progress in regulatory harmonisation in gas (and electricity), undeniably contributing to the diffusion of common regulatory ideas, practices and ideals in the Community.

Public policy convergence

Another observable converging dynamic, initiated by the EU Commission, is the new way the Commission is designing the security of gas supply policy, in particular the way the Commission links the establishment of the competitive internal gas market and the short and long-term security of gas supply. According to the Commission, the proposals on European energy infrastructure and on the security of supply issue are necessary because of the new competitive-based internal gas market. In the proposals it is stated that the new legal order in European gas urges for reconsideration and calibration of current responsibilities to secure continuation and security of gas supply in Europe.[3] In the proposal for a directive on security of supply it is acknowledged that: '(...), market opening and sufficient physical interconnection of markets are not in itself a guarantee of a liquid market based on secure gas supplies from both indigenous and external sources. It is therefore important that the current profound changes in the market and the transition towards a fully open internal energy market are supplemented by transparent and unambiguous new policies defining the overall framework as well as clear roles and responsibilities of the different market players with regard to security of supply within the new market context' (p. 2 of the proposal for a directive).

Furthermore, on Page 4 of the proposal it is stated: 'Liquidity forms an indispensable ingredient of a properly functioning EU internal gas market. Measures aimed at enhancing liquidity, such as spot markets, incentives for new gas supplies from internal and external sources and non-discriminatory authorisation procedures for building storage and LNG facilities, should, among other things, be fully taken into account by security of supply policies in a competitive market environment'.

[3]See Com (2001). 775 final on energy infrastructure and the Web page of the Energy Directorate on the proposal for a directive on the security of supply.

Increased future dependence on gas imports from outside the EU-region as well as differences in current access position to gas resources in the EU region, makes, according to the Commission, the establishment of a competitive-based internal gas market a common interest of the EU. The Member States will become more and more dependent on each other to secure their longer-term supply of natural gas.

According to the Commission, the new internal market order requires to reconsider the security of supply policy of the EU, but at the same time a well-functioning competitive internal market is suggested as an important 'instrument' of the security of supply policy. In this way the Commission links gas market reform with security policy and changes the perspective on gas market reform. In the security proposals, mechanisms of competition, such as spot markets and access to storage facilities are defined as security devices and their establishment, a common interest of the Member States' security of supply position. So increased mutual dependence of the Member States to secure gas supply in the longer term might act as converging dynamic in gas market harmonisation.

4.4. Regulatory Convergence across Electricity and Gas

In energy market harmonisation, electricity and gas for long have been considered and treated as obvious two sides of the same coin. Until 1994, regulatory harmonisation of electricity and gas was an integrated theme in the EU debate. But in 1994 the debate split. According to Stern, it was the European gas industry that caused the separation in the harmonisation debate. In 1994 it became clear that the European gas industry refused to accept the idea of TPA to the gas network. In order to make progress in the harmonisation process the track for electricity and gas went different directions (Stern, 1998, 90ff).[4] For that reason, the gas Directive could only be agreed upon some 18 months later than the electricity Directive and according to many, rather weak in content compared to its counterpart for electricity. Especially on TPA the electricity Directive was more straightforward than the gas Directive, which more explicitly accepted and negotiated TPA as an adequate model for the introduction of competition.

[4]According to Stern, the electricity industry was willing to acccept the idea 'that liberalisation could bring efficiencies and advantages to the industry. (...), By contrast, many of the established actors in European gas industries still regarded the introduction of liberalisation as the equivalent of the end of civilisation' (Stern, 1998, p. 91).

Furthermore, the requirements regarding the opening of the gas market were less advanced compared to electricity.

Convergence on regulatory rules and institutions

In 2003 the harmonisation of electricity and gas regulation seem to be back on the same track again. As shown in both empirical regulation chapters, at the legal level many things have changed in electricity and gas market regulation since the inauguration of both Directives. In electricity as well as in gas, the regulated third party model has developed as the dominant model and almost all countries decided to open up their electricity and gas markets beyond the thresholds set by both Directives. Almost without exception gas regulators have been established simply by extending the regulatory mandate of the already operational electricity regulator. Both in function and to a less extent also in capacity, national regulators for electricity and gas are quite similar. At the end of 2002 only Austria had different types of regulatory interventions for TPA to the electricity and gas grid. For electricity, the regime is based on the ex ante and for gas on the ex-post approval model. All other countries adopted similar regulatory models for electricity and gas.

The regulatory convergence between both energy sectors continues as shown by the 2003 electricity and gas Directive 2003/54/EC. Again, it should be emphasised that it is convergence at the legal level, but the similarities in the proposals for electricity and gas are nevertheless remarkable. With this new Directive, it is clear that the Commission is heading for a complete regulatory harmonisation between electricity and gas. Especially on the issue of TPA, the progress made in ideas is quite remarkable. Regulated third party access is suggested as the future standard for the EU-region in both electricity and gas. The new Directive furthermore, harmonises the functions and capacities of the sector regulators for electricity and gas. The regulator must be independent and in a position to approve ex ante access conditions and tariffs for electricity and gas transmission. With this change Germany is brought back into the majority regulatory model of the EU region with sector regulation in electricity and gas.

So in terms of the establishment of the regulatory *function,* it seems that the convergence across electricity and gas regulation is stronger than the diverging dynamics. Despite the differences between electricity and gas (see the concluding chapter on markets), its regulatory function has matured in rather similar ways and is facing rather similar challenges. These challenges are basically technical in nature and relate to the technical management and co-ordination of the networks, the transparency of access conditions, transmission capacities and the management of (cross-border) congestion. As indicated in the concluding chapter on markets the current electricity networks in Europe have been constructed as national networks with international connections basically serving foreign assistance in case of emergency and shortage. In gas, the pipeline infrastructures clearly have been constructed with a European dimension due to the concentrated location of gas fields. But despite these differences, in gas as well as in electricity, improvement of the transparency, accessibility and management of the networks is currently dominating the regulatory agenda everywhere in Europe. This irreversible tendency towards further harmonisation of the regulatory function is also visible in the 'Europeanisation' of the national idea of Public Service Obligations. The concept of 'public services' has progressively gained importance at the European level over the years. Divergent national traditions pertaining to public services objectives (PSOs) have often constituted obstacles on the path to greater liberalisation. Very often the definition of the content of PSOs, their financing mode and the organisation of their production are the duty of the national political authorities, the EU has progressively gained influence on these matters over the years, through the gradual and slow apparition in the EU legal order of concepts such as 'General economic interests', 'universal' of 'public' services. The emergence of this vocabulary was made necessary by the wide differences between definitions among countries, which used this fuzziness strategically in order to oppose or wrongfully interpret the EU intervention and actions. In the latest amendment proposals these PSOs are harmonised for electricity and gas.

Convergence on the PSOs

Aside from the converging dynamics inherent in the process to mature the regulatory function in gas and electricity, the strategic EU energy policy proposals might also release converging dynamics. Energy is one of the pillars of the European construction: the European Coal and Steel Community and Euratom were designed to lay the foundation of a single or coordinated European energy policy. The creation of an internal energy market quickly became a secondary objective of the completion of the Single European Market. The dependence of a European Energy Policy upon the creation of a corresponding internal market has largely determined the shape and content of this policy. This policy is founded today on four pillars (Matlary, 1996) – price transparency, the development of energy infrastructures (Trans European Networks, TEN), tax policy and the European Energy Charter on investments. In this context, it is possible to consider that the Commission will use the TEN program to build new cross-border transmission capacities, in order to attain competition and convergence objectives. In other words, the Commission as a 'policy entrepreneur' might seek to use the energy policy as a levy towards the completion of internal markets and greater integration of markets and regulatory frameworks. Therefore, the increasing common interest of the Member States to support a European energy policy could play a significant role in the convergence of the regulatory function between the Member States and between electricity and gas. The emerging importance of natural gas for electricity production clearly will strengthen this tendency (see also the concluding chapter on markets).

Overcoming variety of regulatory capacities

Whereas the converging dynamics in maturing the regulatory function in electricity and gas are quite strong thus far, convergence is less strong in the maturation of the regulatory *capacity* across Europe. As shown in the empirical chapters, on the institutional level, differences in regulatory frameworks, such as the powers, missions or the main regulatory authorities and the remaining role of governmental and judicial authorities are still important. There is a tendency towards the independent regulatory model, but within this model the variety in terms of regulatory capacity is still quite substantive across Europe. Some countries do have independent regulators for electricity and gas, but mandated them only with advisory competencies. Other countries did establish already more advanced regulatory bodies and put them independently and autonomously in charge with the regulation in electricity and gas, at distance of the supervising national government. This independent institutional position of the regulatory bodies is not yet common in Europe, but for several reasons, explained in the previous chapters, the most adequate institutional model. Moreover, independent regulatory authorities might induce inherent converging dynamics resulting from the organisational logic pushing them to use their resources and asymmetrical position versus their governmental principals in order to move up the scale of the regulatory powers from simple consultative body to independent regulatory bodies equipped with autonomous regulatory competencies and powers.

In sectors where technology and market mechanisms are dominant features as it is in the energy sectors, flexibility, responsiveness, credibility and professionalism are key elements for efficient and effective regulatory processes. Regulators are both for institutional and organisational reasons better equipped than administrative or political authorities. Moreover, independent bodies are very often precisely designed to increase the credibility, the legitimacy and the efficiency of a specific policy process (Majone, 1997) characterised by uncertainty and a high degree of technical and scientific qualifications.

Supranational move of regulatory powers

The EU is clearly heading in this direction given the new electricity and gas Directive on this point. But the Commission is also looking further ahead when it comes to sector regulation in European electricity and gas. In a harmonised internal electricity and gas market the next step in regulation will be the upscale of regulatory bodies and competencies from the National to the European level. With the creation of a 'European Regulators Group for Electricity and Gas' on 11 November 2003, the Commission indicated the need for a pan-European regulatory supervisor. This group, exclusively composed of regulatory authority unlike the existing forums (Madrid, Florence), was created to 'facilitate consultation, coordination and cooperation among the regulatory bodies in the Member States, and between these bodies and the Commission'. It formally acts as an advisory group for the Commission.

For the short-term institutionalisation and harmonisation of the regulatory authorities, the Commission is relying strongly on informal contacts between national regulators in forums as the Council of European Energy Regulators (CEER), the Madrid and Florence forums. The CEER created in March 2000, brings together 16 European energy regulators and acts as a co-operation platform between national regulatory authorities. These same regulators also take an important part in the Florence and Madrid Forums which brings together since 1998, electricity and gas regulators, the European Commission, transmission system operators, electricity and gas traders, consumers, network users and power exchanges. These two types of 'informal' organisations work outside the direct control of national governmental authorities and have substantially contributed in the preparation of the new European Directives for the European Commission. They also helped through the organisation of discussion and the exchange of information and expertise to resolve practical regulatory problems such as cross-border tariffs in a depoliticised context. These informal entities are also functional answers to the growing complexity of public policies formulation and regulatory task in electricity and gas sectors and encourage the search for optimal solutions outside the national context. These forums or information and expertise platforms represent the ideal infrastructure to initiate policy learning and emulation process, in which actors learn from experience and best practices. Coercion is not the sole driver for policy transfer, the perception of the necessity of a transfer and lesson-drawing are also powerful instruments (Dolowitz and March, 2000: 13). By organising these platforms, the Commission suggests best practices, models and original solutions (Radaelli, 2000: 26), although indirectly since it is either the experts or the actors of the forums that build together the outcomes. In this way these informal mechanisms facilitate the Commission's goal of internal market creation and regulatory convergence.

4.5. Conclusion: The Constitution of a Multilevel Technocratic Regulation

Although liberalisation and regulation goals still differ among countries in terms of scope, scale and focus, as Chapters 2 and 3 show, common regulatory objectives can be identified. The creation of effective gas and electricity markets is the core feature of liberalisation and regulation in all the countries. With its willingness to deepen and accelerate liberalisation, the Commission contributes to reaffirm the rationale for the reforms and confirm their strategic importance for European integration. Convergence on PSOs might be slower and less intense, due to the highly political dimension of the issue at national levels and the important divergence that still exists in national tradition. However, the slow emergence of a public service doctrine at the EU level and the new amendments of the electricity and gas Directives can be considered as a powerful driver for convergence on minimal standards for the definition of European public services obligations. The convergence of energy policies is however more uncertain. National traditions and institutional divergences constitute heavy obstacles. It is however not impossible that market dynamics, and especially markets integration, could affect the pertinence and effectiveness of existing energy policy instruments.

This raises the issue of policy instruments. Under this category, we must distinguish two things: policy tools and the institutions that are supposed to use them. As for the tools to liberalise and especially to regulate energy sectors are already quite similar across countries and sectors. Competition regulation is conducted through M&A and through antitrust tools. More precisely, the development of incentive remedies, such as divestiture or sale of assets tend more and more to replace structural remedies. As in sectoral regulation the rTPA model is already almost universal. In environmental matters, certification, levee and taxes to encourage renewable energies are already widely used and new instruments might emerge together with the integration of markets. As for institutions and more precisely regulatory authorities it can be expected that although convergence is due to happen it will be on a slower pace and on a smaller case. Institutional dynamics and political inertia will certainly limit the degree to which regulatory authorities will become alike.

However, three types of actors have appeared as credible convergence dynamics bearers: the EU Commission, the informal institutions at the EU level and the national regulatory authorities. The statutory powers of the Commission and its central situation in the EU energy regulatory regimes place it in a strategic position. Initiator and supervisor of the liberalisation and re-regulation processes, the Commission can follow various strategies to achieve its goal of internal market creation and harmonisation. If the legislative and the competition regulation paths remain at the centre of its intervention, it has the opportunity to progress without the burden of political bargaining by relying on the many informal institutions that bring it expertise, legitimacy and credibility. These same informal institutions are expected to gain more power and centrality, especially as long as the legislative path is blocked at the EU level and act as heavy convergence drivers both at the EU and national levels. Finally, national regulatory authorities and especially independent regulatory authorities will most certainly give greater momentum to convergence dynamics that originated at the supra-national level. Therefore, an important conclusion regarding convergence in energy market regulation is that convergence could be achieved less through the political process, than through an alternative technocratic and network shape-like process integrating national and supranational levels in one emergent but coherent governance mechanisms. We believe so for the following reasons.

The actions of the Commission – the use of competition regulation to intervene in sectoral regulation, the use of energy policy, the use of regulation by information – the growing importance of European law and the actions of the national regulatory agencies as relay and support for the Commission are, as we saw above, important drivers for harmonisation. But they are also elements of a wider process of technocratisation of regulatory intervention based on expertise and independence. What is new is the nature of this process integrating multilevel regulatory intervention. This new regulatory system, parallel to the political process, includes national and supra-national actors and deals with both national and supra-national issues. Although there is still no clear empirical evidence that this process will lead to a full-scale technocratic multilevel regulatory system, there is substantive indication that, at least, this is a credible hypothesis.

As we saw above, the emergence of these informal institutions and processes is a functional answer to the growing need for harmonisation, a need which is itself largely created by markets dynamics and business strategies. In other words, what the political process, for structural and institutional reasons, could not deal with, for instance the issue of cross-border tariffs, was simply transferred to a new nascent informal multilevel arena. We think that this situation will remain pretty much identical in the long term and that the power of this new regulatory process will develop in the future for at least three reasons.

1. The process is favourable to the industries because in their perception a depoliticised arena is better able to address the regulatory issues at stake, and favourable to the Commission, whose objectives are to achieve the internal market.
2. The process is in tune with the general evolution of public governance, where non-political arenas and technocratisation of public intervention develop according to delegation theories (Majone, 1997) and new governance approaches (Kickert and Koppenjan, 1997).
3. The globalisation or at least the supra-nationalisation of regulatory problems is a reality, which somehow needs to find adequate regulatory mechanisms to deal with them.

Although these evolutions raise serious questions pertaining to public governance in general, and energy market regulation in particular, for instance in terms of democratic accountability, we do not think that this is just a temporary development. It is here to stay. Therefore, the emerging political question is how to design the most efficient, credible and accountable regulatory system.

References

Bennett, C. (1992). What is policy convergence and what causes it? *British Journal of Political Science* 21(b): 215–233.

Bernstein, M.H. (1955). *Regulating Business by Independent Commission.* Princeton University Press, Princeton.

Commission of the European Communities (2002). *First Benchmarking Report on the Implementation of the Internal and Gas Market.* Bruxelles, p. 138.

Commissions de Régulation de l'Electricité (2001). *Communication de la CRE relative à la liberté du négoce de l'électricité du 6 septembre.* CRE, Paris, www.cre.fr.

Crozier, M. and Friedberg, E. (1977). *L'acteur et le système.* Seuil, Paris.

Dehousse, R. (1997). Regulation by networks in the European Community: The role of European agencies. *Journal of European Public Policy* 4(2): 246–261.

Dolowitz, D. and Marsh, D. (2000). Learning from abroad: The role of policy transfer in contemporary policy-making. *Governance* 13(1): 5–24.

Eyre, S. and Lodge, M. (2000). National tunes and a European melody? Competition law reform in the UK and Germany. *Journal of European Public Policy* 7(1): 63–79.

Freeman, G.P. (1985). National styles and policy sectors: Explaining structured variation. *Journal of Public Policy* 5(4): 467–496.

Kickert, J.M. and Koppenjan, J. (1997). *Managing Complex Networks, Strategies for the Public Sector.* Sage, London.

Levi-Faur, D. (1999). The Governance of competition: the interplay of technology, economics, and politics in European Union electricity and telecom regimes. *Journal of Public Policy* 19(2): 175–207.

McGowan, F. and Wallace, H. (1996). Toward a European regulatory state. *Journal of European Public Policy* 3(4): 560–76.

McGowan, L. and Cini, M. (1999). Discretion and politicization in EU competition policy: the case of merger control. *Governance: An International Journal of Policy and Administration* 12(2): 175–200.

Majone, G. (1996). *Regulating Europe.* Routledge, London.

Majone, G. (1997a). From positive to the regulatory state: Causes and consequences of changes in the mode of governance. *Journal of Public Policy* 17(2): 139–167.

Majone, G. (1997b). The new European agencies: Regulation by information. *Journal or European Public Policy* 4(2): 262–275.

Majone, G. (2000). Innovations politiques et retards institutionnels. In J. Vandamme, and F. van der Mensbrugghe (dir.). *La régulation des services publics en Europe,* Editions ASPE, Paris, pp. 355–375.

Marin, B. and Mayntz, R. *Policy Networks, Empirical Evidence and Theoretical Considerations* Campus Verlag, Frankfurt.

Matlary, J.H. (1996). Energy policy: from a national to a European framework? In H. Wallace, and W. Wallace, Eds., *Policy Making in the European Union,* Oxford University Press, Oxford, pp. 257–277.

Neven, D., Papandropoulos, P. and Seabright, P. (1998). *Trawling for Minnows, European Competition Policy and Agreements Between Firms.* CEPR, London.

Pelkmans, J. (2001). Making EU network markets competitive. *Oxford Review of Economic Policy* 17(3): 432–456.

Phillips, C.F. (1985). *The Regulation of Public Utilities, Theory and Practice.* Public Utilities Reports, Inc., Arlington, Virginia.

Radaelli, C.M. (2000). Policy transfer in the European Union: Institutional isomorphism as a source of Legitimacy. *Governance* 13(1): 25–43.

Rodrigues, S. (2001). *La nouvelle régulation des services publics en Europe. Energie, postes, télécommunications et transports.* Editions TEC&DOC, Paris, p. 694.

Schmidt, S.K. (1998). Commission activism: Subsuming telecommunications and electricity under European competition law. *Journal of European Public Policy* 5(1): 169–184.

Schmidt, S.K. (1997). Sterile debate and dubious generalisations: Europeans integration theory tested by telecommunications and electricity. *Journal of Public Policy* 16(3): 233–271.

Scharpf, F. (1993). *Games and Hierarchies in Networks. Analytical and Empirical Approaches to the Study of Governance Institutions.* Campus Verlag, Frankfurt.

Scharpf, F. (1999). *Governing Europe: Effective and Democratic?* Oxford University Press, Oxford and New York.

Stern, Jonathan P. (1998). *Competition and Liberalization in European Gas Markets. A Diversity of Models.* The Royal Institute of International Affairs, London.

Tuot, T. (2001). La loi du 10 février 2000 un an après l'installation de la régulation du marché électrique. *Cahiers juridiques du Gaz et de l'Electricité* 573, pp. 51–57.

SECTION II
Market

Chapter 5
Electricity and Gas Markets in Europe: Competition and Integration Analytical Framework

JEAN-MICHEL GLACHANT AND DOMINIQUE FINON

5.1. Introduction

This part deals with the characterisation of the different national electricity and gas markets in terms of competitive intensity after the first waves of the reforms, and the chances of integration of these markets in two single European electricity and gas markets.

Competition reforms shake as never the European electricity and gas industries. Under the momentum of the European liberalisation process, radical changes occurred and are occurring, but in very different ways. At the European level the policy principle of economies' integration by competitive exchanges justifies the suppression of barriers to electricity and gas trade between national markets and the mandatory introduction of provision of third party access in national legislation and the suppression of the monopolies to supply, import and export. But each national industry follows its own track in the framework defined by the two first European electricity and gas Directives, some of the countries which have been exemplified having besides pioneered the liberalisation movement (Gilbert and Kahn, 1996; Midttun, 1997, 2001; Bergman *et al.*, 1999; Arentsen and Kunnecke, 2003; Glachant and Finon, 2003).

So the two respective European Directives on electricity and gas voted in 1996 and 1998 and the reforms that they brought about in the 15 EU Member States and the Associate States have not promoted a homogenous competitive model and a fortiori a single integrated European market up to 2004. This institutional variety limits the scope of integration of national markets. The European electricity and gas markets remain multidomestic but with some increasing elements of integration between them at the level of the wholesale market. And each of the two will remain so up to the concretisation of the second wave of reforms which will occur with the transcription of the 2003 Directive on electricity and gas markets which mandates a deepening of the liberalisation reforms and creates the conditions of institutional and regulatory convergence.

Market integration is still not the main determinant of the competition profile of each national market, given the remaining market rules and technical limitation on inter-countries trade. Indeed, the technological dependency upon a fixed infrastructure for trade in the two industries and the need for a centralised technical co-ordination complexify significantly the conditions allowing hourly or daily wholesale markets integration. It needs in particular a set of harmonised rules and balancing mechanisms between the different countries which is a necessary condition for the development of a liquid spot market. So the competitive profile of each national market will have to be independently identified, including elements of integration with other markets as one aspect of the competitive forces.

5.2. Market Theories

The analysis of the competition profile of each national market is based on the market theories. We make the hypothesis that basic conditions (technical infrastructures for the trade, regulation and market rules) and industrial structures determine the effectiveness of the competition and the market efficiency, via the possibility of balancing of dominant firms' market power in the oligopolistic game. But we have to refer to the two concurrent theories to characterise the competition profile of the different national markets: the standard market theory and the contestable market theory which insists upon the technical and jurisdictional conditions of credible competitive threat and virtual competition rather than the structural conditions.

Standard market theory considers that competition must imply a number of producers and suppliers in competition on the different levels of the value chain, under the hypothesis that the structure determines the players' conducts and the efficiency of the markets (Armstrong *et al.*, 1994 ; Newbery, 2001). So in the electricity and gas industries conditions are the following :

- The competitive activities – power generation, gas bulk purchase, import on the upstream side; wholesale and retail supply on the downstream side – must be horizontally de-integrated among a significant number of market players.

- The organisation of the interface between the regulated and competitive parts of the chain of activities must limit the exercise of market power by incumbents. Given the technical peculiarities of the energy networks and the consecutive necessity of technical coordination by the system operator, the risk of market power exercise by the operator suggests to separate clearly the competitive activities from the transmission system operation, that it is not the case in telecommunications where such a technical source of market power does not exist.
- Vertical integration between generation (or gas bulk purchase) and supply under hierarchical structure (or long-term contract) must be limited in order to reduce entry barriers.

The concurrent model of virtual competition, proposed by the theory of contestable markets (Baumol *et al.*, 1982), opposes its conclusion to the competitive model by considering that the main objective of competition could be simply reached by suppressing the legal and technical barriers to entries. Credible threat by entries would be sufficient to incite incumbents to productive and allocative efficiency. In other words industrial structures might be preserved. They do not determine a priori inefficient behaviour in vertically integrated industry if the conditions of contestability are respected. It means that horizontal concentration and vertical integration between upstream and downstream activities could be preserved if non-discriminatory access to the respective grids is guaranteed to the incumbents' potential competitors with a complete unbundling of the networks which could be justified in this sense. From this point of view, in the electricity and gas industries, it would be sufficient to have credible threat exerted under these conditions by the foreign competitors which are themselves incumbents in the adjacent national markets, that avoid divestures of the incumbents' production assets.

The menu of various provisions defined by the electricity and gas Directives and transcribed in national laws is set at the confluence of these two models which reflect the respective references of the proponents and the opponents of the straightforward market liberalisation. This inspires our analytical framework in two directions:

- the focus on the technical, institutional and regulatory conditions of competition in each country to apprehend the conditions of market contestability,

- a just as much important focus on the industrial structures and the capital markets.

Empirically, indeed, vertical and horizontal integration has been observed as one of the main barriers to the development of competition in national markets. (Bergman *et al.*, 1999; EC-DG TREN, *Benchmarking Report*, 2002). Given the initial situation characterised by the presence of national and regional monopolies and the (quasi-) vertical integration of the two value chains in gas and electricity, the main barrier to the establishment of competition appears to be the presence of incumbents. The development of active wholesale and retail markets is not at all guaranteed by the simple enactment of a third party access provision and opening up of final markets. A complementary set of market rules, market institutions and re-structuring measures is necessary to allow effectiveness of competition, development of entry opportunities, and trade development in the two industries.

5.3. Competition Intensification and Market Integration

European market integration will contribute to the effectiveness of competition on each national market. The main factors of power trade between countries are the differences of cost structures of electricity generation of the companies along the successive hourly and weekly periods of the year, which are supposed to be reflected in the spot price making. In the gas market the differences in production and import costs and short-term wholesale prices are also the drivers of the intra-European trade. The playing field which is the first in time in the process of market integration is the wholesale trade (bilateral contracts, daily arbitrage on marketplaces) which concerns large buyers: distributors, suppliers, traders and very large industrial consumers. To levelise the playing field implies several institutional conditions on which the European Commission is acting up:

- To lower technical, regulatory and informational barriers to trade,
- To harmonise market rules and institutions, in particular the transmission access.

Beyond the economic incentives to trade, the different characters of national markets will determine entrants' industrial strategies and will have some influence on inter-country trade in the competition. In the European Union few electric markets are highly dynamic, having a slow growth despite some new uses because of the maturity of the majority, and most of the gas markets are so, with only dynamism on the segment of power generation, at the exception of the recent Iberian and Irish gas markets. So we cannot wait the same rush in competition by attraction of foreign entries and steady competitive sales from abroad than in other dynamic sectors such as telecommunications. Moreover, the merger of the national markets in a single European electricity or gas market is not only a matter of trade. It implies the opening up to players other than the incumbents, in particular foreign producers, suppliers or traders on the respective wholesale markets, and entrants which seize opportunities to acquire companies or assets by the help of mandatory divesture or privatisation process.

The comparison of national prices on different market levels will be indicative of the reality of market integration between countries. Competition with market integration is supposed to make prices converge. This convergence in a period of overcapacity exerts its effects by the price decrease towards short-term marginal costs, in the limit of the technical capacity of trade. In a period of mid-term anticipation of capacity deficit in different countries, the price convergence would occur on another basis towards the long run marginal cost, in a situation which implies less exchanges between countries but maintains the mutual influence between markets. But the divergence of prices or the lack of convergence at a certain moment will reflect the existence of remaining technical and regulatory barriers.

5.4. Characterisation of the Competition Profiles of National Markets

In order to characterise the competitive profile of each national market with its potentialities of exchanges with the other markets in its European market environment (see Table 5.1), we refer to the concept of 'market attractiveness' of the Mackinsey matrix of strategy analysis for focusing on the potential of attractiveness for entrants and the set of barriers on the competitive game.[1]

Mackinsey's concept of attractiveness is characterised by different parameters: market size, market growth, price level (which determines

[1]The MacKinsey consulting group has proposed a pioneering framework in the sixties to assess the opportunity of investment in various field of activities by crossing the characteristics of markets in these fields in terms of attractiveness and the trump cards of the concerned enterprises market share, competitiveness in terms of price/quality ratio, trademark, etc.

Table 5.1. Characters of the competitive profile of national electricity and gas markets.

Attractiveness		Accessibility	
Size	Overall size, segment size	Technical accessibility	Interconnections, Internal congestions
Growth	By segments: industrial, commercial, power generation, etc.	Regulatory accessibility	Degree of openness, Type of TPA, System operation unbundling, Transmission pricing, Regulatory institutions
Profitability	Price level	Commercial accessibility	Wholesale price index, Exchanges (burses, hub) or pools, Balancing market
		Industrial and capital-market accessibility	Horizontal integration Vertical integration (Production/import – supply) Regime of ownership

the incumbent's profit level), but also the intensiveness of the competition and the market barriers, in particular the market structure (market shares of the main competitors), the vertical integration (including long-term contracts), and the role of the technology in the competition. In order to upgrade the concept of attractiveness, we shall distinguish in a narrow sense the market attractiveness which characterises their commercial potential (market size, growth, profitability), and the market accessibility which defines the various barriers to entries in different dimensions: technical, regulatory, commercial and industrial.

So, in the two next chapters of this section, the characteristics of the respective national markets and their degree of integration with the other markets are analysed in terms of competitive potentialities. Recognising inherent differences between the electricity and gas markets, in part due to the specific allocation of gas resources, the section starts out with a two-level market perspective on the gas market, while the electricity market is discussed in a more traditional one-level analysis. The two levels refer to the trade and distribution of gas domestically at the national level, and the two-sided oligopolies that balanced between the major producers and major national companies at the international level, with strong geopolitical overtones.

The final chapter analyses the potential for European unification of electricity and gas markets and for the so-called electricity and gas convergence at the national level as elements of development of competitive forces.

References

Arentsen, M. and Kunnecke, R., Eds. (2003). *National Reforms in European Gas*. Elsevier, Amsterdam, pp. 245–282.

Armstrong, M., Cowan, S. and Vickers, J. (1994). *Regulatory Reform: Economic Analysis and British Experience*. The MIT Press, Cambridge (Mas.).

Baumol, W., Panzar, J. and Willig, R. (1982). *Contestable Market and the Theory of Industry Structure*. Harcourt, New York.

Bergman, L., Brunekreeft, G. and Doyle, C. et al. (1999). *A European Market for Electricity? Monitoring European Deregulation 2*. Center for Economic Policy Research, Londres.

European Commission EC-DG TREN. (2003). *Second Benchmarking Report on the Implementation of the Internal Electricity and Gas Market*. European Commission, Brussels.

Gilbert, R. and Kahn, E., Eds. (1996). *International Comparisons of Electricity Regulation*. Cambridge University Press, New York.

Glachant, J.M. and Finon, D. Eds. (2003). *Competition in European Electricity Markets: A Cross-Country Comparison*. Edward Elgar, Cheltenham.

Midttun, A., Ed. (1997). *European Electricity in Transition*. Elsevier, Amsterdam.

Midttun, A., Ed. (2001). *European Energy Industry Business Strategies*. Elsevier, Amsterdam.

Newbery, D. (2001). *Issues and Options for Restructuring the Electricity Supply Industries*. Cambridge University, Working Paper, Department of Applied Economics.

Chapter 6
European Electricity Markets: Variety and Integration

JEAN-MICHEL GLACHANT[1]

The idea of constructing an internal European market for electricity was launched about fifteen years ago as a consequence of the European 'Single Act'. However, there are still enough differences and separations between the Member States as between the non-Member States (like Norway or Switzerland). In fact, six years after the adoption of the Directive 96/92, the European 'internal market' for electricity is neither unified nor uniform. Even within the EU, each country appears to be cultivating its own preferred variant on the 'Single Market', and no one can accurately predict when and how the variety of national electricity markets may vanish into the rising tide of pan-European market rules harmonisations and convergences.

In this chapter we contend that convergence of Europe's electricity markets has not yet occurred, and is not imminent except for the creation of a loose continental market integration. This assertion is organised into two parts. The first part focuses on *'attractiveness'* of the European electricity markets before 1999 (deadline for implementing the European electricity directive), the second part on *'accessibility'* of the markets created by then. By *attractiveness* of electricity markets we mean their business potential in static terms (market size) and dynamic terms (market growth, price differential). We will see few highly dynamic markets in terms of growth and development of new uses: the electricity industry is a very mature one, so we cannot expect the same rush in competition and new entry like in the telecoms industry. Attractive or not, none of these markets will merge into a 'single European electricity market' unless it opens up to players other

[1] I thank Nicolas Chauvet (now in the French Electricity Regulatory Commission Staff) for the work of collecting and harmonising data.

than its national incumbent operators. That is the *accessibility* dimension of our typology. The opening of some access to European electricity markets started in every country but it is still far from being finished or harmonised.

In the case of the electricity sector, it is not enough to declare it 'open' to competition for desirable competitive mechanisms to appear in all dimensions of the reform (Glachant, 2002, 2003a; Holburn and Spiller, 2002; Joskow, 2002). On the one hand, in the absence of alternative infrastructures for the transmission and distribution of electrical energy, this industry retains strong 'imperfection' characteristics associated with traditional network industries: natural monopoly (owing to economies of scale and economies of scope) and asset specificity (in the sense of Glachant and Finon, 2000; cf. Williamson, 2000, 2002). On the other hand, the impossibility of stockpiling electrical energy, and the 'real time' loop flows externalities (congestion, counterflows) between the networks' users (producers and consumers), combine to create an economic rationale for maintaining central coordinating authorities (Smeers and Boucher, 2001; Glachant, 2002; Hogan, 2002). Finally, owing to the importance that the incumbent electric concerns assigned to vertical or horizontal, special or exclusive, relationships among the industry's players (production, transmission, distribution, sales), and with their primary energy sources (gas, coal, nuclear), this sector contains many forms of market power and strategic asymmetries between the various industrial and financial players (Newbery, 2000, 2002a,b).

The experience of the reforms has revealed[2] that creating competitive electrical systems is not a linear process, and that the reforms take place in a three-dimensional space (Glachant, 1998; Glachant and Lévêque, 1999; Glachant, 2000a,b). A first dimension of competitive reform involves its legal and regulatory aspects, amending laws and regulations governing the practice of electrical businesses and establishing rights to enter markets and access networks. A second dimension, equally important, pertains to the operational development of technical and commercial infrastructures for these competitive businesses, such as the organised markets. Finally, a third dimension (pre-eminently strategic) concerns industrial assets – the transformation of business structures and portfolios, as well as forms of corporate ownership. In practice, the

[2]As of 1995 I have successively studied the British, Norwegian, Swedish, Californian and German electricity reforms in detail. For each country I combined field research with collecting essential documents (Glachant, 2002a).

actual form assumed by competitive electricity reforms is uniquely determined by the combination of these three dimensions in a given country or area of Europe. This is why there is so much potential for diversity in electricity markets: in essence, the competitive processes are the outcome of interactions between three different dimensions of transformation. On these criteria, we find that, between 1999 and 2003, there is no 'single European electricity market' covering the entire EU. In reality, the European electricity markets are still neither very open nor unified in 2003, despite of progress in the implementation the first directive, and the largest national markets could be categorised in 'partially open' markets – 40% of total EU consumption with France and Germany – or 'not very open' – 20% of the total with Italy and Spain –, at the exception of the UK market (11%). In the end, we find none of these 16 national markets (EU markets and Norway) to be both fairly attractive and very open.

6.1. Limited Attractiveness of European Electricity Markets

The *attractiveness* of the Member States' electricity markets before the implementation of the European directive depends on their business potential in static terms (market size) and dynamic terms (market growth, price differentials). The concept of 'attractiveness' describes the relative attraction of the national electricity markets compared with one another on the basis of a handful of variables that measure their business potential. In static terms, we can reasonably measure attractiveness from the volume of domestic consumption. The dynamic measure of attractiveness will be provided by market changes, which we can measure by the physical growth in consumption volume. But a better measure of business dynamism is provided by the change in the real value of electricity sales which combine price level and size of the market. Accordingly, we have deflated the changes in the nominal value of electricity sales by national price indexes (producer prices or consumer prices). The combination of size and growth criteria gives us the attractiveness of the European markets. This attractiveness is then complemented by a profit sustainability criteria at the level of electricity prices (Glachant, 2000). As we consider the situation during the post-Directive period (1998–2003), the market characterisation will be made on the basis of the pre-reform situation.

Three segments of markets have to be considered, each with its very different institutional movements and business characteristics: industrial and transportation customers; commercial and tertiary customers; residential customers. Distinguishing these three market segments (Industry, Services and Residential) in the 15 EU countries plus Norway will make a total of 48 elementary electricity markets.

- The industrial market makes up the core of the eligible customer base defined in Directive 96/92/EC. While the largest in size (930 TWh in the EU in 1998), this market segment is already shrinking at the European level, having lost 9% in constant value between 1990 and 1998.
- The commercial market depends mainly on voluntary national measures for eligibility (until the next European Directive to come in July 2004). This is the smallest of the three markets, with 500 TWh consumed in 1998. But it is also the fastest growing, with a 22.3% increase in constant value in 1990–1998.
- Residential customers, who were eligible in 2003 in only six countries: Austria, Finland, Germany, Norway, Sweden and the United Kingdom. At 630 TWh, the market is mid-way in size between the two others. Its growth rate has been weak. Between 1990 and 1998, the increase was only 12.6% in constant terms – i.e., an annual average of less than 1.5%.

6.1.1. Markets size and maturity

The four largest markets – Germany (DE), France, the United Kingdom and Italy – make the core of European electricity, each accounting for between 12% and 22% of the single EU market (2100 TWh in 1998). These 'big four' represent a combined two-thirds of the total EU consumption. The next six mid-sized markets together represent one-third of the total EU consumption. By decreasing size, they are: Spain, Sweden, Norway (not an EU member), The Netherlands, Belgium and Finland. Then the seven smallest markets, each accounting for between 0.2% and 2.3% of the single EU market, add up to 10% of the total EU consumption. By decreasing size, they are: Austria, Switzerland,[3] Greece, Denmark, Portugal, Ireland and Luxembourg (Fig. 6.1).

In terms of maturity, in the period before the implementation of European electricity directive, there is a wide difference between the highest and lowest European market growth rates. At the aggregate

[3]Switzerland (CH in Fig. 6.1), which is closely tied to the operation of other EU continental markets, is added.

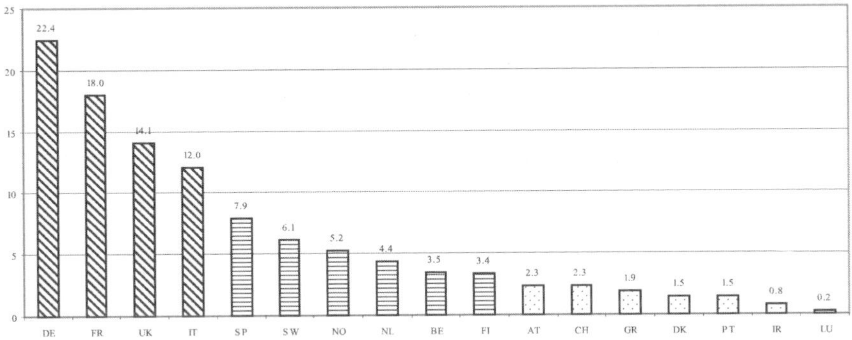

Fig. 6.1. Size of European electricity markets before the reforms (1998). *Note*: by volume and as% of the single EU market. *Note*: The total of the three categories exceeds 100% of total EU consumption, as two countries are not EU members.

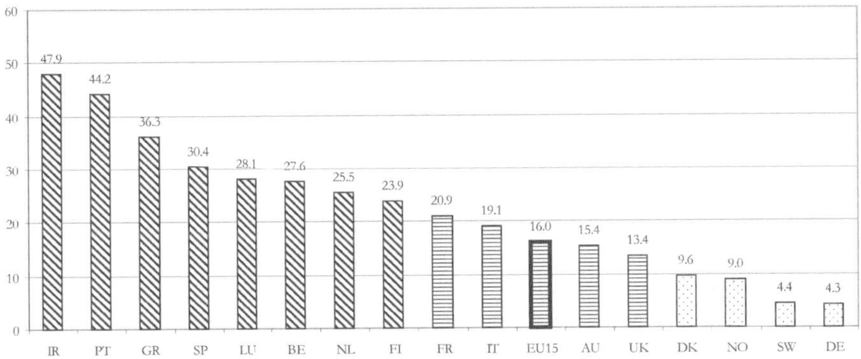

Fig. 6.2. Growth in electricity consumption by volume during the pre-reform period 1990–1998. *Sources*: Eurostat et Unipede, 1999.

EU level, the electricity industry has definitely reached maturity: in eight years, it has grown by an annual volume average rate of 1.9%. We can nevertheless classify the markets into three categories, both for volume growth (Fig. 6.2) and for constant value growth (Fig. 6.3, where the growth in electricity sales is adjusted by national price indexes).[4]

[4]National volume series are based on the data from the Electricity Information yearbooks of the International Energy Agency (IEA) (1996–1998 eds, OECD, Paris), except 1998, which we extrapolated from the national volume indicators provided in the UNIPEDE's Electricity Outlook 1999 (UNIPEDE, Brussels, 1999). The constant value indexes were computed by applying to each country the real national electricity price index published by the IEA in its Electricity Information yearbooks (European indexes of industrial consumption and residential consumption).

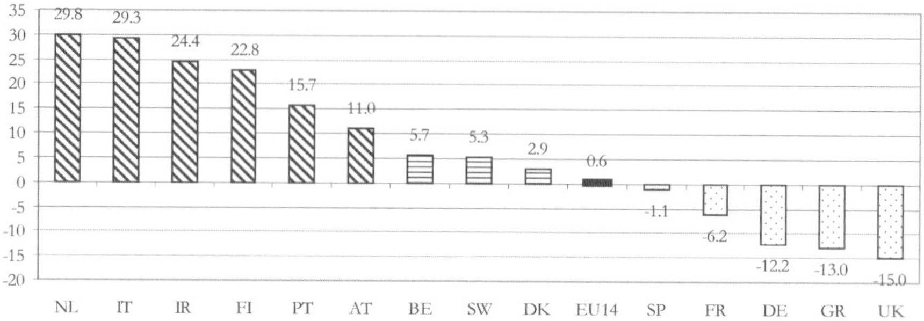

Fig. 6.3. Growth of electricity consumption in constant value terms, on the pre-reform period 1990–1998 (in% of 1990). *Sources*: AIE, Eurostat, Unipede, 1999.

By volume, there are three groups.

1. Eight high-growth countries (with rates ranging from 23.9% to 47.9% in eight years, i.e., annual rates of 3%–5%) with the countries in economic catching up Ireland, Portugal, Greece, Spain, with some more mature economies, Belgium, The Netherlands, Luxembourg and Finland.
2. Four medium-growth countries (from 20.9% to 13.4% in eight years, including the EU average of 16%, or an annual rate of 1.6% to 2.4%). By decreasing order, they are: France, Italy, Austria and the United Kingdom.
3. Four countries with distinctly weaker growth rates (from 4.3% to 9.6% in eight years, or an annual rate of 0.5% to 1.1%). By decreasing order, they are: Denmark, Norway, Sweden and Germany. Germany is thus the largest market, but the lowest-ranked of the 17 countries in terms of volume growth.

By adjusting value growth in electricity sales by national price indexes we calculate 'constant value growth'. In constant value terms, most of the growth rates drop significantly with the exception of Italy, The Netherlands, Sweden and Finland. The country breakdown is quite different. The EU14[5] constant value growth is not really different from zero (+0.6% after eight years). There are three categories of countries there. 1° six medium-growth countries: The Netherlands, Italy, Ireland, Finland, Portugal and Austria; 2° three small-growth countries: Belgium, Sweden and Denmark; 3° five negative-growth

[5]European Union less Luxembourg (data unavailable).

Table 6.1. Attractiveness of European electricity markets before the reforms (1998).

	Negative growth (from −15.0% to −1.1%*)	Medium growth (from +3.0% to +5.7%*)	High growth (from +11.0% to +29.8%*)
Large Size (12%–23 % of EU15 market)	France Germany United Kingdom		Italy
Medium Size (3.4%–8% of EU15 market)	Spain	Belgium Sweden	The Netherlands Finland
Small Size (0.2%–2.3% of EU15 market)	Greece	Denmark	Portugal Ireland Austria

*Cumulated growth on the period 1990–1998.
Sources: Eurostat, Unipede, 1999.

countries: Spain, France, Germany, Greece and United Kingdom with three of the four biggest EU countries.

Overall, the EU markets have already entered an advanced maturity phase, in which the value added by the electricity industry is nearly stagnant. It is revealing that, on the German market, the largest one, the value of electricity sales in real terms fell 12.2% between 1990 and 1998, despite a rise of more than 4% in the volume of electricity consumed. In most European electricity markets there is a future of slow but steady growth. Electricity is not the gas or the telecoms industry.

The relative attractiveness of each electricity market is characterised by cross-tabulating the size and growth indicators (Table 6.1).

In most cases, the effects of the three attractiveness factors – size, volume growth and value growth – are scattered. Three large-sized countries (Germany, France and the United Kingdom, whose combined consumption equals 55% of the EU total) exhibit negative value growth rates. Conversely, six countries with high value growth (Austria, Portugal, Finland, Ireland and The Netherlands) have small or medium-sized markets; their combined consumption represented 12% of the EU total. So it results that only five countries can be considered as more attractive than the others, since they combine either a large or a medium size with a medium or a high growth. Of which, Italy is the only country characterised by a large-sized market (12% of EU) and a high value growth (+29.%).

The four other countries – Spain, Belgium, Finland and The Netherlands – have medium sizes between 3.4% and 7.9% of EU. Spain and Belgium presented low value growth rates (−1% and + 6% respectively) with high volume growth (+27% and +30% respectively).

Only Finland and The Netherlands experienced both high value growth (+22.8% and +29.8% respectively) and high volume growth (+23.9% and +25.5% respectively).

6.1.2. Size, maturity and profitability of market segments

Given the limited attractiveness of the European electricity markets, we have to analyse it more precisely by market segments and by differentiating the situations between countries. There is a considerable difference in the growth rates of the three segments. Notably when the value of the Industry segment was on the decline (−9%) that of Services was still growing (+14.6%). There is also a great variety for each segment between countries in terms of size, growth and price.

With such price and growth differentials in the different market segments opened up to the competition, we could expect hard pressures on the incumbent companies in many countries and a fierce competition on this market field. In particular, the niche segments with high national prices and steady growth, or medium national prices plus positive growth, would be targetted by entrants coming from other countries or by national ones. But the opportunities and the constraints are not the same in the industry segment as in the service and residential segments.

In the commercial and residential segments there are several elements of differentiation which give advantages to the incumbent because they constitute barriers, costs of entry and elements of entry deterrence. The penetration on commercial markets and a fortiori on residential ones is not easy to manage for foreign enterprises, as for local outsiders. The price is not the unique dimension of competitiveness. First, competition includes the marketing responsiveness of the entrants, the technical treatment of the customer (customer file management, metering, billing), the range of services offered to customers (like multi-sites billing, multienergies offer), and the economies of scale and scope associated with these activities. Second, in technical terms these customers correspond to hundreds of connection points and meters for a supplier, and then the process of analysing, scheduling, balancing, metering and billing them is much more complicated and more hazardous than for big industry customers. Foreign or local entrants have to develop local marketing business for penetrating this segment. Third, the incumbent benefits from advantages such as the knowledge about the customers and the trademark effect.

Will these entries and their consolidation be easy to finance and to realise? Are their competitive advantages strong enough to finance the costs of entry and the marketing strategy to attract customers linked by loyalty to the incumbent? These questions have to be answered by business strategists. The answers will depend pretty much on the actual accessibility of each national market.

Industry market

The segment (930 TWh in the EU in 1998) displays a strong overall downtrend, with an already heavy decline in certain countries. In eight of the fifteen countries, the real value of electricity sales fell between 1990 and 1998, with a loss of 9.0% for all the EU14 markets combined. The 'negative growth' zone in the Union (France, Sweden, Portugal, Spain, Germany, United Kingdom and Greece), represented more than 70% of EU industrial consumption. The main reason is the combination of slow growth and improvement of the price level for the buyers. This regression of the industrial electricity markets is all the more remarkable as it predates the transcriptions of the European Directive, and therefore the steep industrial price falls of 1999–2000 (estimated at between 10% and 20% depending on the country and the type of supply) (Fig. 6.4).

For the five last countries (Portugal, Spain, Germany, United Kingdom and Greece), the value shed between −9.0% and −34.4% in constant value terms in 1990–1998.

Only three large or medium-size countries seem relatively more attractive than the rest (Italy, The Netherlands and Finland, totaling 23% of EU industrial consumption), with growth rates from 17.2% to 38.3% in 1990–1998. One of the reasons is the high or

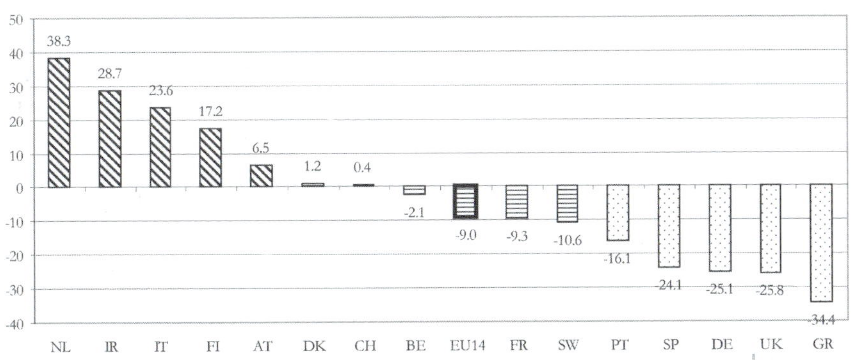

Fig. 6.4. Growth of industry markets on the pre-reform period (1990–1998). *Note*: Growth in constant value terms between 1990 and 1998.

Table 6.2. Attractiveness of industry markets on the pre-reform period 1990–1998.

	Negative growth (from 34.4% to –16.1%*)	Small negative or positive growth (from –10.5% to 1.2%*)	High growth (from +6.5% to 38.3%*)
High price €/MWh (80.5–63.6)	Portugal Germany		Italy Austria
Medium price €/MWh (57.3–49.4)	UK Spain	Belgium Denmark	The Netherlands Ireland
Low price €/MWh (45.6–30.1)	Greece	Sweden France	Finland

Note: Growth of EU industry market in value terms on 1990–1998: –8.88%.
Sources: AIE Eurostat, Unipede, 1999.

medium industrial prices which reflected some inefficiencies in their electric supply industries before the reform (Italy, The Netherlands) (Table 6.2.).

By referring the average European price of 54.4€ for one MWh (= 100%), actual prices for industry range in continental Europe from 80% in France to 166% in Switzerland (CH). Absolute best prices appear in Nordic countries like Finland (FI) and Sweden (SW), the cheap Swede (55%) being two times less than the European average and three times less than the expensive Swiss (Fig. 6.5).

With such price and growth differentials in the industry markets we may expect hard pressures on the incumbent companies in many

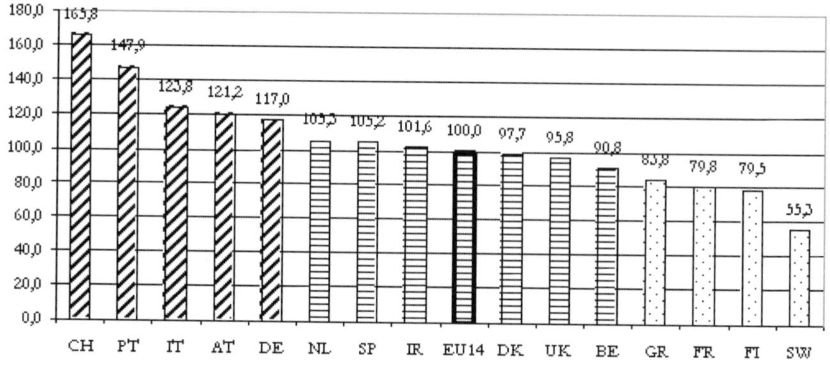

Fig. 6.5. Prices of electricity in the industry markets before the reforms (1998). (All taxes excluded; EU average price: 54.4€/MWh).

countries and a fierce competition on the industry market battle field. In particular, the six favourable niches of high national prices (Portugal, Germany, Italy, Austria), or medium national prices plus positive growth (The Netherlands, Ireland), would be tempting targets for the European competition.

Commercial markets

This market segment (500 TWh in the EU in 1998) displays a more growing profile than the Industry segment. The four top-ranking countries are Italy, Portugal, Ireland and The Netherlands, representing 20% of total EU commercial consumption. Their growth in real value terms was strongly positive, ranging from 30.3% to 52.2%. This compares with a EU14 average of 14.6%. The six middle-ranking countries are Belgium, Spain, Finland, Greece, Austria and Denmark, accounting for 18% of the EU14 total. Their value growth lies between 15.5% and 24.0%. Four countries performed badly: Sweden, United Kingdom, Germany and France (62% of EU in size), experienced low growth of 5–6% and the only negative growth with –3.3% in France (Fig. 6.6).

On the basis of these positive value trends, three markets emerge as more attractive than the rest: Italy, The Netherlands and Spain (24% of EU total) with 1990–1998 growth rates of 22%–52.2%. Italy can be (again) considered as the most attractive country with a large size (11.2% of EU) and growth rate of 52.2% (Table 6.3).

With an average European price of 77.2€ for one MWh (= 100%), actual prices for services and institutional range in continental Europe

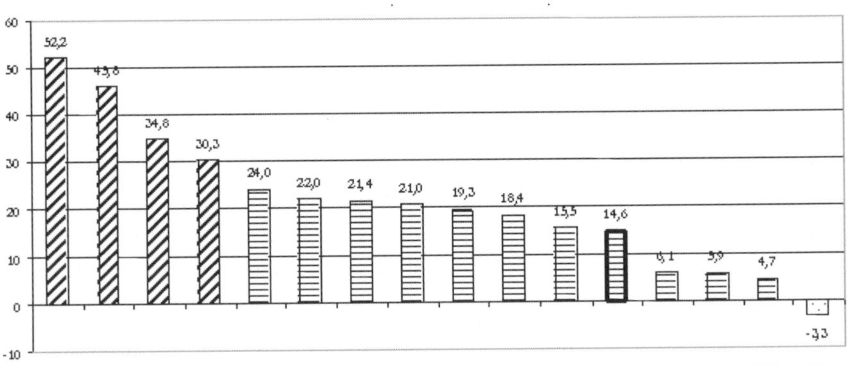

Fig. 6.6. Growth of commercial markets (pre-reform situation on 1990–1998). (In constant value terms).

Table 6.3. Attractiveness of commercial markets in the pre-reform period 1990–1998.

	Negative growth (−3.3%)	Medium growth (from 4.7% to 24%)	High growth (from +30.3% to 52.2%)
High price €/MWh (84.6–104.1)		Spain Germany Austria Belgium	Italy Portugal
Medium price €/MWh (71.5–77.6)		UK	The Netherlands Ireland
Low price €/MWh (43.6–65.1)	France	Denmark Finland Sweden Greece	

Sources: AIE, Eurostat, Unipede, 1999.

from 63% in France to 135% in Portugal (P). Absolute best prices appear in Nordic countries like Finland (FI) and Sweden (SW), the cheap Swedish price being two times less than the European average and two times and a half less than the expensive Swiss or Portuguese one (Fig. 6.7).

Residential customers

The opening up of the residential market concern in 2004 six countries: Norway (1991), Sweden (1996), the UK (1998), Germany

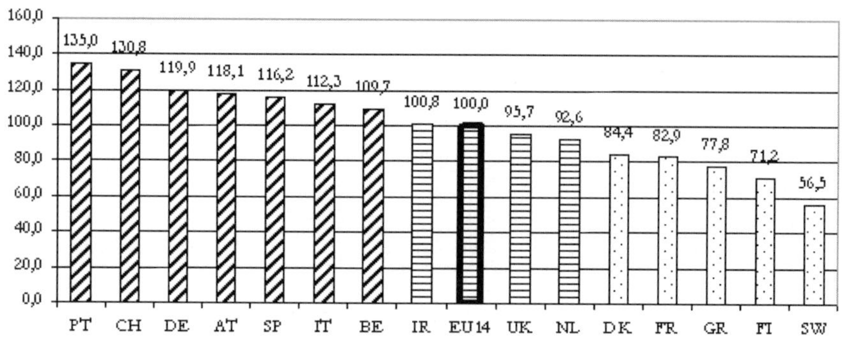

Fig. 6.7. Prices on commercial markets before the reforms (1998). (All taxes excluded; EU average price = 77.2€/MWh = index 100).

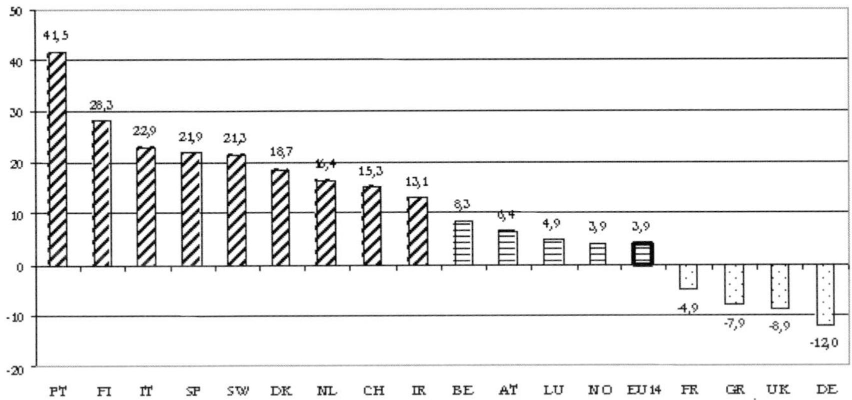

Fig. 6.8. Growth of residential markets on the pre-reform period 1990–1998. (in constant value terms).

(1998), Austria, Finland and Spain (all in 2003). Despite that the technical and commercial conditions of competition are still more stringent than on the commercial segment and that the opening up is more differentiated and slower than for the services segment, size and maturity of the markets give another element of appreciation of the competitive stake for the market players. The residential segment (620 TWh in the EU in 1998) displays a positive growth trend in constant value term, but more moderate than the services segment and with some countries with negative growth rates. The EU14 average growth is 3.9% between 1990 and 1998 but with a quite large range from a group of nine countries with a growth of more than 13%. A group of four countries – France, Greece, Sweden and Germany (61% of EU residential consumption) – posted a negative growth of −4.9% to −12.0% (Fig. 6.8).

On the basis of these trends, only four large or medium-sized markets can be considered as more attractive than the rest: Italy, Spain, Sweden and The Netherlands, which make up 26% of total EU residential market. Their 1990–1998 growth rates range from 16.4% to 27.9% (Table 6.4).

With an average European price of 99.9€/MWh, actual prices for residential range in continental Europe from 77% in Denmark to 121% in Germany (DE). Absolute best prices appear in Nordic countries like Finland (FI), Sweden (SW) and Norway (NO), the cheap Norwegian (43%) being two times less than the European average and three times less than the expensive German price (Fig. 6.9).

Table 6.4. Attractiveness of residential markets in 1998. (Growth of EU residential market in value terms, 1990–1998 = 3.9%).

	Negative growth (from −12% to −4.9%)	Medium growth (from 3.9% to 8.3%)	High growth (from +13.3% to 41.5%)
Large size (10% to 20% of the EU market)	France Germany United Kingdom		Italy
Medium size (3% to 10% of the EU market)		Norway Belgium	Sweden Spain The Netherlands
Small size (0.1% to 2.9% of the EU market)	Greece	Austria Luxembourg	Finland Portugal Ireland Denmark

Sources: AIE, Eurostat, Unipede, 1999.

Ignoring the technical and marketing constraints, could we expect hard pressures from foreign and local entrants on the incumbent companies with such low price and growth levels in the residential market in many countries? On the contrary, will the markets with high national prices and medium growth, or those with medium national prices plus positive growth, be targets for the European competition?

The answer cannot be the same as in the services markets, a fortiori as in the industry markets. Residential market is a mass retail market

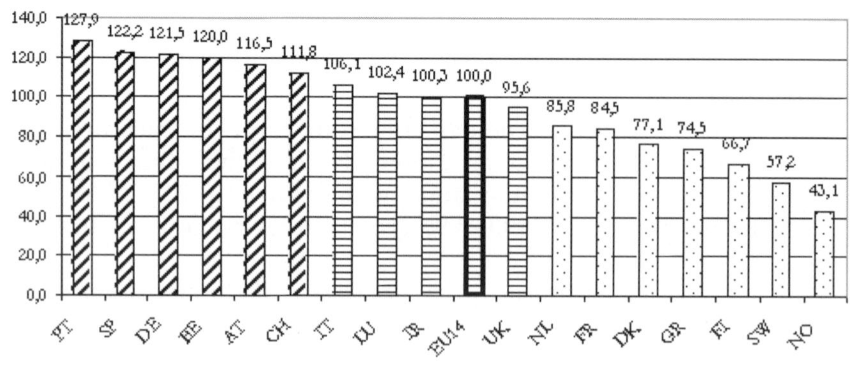

Fig. 6.9. Prices on residential markets before the reforms (1998). (All taxes excluded; Index 100 on EU average price: 99.9€/MWh). *Sources*: AIE, Eurostat, Unipede, 1999.

and it is considered as a very specific electricity market. First, price competition cannot be imported from abroad by foreign enterprises having no local organisation like in the services markets. The price of wholesale energy is not the core dimension of competitiveness. Competition in this segment is mainly on the efficiency of the management of the customers (management of sales, information technologies, call centers, etc. and their economies of scale and scope) and on the appeal of the marketing and the branding of each competitor.

Second, crucial barriers arise, in particular the financing of entry investment easy with large uncertainty on the profitability because of the low expected margin by customers. Are their competitive advantages strong enough to finance the costs of entry and to deter reactions of incumbents? Are the targeted customers willing to change their suppliers for new foreign entrants? All questions haunting business strategists' minds: years before being bankrupted the flamboyant Enron failed its entry into retail competition in California in 2000. In fact foreign competitors as local candidates to entries on this segment have to buy or ally with local distributors in electricity or gas before penetrating this segment.

Main traits of attractiveness

From this analysis of market attractiveness we could draw three main conclusions.

First, at the end of the 90s, there are few European markets being obviously attractive with good growth and high prices. Only two countries (Italy and The Netherlands, 16% of the total EU market) qualified three times as attractive countries on each segment. Only two others (Spain and Sweden, 14% of the total EU market) qualified two times as attractive countries.

Second, we cannot infer the results of integration effects from the ex ante observation of the respective European markets attractiveness and their segment profitability. There are at least three types of markets and of European price competition. We may consider the industry markets to be the most sensible to pure price offers and then the most open to foreign attacks in case of high price differentials between neighbouring countries. In the services markets we may consider competition as being a mix of price and service offers. Then price differential plays only when additional services are offered on the same footing by foreign entrants and existing incumbents. This implies that competition from abroad has a severe handicap vis-à-vis competition from incumbents on the services and residential markets.

These cannot be served from abroad, and international energy prices differential is not the key for entering. Entering this mass market necessitates a local organisation as well as the knowledge of the retail business which is far to be a pan-European common knowledge. For the moment in 2004 only England & Wales and some Scandinavian countries really know what open retail competition on residential market could be, as the low rate of switchings reveals in other concerned countries (Germany, Denmark, Austria and since 2003 Spain). In 2002 the rate of switchings was less than 5% in Germany while it was established at the level of 12%–14% in UK, Norway and Sweden (European Commission, 2004).

Third, prices differential may provoke intensive foreign competition only if energy can flow easily between countries. Limited interconnections between some European countries, like the so-called peninsulas such as UK, Iberian countries and Scandinavia with the continental markets, or larger connections too much congested for short-term transactions because of the existing pre-reform long-term contracts, like with The Netherlands or Italy, isolate the national markets from foreign price competition. For the moment, the connection infrastructures for creating a pan-European wholesale market exist only in the so-called Continental Europe: from West Denmark to Switzerland, and from France and Belgium to Austria and Poland. So accessibility of European markets is as important as attractiveness in shaping the national markets' dynamics.

6.2. Accessibility of European Electricity Markets

Competitive reforms are affecting not only the fifteen countries of the European Union, but extending beyond them to nations like Norway and Switzerland, as well as to the ten new members in the EU (Poland, Hungary, the Czech Republic, etc.). Nevertheless there is no convergence towards a single model and a significant diversity persists. The attractiveness of national electricity markets in the EU has given us a relative gauge of their business potential. But a market's attractiveness does not necessarily define its accessibility to other players than its incumbent operators. Accessibility of these markets is at the core of the remaining diversity. By accessibility we mean actual openness of European electricity markets: technical and inter-trade openness; legal and regulatory openness; physical and commercial openness; industrial and financial openness (Glachant, 2003a,c).

We successively characterise these four parameters of accessibility before concluding on the convergence of European electricity reforms and the reality of market integration by examining the evolution of wholesale and retail prices. The third character – the physical and commercial openness of markets – will be identified by the capacity of interconnections with other countries, access rules during periods of congestion, and the rules of wholesale and retail markets. The fourth character presents the different countries' approaches to the access of industrial assets and capital stock of the national companies. We emphasise in particular the technical and commercial fragmentation of the Electric Europe, affecting not only islands and geographic peninsulas, but even the 'European copper plate' (Benelux, Germany, France, Austria).

6.2.1. *Initial situation of technical and inter-trade opening*

We need to assess the effective opening of these national markets to European competition.[6] We perform two successive assessments of national market opening: ex ante (before the transposition of Directive 96/92/EC) and ex post (after the transposition).

The openness of the national markets before the Directive transposition is evaluated on two criteria: 'technical opening' and 'inter-trade opening'. Technical opening depends on the capacity of cross-border transmission of electricity flows, and thus on the density of interconnections between the national grids and the grids of other European countries. Inter-trade opening is expressed by the actual volume of electricity supply purchases from other European countries. The degree of inter-trade opening in the national markets is measured from import statistics, by calculating gross electricity imports as a percentage of total domestic consumption. The combination of the two measures – technical and intertrade – provides a gauge of the national markets' ex ante openness.

Technical opening

A cross-border connection capacity equal to about 10%–12% of the national peak power demand seems to open a national market enough to allow a strong mutual influence with the adjacent wholesale markets and even to allow the merger of its wholesale market with these ones. This is effectively the case between Norway and Sweden, which have been operating a joint wholesale electricity market

[6]See our contribution to the retail competition measurement in the OXERA report to the European Commission DG TREN (OXERA, 2001) which was the basis of the *First Benchmarking Report* (European Commission, 2001).

Table 6.5. Ratios of interconnection capacity on total installed capacity in EU countries.

A	B	DE	DK	Finl.	F	Gr	Ir.	I	Neth	P	Sp	Sw	UK
21%	25%	9%	29%	19%	12%	11%	5%	8%	21%	8%	4%	21%	3%

Source: European Commission (2003).

(NordPool) since January 1996, joined later by Finland and Denmark (Table 6.5).

This technical opening criterion divides the EU national markets into three different geographic markets:

- the isolated markets (United Kingdom, Ireland, Greece) covering 17% of the EU market;
- the regional markets (Portugal–Spain inside the Iberian area; Norway–Sweden–Finland–East Denmark inside the Scandinavian area) representing 25% of the EU market;
- the continental markets (Denmark West, Germany, The Netherlands, Belgium, France, Italy, Austria) covering 63% of the EU market.

In other words, the heart of the European electricity market actually consists of an aggregation of markets on the so-called Continent into one partial but large pan-European electricity market. To this group we should add Switzerland, which has a function of hub for the pan-European electricity flows. But we have to subtract Italy and The Netherlands due to their cross-border congestions resulting from pre-reform long-term contracts.

The initial degree of inter-trade intensiveness

This criterion ranks EU countries by the size of their cross-border electricity flows. The European markets in 1997 displayed an average inter-trade opening of 8%, measured by the ratio of electricity imports to domestic consumption. From this average, we can distinguish three types of national markets.

- Three countries with an inter-trade opening close to the average 8% value: these include the two European countries that have already merged their wholesale electricity markets – Norway and Sweden – plus Germany. The three represent a combined 34% of the EU single market.
- Eight countries with an above-average trade openness, from 12% to 40%: Portugal, Finland, Switzerland, Austria, Italy,

The Netherlands, Belgium and Denmark. They account for 30% of the EU single market.
- Five countries with a below average trade opening, from 0% to 5%: United Kingdom, Ireland, Greece, Spain and France which is a major net exporter, but a small importer. Their aggregate size is equal to 43% of the EU single market.

By combining technical and trade openness, we can distinguish between three zones of different openness and size:

1. the 'very open' European zone is in fact the smallest of the three. It includes the six countries that combine above average trade openness and involvement in the emerging pan-European continental market: Switzerland, Austria, Italy, The Netherlands, Belgium and Denmark. Together, they equal 26% of the EU single market.
2. the 'partly open' European zone is considerably bigger. It includes four countries located in regional markets (Portugal, Finland, Norway and Sweden) and a large country of the pan-European continental bloc (Germany). The five represent 39% of the EU single market.
3. the 'little or not open' European zone is the largest of the three. It comprises five countries (United Kingdom, Ireland, Greece, Spain and France), accounting for 43% of the EU single market.

The 'very open' European zone before any implementation of the European directive is only a quarter of the EU market. This opening does not preclude that cross-border congestions then restricted access after the implementation of the European 1996 directive (Italy, The Netherlands, Belgium), or that legal and regulatory European opening does not apply (Switzerland). After taking into account these restrictions the 'very open' European zone will decline to only 6% of the EU total.

6.2.2. Legal and regulatory openness

The compromises incorporated in the European Directive 96/92 in 1996 reflect the level of disagreement expressed during the many years of negotiation between the European Commission, the member countries and the European Parliament. Remember that the Commission's *Green Book* that initiated the reforms dated back to 1989. This is why the Directive 96/92 has made ample provision for many different implementation modes for several key elements of the competitive reform. A priori, with so many available options, it appears that the potential for variations available to the 15 Member States was extremely high (Hancher, 1997). Nonetheless, in spite of skeptical predictions, national implementations have generally converged towards the same options, those considered the most competitive (Bergman, 2000; Glachant, 2000a, 2003a) for entry in generation, third party access to the network and consumer eligibility. It has been helped by the stimulation impulsed by the rule harmonisation process that the Commission is legally in charge to lead. The process was paved by successive declarations of Energy Council of Ministers or European Summit (Stockholm in 2001 and Barcelona in 2002) in favour of this alignment on the more competition-oriented rules. It brought negotiations to the new Directive 2003/54 of completion of market rules which has been voted on June 2003 and has been transcribed in national laws in 2004.

So at the beginning in 1998, it was predicted that at least six countries of the EU (with a consumption of 650 TWh) would select tendering instead of authorisation; in fact, by 1999, only one country had, Portugal (with 33 TWh). As to the conditions of third party access to the network (TPA), in 1997, at least six countries (with 930 TWh of consumption) were expected to prefer negotiated TPA over TPA regulated by a public authority – in fact, by 1999, there were only three (Germany, Greece and Denmark: 580 TWh). Moreover, in 1997, a single-buyer model was foreseen for at least six countries (1280 TWh domestic consumption, more than half the Union's consumption). In 1999, only three countries (Germany, Italy and Portugal: 800 TWh) have partially implemented the single-buyer mechanism for the distributors and it has been not really used. Overall, among the nine large and medium-sized countries in the EU (domestic consumption greater than or equal to 70 TWh), only Germany with a negotiated-TPA regime and a single-buyer at the local level, is at variance with the de facto European standard, i.e., the combination of authorisation for generation and regulated TPA for network access (Box 6.1).

As to the degree of openness of retail markets, most countries chose to exceed the minimum for eligible clientele established by the directive (26% in 1999, 29% in 2000 and 33% in 2003). Five EU countries (Great Britain, Sweden, Finland, Germany and Austria: 1100 TWh) already report 100% eligibility, as does Norway outside the Union as early as 1991. Three did it in 2003 (Denmark, The Netherlands and Spain: 330 TWh). Three other countries have either exceeded the minimum as of 1999, or will by 2003 (Belgium, Italy, Luxembourg: 380 TWh). Of the large and medium-sized European

Box 6.1
Definition of legal terms in the European legislation

At the level of generation the *tendering* process allows public authorities to dictate the evolution of the nation's stock of generating stations using a mechanism to elicit competitive bids (analogous to the call for tenders in government procurement). Conversely, the *authorisation* process leaves the initiative and responsibility for building new generating stations in the hands of operators, subject to general conditions established ex ante by public bodies that issue the authorisations for construction and operation.

The *single-buyer* procedure of the European directive places an intermediary between producers and sellers outside the network and consumers having access to a competitive electricity market. However, this intermediary is obligated to perform all supply contracts concluded directly between the third parties to the network. At the level of the network, the expression *Third Party Access to the Network* or TPA, designates the right of those operators (producers or resellers) who do not own transmission networks (long-distance, high-capacity power lines) or distribution networks (lower-capacity power lines that reach the final consumer) to access existing networks to supply their own clientele. Access is *regulated* when the main contractual and rate-setting provisions are set by a supervisory public body. Access is *negotiated* when the conditions are directly determined by the users of the network.

Clients are *eligible* if they have the right to freely select their supplier on a competitive market.

countries, only France (450 TWh) is emerging as a laggard in terms of the degree of openness (maintained at a minimum). The French implementation has not respected the deadline of February 1999, and the enabling legislation was only enacted in February 2000, while an important decision like the new tariff for transmission waited until Summer 2002.

However, the European directive provides further options for national governments wishing to subvert the impact of competition and undermine the uniform operation of the European market. But it has been used only by Germany and Spain. These options include

Table 6.6. Diversity in Implementation of reforms of EU state members in 2002.

	Market opening rate (%)	Full opening date	Unbundling of the TSO	Type of TPA	Regulation	Network tariffs 2001	Subsidies and stranded cost
Austria	100	2001	Legal	rTPA	Ex ante	high	
Belgium	35	2007	Legal	rTPA	Ex ante	medium	
Denmark	90 (distrib.)	2003	Legal	rTPA	Ex ante	low	
Finland	100	1997	Own.	rTPA	Ex ante	low	
France	30	none	Manag.	rTPA	Ex ante	medium	
Germany	100	1998	Account.	nTPA	Ex post	high	yes
Greece	30	none	Manag.	rTPA	Ex ante		
Ireland	30	2005	Legal	rTPA	Ex ante	medium	
Italy	45	none	Legal	rTPA	Ex Ante	medium	
The Neth	33	2004	Own.	rTPA	Ex ante	medium	
Portugal	30	none	Legal	rTPA	Ex ante	high	
Spain	54	2003	Legal/Own.	rTPA	Ex ante	high	yes
Sweden	100	1998	Own.	rTPA	Ex ante	low	
UK	100	1998	Own.	rTPA	Ex ante	low	

Source: European Commission, *Second Benchmarking Report on the Implementation of the Internal Electricity and Gas Market*, March 2003.

support for domestic energy sources such as hard and brown coal in Spain and Germany (with 7 billions DM/year) or assistance for stranded costs (nearly 8 billions€ announced in Spain).

On the whole, if we broaden the comparison between countries to cover other important aspects of the reforms, the diversity of national approaches again appears quite pronounced. In particular it was the case for the provision for unbundling competitive activities from natural monopolies (in particular the ownership of the network), the regime governing the use of transportation networks, or the nature of the public authority responsible for overseeing electricity reform (Table 6.6).

We can view the first three countries to reform as trailblazers. They conceived their reforms well before the vote of the Directive 96/92 and implemented them voluntarily, free of any obligation ensuing from that Directive (cf. Group 1 in Table 6.7). These countries are England[7]

[7]The competitive electricity reform of the 1990s in the United Kingdom was not uniform. In particular, Scotland and Northern Ireland were not subject to the same regime as England & Wales. Consequently, the famous competitive wholesale market, called the Electricity Pool, was only operational in England & Wales. Scotland retained a vertically integrated system without an organised wholesale market, and Northern Ireland applied a 'strict' single-buyer system (with purchasing monopoly and a single reseller) that was finally disallowed by the 1996 European Directive.

(reform in April 1990), Norway (reform in January 1991) and Sweden (reform in January 1996). We note that these pioneering countries, all early and eager reformers, feature similar provisions for most of the main issues: eligibility regime, authorisation for generation, network access with the regulated TPA, legal and ownership unbundling combined with the transmission system operator (TSO), and an independent regulation authority responsible for regulating access to the network (Table 6.7).

The second wave of reforms concerns Spain, Germany and The Netherlands (cf. Group 2 in Table 6.7). These reformers are motivated and quite diligent. They began their national reforms just after legal publication of the European Directive (in February 1997), not waiting for the two-year deadline to expire. However, not one of these three countries has fully replicated the reform programme already in effect among the trailblazers. These three countries also differ amongst themselves – The Netherlands and Germany contrast with Spain in terms of the eligibility regime, TPA, unbundling, the constitution of the TSO and the body responsible for network access.

Finally, the third wave concerns the latecomers. Belgium, Italy and France did not manage to reform before the official deadline of February 1999 had expired (cf. Group 3 in Table 6.7). None of these belated candidates has replicated the full reform agenda of neither the trailblazers nor any of the three variations in the second wave. On the whole, only a single element of electricity reform has been unanimously accepted by all the nine countries the authorisation regime for generation. But the regulatory TPA has been chosen everywhere at the exception of Germany.

6.2.3. Technical and commercial openness

If the new legal and regulatory provisions establish the general rules of the competitive reform, the development of both technical infrastructures and commercial ones with creation of organised markets creates the operational means. In practice, effective entry into a domestic electricity market directly depends on the conditions of physical access (the existence of network interconnections between countries; the treatment of access during congestion periods on interconnections) and on the conditions of commercial access rules of wholesale and retail markets).[8]

[8]See Glachant and Lévêque (1999); Smeers and Boucher (2001); Glachant and Pignon (2002); Stoft (2002); Hunt (2002); Glachant (2003b).

Table 6.7. Map of the Competitive Reforms in 2000 and 2003.

	Eligibility in 2000 (%)	Eligibility in 2003 (%)	Generation	Network access	Unbundling	Transmission system operator	Regulation body of network access
Group 1	100	100	Authorisation	rTPA	Complete (T)	Integrated	Independent body
England					Legal (D)	Own-Operate	
Group 1	100	100	Authorisation	rTPA	Complete (T)	Integrated	Independent body
Norway					Legal (D)	Own-Operate	
Group 1	100	100	Authorisation	rTPA	Complete (T)	Integrated	Independent body
Sweden					Legal (D)	Own-Operate	
Group 2	54	100	Authorisation	rTPA	Complete (T)	Integrated	Independent body
Spain					Legal (D)	Own-Operate	
Group 2	100	100	Authorisation Single Buyer (distrib.)	nTPA	Accounting (T)	Regional	Competition body
Germany					Accounting (D)	Owned by producers	
Group 2	32	63	Authorisation	rTPA	Legal (T)	Operat. Indep	Special department
The Netherlands					Legal (D)	Owned. by producers	
Group 3	33	80	Authorisation	rTPA	Legal (T)	Operat. Indep.	Five independent bodies
Belgium					Accounting (D)	Owned. by producers	
Group 3	35	66	Authorisation	rTPA	Legal/Complete (T)	Operat. Indep.	Independent body
Italy					Legal (D)	Owned. by producer	
Group 3	30	37	Authorisation	rTPA	Management (T)	Operat. Indep.	Independent body
France					Accounting (D)	Owned. by producer	

Abbreviations: T – transmission; D – distribution; Own – ownership, Operat. – operation.
Source: European Commission, *Second Benchmarking Report 2003*.

Consequently, the implementation of competitive reforms on a pan-European scale primarily requires the physical opening and increasing of domestic transmission networks. Of course, harmonisation of the principles underlying access rules, rate structures and levels for using the networks is as important as physical connection. Such a harmonisation is currently being finalised by the coordination of the TSOs (the European ETSO Association) in collaboration with the European Commission, and the European Regulators Association (CEER) under the auspices of the so-called Forum of Florence. The European Commission is pressing for having a unified European framework for cross-border trade. A common and mandatory European regulation, debated since March 2001, has been adopted by the European Council in June 2003 in this direction.[9]

The physical constraints

A purely physical constraint remains: the transmission capacity between member countries. Practically, because of inadequate interconnections, Europe will never be a totally unified 'internal market'. As seen in Section 6.2.1, seven members of the Union are almost separated from the 'continental plate' to form five physically autonomous subsystems: Ireland (17 TWh), the United Kingdom (333 TWh), the Iberian peninsula (200 TWh), Scandinavia (325 TWh) and Greece (40 TWh). On the continental plate (nine countries with an overall demand of 1500 TWh) interconnection capacity is rather high (at least more than 15% of each national peak load). By comparison, Norway and Sweden were able to merge their domestic wholesale markets into a single wholesale electricity market with a capacity connection ratio of around 15% of their peak loads. Nevertheless large importers like Italy and The Netherlands are still too congested at their connexion points partly because of the reservation of capacity for pre-reform contracts for working smoothly into a pan-European single wholesale market. For less connected countries (like at the Spain–France border) the European policy is focused on the increase of connection capacities. The Barcelona European Summit in spring 2002 added to the electricity reform goals the building of a minimum of 10% connection capacity within the European Union.

[9]European Commision – Regulation on conditions for access to the network for cross-border exchanges in electricity. Regulation (EC) No 1228/2003 of the European Parliament and of the Council of 26 June 2003.

Commercial infrastructures

Competition reforms also require commercial infrastructures. New intermediation infrastructures are required to bring together many sellers and buyers around various types of trade. The purpose of these commercial infrastructures is to handle new types of supply which emerge in competitive electricity markets.

A first kind of required commercial infrastructure is the multilateral power exchange, which organises the recurrent and transparent meeting of supplies and demands of large quantities of a homogeneous product on a hourly basis.[10] This new type of commercial infrastructure allows to expand competition by introducing uniformity, giving each stakeholder identical information about volumes and prices on a daily basis. It also provides for easy and equal access to daily transactions concluded for different maturity dates. In addition to these energy spot markets, like the day-ahead market on the day D–1 and the hour-ahead market, or balancing market, on day D, power exchanges may provide facilities to hedge against price risks with future contracts on a specific market. Moreover, on the fringe of these markets, the traders (the best known in Europe has been the American Enron before its fall in 2001) can intermediate between electricity producers and buyers, expanding competition between energy suppliers beyond the small club of producers (with familiarity with speculative purchases and sales and risk management of energy portfolios). Finally, brokers can help to match supplies to demands on a bilateral basis of the unorganised markets with Over The Counter (OTC) transactions. Organised wholesale markets (pools, power exchanges) exist for several years in two European regions in England and Scandinavia. It has been adopted in Spain in 1998 and then in different countries of the Continental Plate. First it was the APX, opened in Amsterdam in the summer of 1999 (with large electric concerns from Belgium, Sweden and Spain as shareholders), then in Germany two competing regional power exchanges opened and then merged, EEX in the West (Frankfurt) and LPX in the East (Leipzig) in 2001, and in France an exchange opened in the autumn of 2001. Italy prepared since 2000 a national pool which has been progressively

[10] A clear distinction has to be made between multilateral markets – 'organised' markets on which all supplies and demands simultaneously meet on the basis of a set of detailed rules, such as markets for stocks and bonds – and bilateral contract markets – on which a buyer meets one supplier at a time in direct negotiations, such as the market for architectural or legal services. Between these two extremes we find various types of grey markets.

postponed to 2004. Spain and Portugal share an Iberian PX in 2003. In the absence of this type of organised market in some other regions of Europe, several price indexes indicate trends on more informal grey markets, such as those in central Europe (the Swiss SWEP index, the German CEPI). The most used European grey market indexes are the OTC prices collected and published by Platts.

A second development in commercial infrastructures for competitive electrical systems is the creation of an organised retail market. Indeed, regulatory opening to competition with 100% eligibility does not solve the transactional difficulties facing domestic and small business customers (in Germany only 3% of consumers actually changed supplier, and it appears that none of them selected a foreign supplier). How can these consumers change supplier when they remain physically connected to a monopolistic network belonging to the incumbent operator who retains responsibility for the actual delivery of their energy?

In practice, these organised competitive retail markets function as mechanisms for allocating, measuring and controlling property rights using dedicated information networks that collect, code, exchange and store information on participants. These market information networks establish and disseminate physical and commercial measures of the supply flows, and through them knit ties of commercial and financial responsibility between millions of clients/consumers and dozens of suppliers who keep records of access to electrical networks and to wholesale markets. To date, these organised competitive retail markets only exist in 2004 on the periphery of the European Union (in Great Britain and Scandinavia), where they breathe additional competitive life into electricity reforms. They probably are laying the groundwork for buying electricity on the Internet if retail trade will go in this direction. In Germany, the competitive retail market remains negotiated, thus strongly dependent on the operating decisions made by the 900 different local distribution networks which are increasingly controlled by the new four top national companies.

Comparison of effective openness

The technical infrastructure and the commercial provisions underlying the reforms in the nine largest electrical countries of Europe is assessed by their comparison. In terms of physical interconnectedness, it is noteworthy that the three trailblazers (England, Norway and Sweden) are not well linked to the other European countries. The operation of the UK domestic market and the Nordic one remains physically separated from that of the other European markets. In fact, these pioneers' reforms were self-contained. Furthermore, while Sweden effectively shares the same electricity market with Norway (the Nord Pool), this wholesale market splits into local sub-markets operating under different rules during 40–60% of the time (with redispatching by counter-trading in Sweden and market splitting in Norway) because of congestion on the inteconnections or in the national transmission networks.

Finally, centralised rules governing the wholesale market, such as those applied in England between April 1990 and the end of March 2001, never existed in Norway or Sweden, where all electricity companies retained control of their merit order dispatch. Indeed, in the former centralised British pool, participation was mandatory for all buyers and sellers of energy, and the market manager (NGC) authorised electricity production on the basis of the day-ahead auction's outcome. In voluntary pools, on the other hand, participation in the organised market by buyers and sellers is optional, and winning bidders decide which facilities will produce the electricity. In fact most of the electricity sold to final consumers comes from vertically integrated suppliers (from 80% to 90% on an average) (Table 6.8).

In the second group, that of the diligent reformers, Spain's physical isolation sets it apart as a new case of self-contained reform. The two other countries, Germany and The Netherlands, are connected each another and with other European countries. These two countries are also similar in aspects relating to commercial access (management of congestion, wholesale market). However, from the start, their wholesale markets operated quite differently from those of the three trailblazers and Spain. The Amsterdam APX and Frankfurt and Leipzig exchanges (EEX and LPX) were created as private power exchanges, i.e., they are not harmonised with complete provisions for network management by the TSOs. In the trailblazers (England, Norway and Sweden), the TSOs created and managed the electricity pools, facilitating physical execution on the networks of commercial transactions concluded in the market. But, in The Netherlands, difficulties that arose in 2000–2001 attempting to coordinate the private PX with physical management of the transmission grid, given its capacity limitations and security regulation, prompted the government to nationalise the network. The new public transmission corporation (Tennet) bought out the APX. In Germany, where bottlenecks in the transmission network are less of an issue, several transmission system operators each manage a region, rendering consistent execution of the transactions concluded on the two

Table 6.8. Physical and commercial access to the nine major markets in 2002.

	Interconnectedness	Congestion management	Wholesale market	Retail markets (services and residential)
1. England	Domestic market	Redispatching	Centralised, mandatory (up to 2000)	Regulated competition (1999)
1. Norway	Regional market	Market splitting	Centralised, optional	Regulated competition (1997)
1. Sweden	Regional market	Redispatching	Centralised, optional	Regulated competition (1997)
2. Spain	Domestic market (Regional with Portugal in 2004)	Redispatching	Centralised, partially mandatory	Regulated (2003)
2. Germany	Pan-European market	Redispatching	Bilateral + private PX	Negotiated competition (1998)
2. The Neth.	Pan-European market	Redispatching	Bilateral + PX	Monopoly
3. Belgium	Pan-European market	Redispatching	Bilateral	Monopoly
3. Italy	Regional market	Market splitting	Bilateral + centralised PX in 2004	Monopoly
3. France	Pan-European market	Redispatching	Bilateral + private PX in 2001	Monopoly

Source: European Commission, *Second Benchmarking Report*, 2003.

German exchanges difficult in very practical details (such as 'balancing').[12] Furthermore, Dutch and German trade flows are not coordinated under standardised rules of exchanges like in a unified central marketplace, in particular for allocating access on cross-border connections and for the balancing. The same observation is reinforced for the German retail market, where the rules of trade are not uniform and regulated, but negotiated between sellers and several hundred TSOs and distribution systems operators.

The latecomers in the third group (Belgium, France and Italy) are still managing some provisions of the initial implementation of the European Directive. We can observe that all the three belong to the highly interconnected pan-European market, the so-called Continental market. In the case of Italy, however, the TSO manages congestion by market splitting. On the other hand, these three countries do not wish to replicate either the English or the Scandinavian models, nor do they intend to establish the same wholesale market mechanisms: organised day-ahead and balancing markets[13] are opened or to be opened soon in cooperation with TSOs in France and Italy (with in this country a unique buyer of wholesale electricity on behalf of distributors for their ineligible clients), but not in Belgium, which may integrate into the Dutch or French markets some day.

These initial differences between the latecomers should be exacerbated by the unequal powers of attraction between their respective markets (contracts and PX) and the private exchanges or grey markets operating at their borders (The Netherlands and Germany for the PXs and Switzerland for a grey market) as well as by not harmonising the allocation of capacity on cross-border connections. Nevertheless, the quite synchronised functioning of the German and French PXs since their creation in the autumn 2001 suggests that a continental PX is something reachable in the future, creating a reference continental price which will influence a dozen of national markets in the EU and around (Switzerland and some

[12]The balancing problem is that either the generator or the consumer can have real imbalances from the expected schedule arranged on the wholesale market on Day-1. These imbalances have to be compensated within the rules set by the network operator apart from the market rules working on the wholesale market.

[13]A day-ahead market is a standard organised market (pool or private exchange) in which supplies and demands meet 24 hours before the physical realisation: 'the eve for the next day'. A balancing market is a very special organised market managed by the TSO, its main user, to buy the increases or decreases in production required for the equilibrium and security of the network two or three hours ahead of time.

Applicant Countries like Poland, Hungary, Czech Republic). But the liquidity of such a pan-European PX will probably remain limited to 1–2% of the physical consumptions of this area. It will stay a thin and loose PX, hatched by cross-borders congestions and contained by the self-dealing of overwhelming national vertically integrated electricity suppliers.

6.2.4. Industrial and capital stock accessibility

In tandem with legal and regulatory opening and the development of their technical and commercial infrastructures, two other significant dimensions of the competitive openness impose themselves: industrial openness and the accessibility to the companies capital stock (Glachant and Finon, 2003). This matters for two principal reasons. On the one hand, the effectiveness of electricity competition depends just as directly on industrial and financial opening as it does on market rules and technical and commercial infrastructures. On the other hand, for reasons that are as much economic as commercial, electricity companies do not wish to establish a permanent presence in foreign markets in which they do not actually own production or marketing assets. Those specialised in consulting and trading, as was initially thought to be the exception with the flamboyant example of the Enron entry. But after its bankruptcy and the American traders' retreat from the EU electricity markets in 2002 the pure trader model lost its credibility in the EU as well as in the US (Table 6.9).

Few of the European industrial manoeuvres inside and between countries would have occurred had the firms' capital remained in the hands of the public authorities. Indeed, none of the 17 countries featured an entirely private electricity sector before the 1990s. Thus, it was primarily privatisation that opened up the capital market that drives the current sectorial and geographical diversifications and the horizontal and vertical integrations based on mergers and acquisitions.

As to vertical and horizontal integration of electric concerns, two different types of industry existed in the 90s before the reforms. To the West and South of a line connecting Edinburgh–Brussels–Rome–Athens, eight countries had the highest level of vertical and horizontal integration of the electric industry in Europe, with in some cases a unique firm under public ownership. Among these countries only two have chosen to break up the incumbents (England and Italy). Another electrical Europe existed to the North and East of the Edinburgh–Athens line in nine countries with much less concentrated

Table 6.9. Industrial and capital access to the nine major markets in 2003.

	Concentration of production* (%)	P-D integration	Dominant ownership	Production by private firms (%)
England–Wales	37 (C_3)	Restricted (1990–1998)	Private (P-T-D)	>90
Norway	24 (C_3)	Restricted	Govt. public (P-T), Munic. public (D)	<10
Sweden	50 (C_3)	Unrestricted	Govt. public (P-T), Munic. public (D)	<10
Spain	79 (C_3)	Unrestricted	Private (P-T-D)	>90
Germany	61 (C_3)	Unrestricted	Private or mixed (P-T), Munic. public. (D)	>50
The Netherlands	33 (C_3)	Unrestricted	Private (P), Public (T) Munic. (D)	>75
Belgium	60 (C_1)	Unrestricted	Private (P-T), Munic. or mixed (D)	>90
Italy	72 (C_3)	Restricted (since 2001)	Mixed (P-T) Public or munic. (D)	>33
France	80 (C_1)	Unrestricted	Govt. public (P-T-D)	<10

P : Production; T : Transmission; D : Distribution.
*Notice that the market share of the Top-3 takes into account the import capacity.
Sources: European Commision, *Third Benchmarking Report*, 2004; Glachant (2000a).

vertically and horizontally (several dozen operators in each one) and a myriad of municipal distribution companies.

In most part of this second Europe, competitive reforms stimulated an unprecedented wave of horizontal and vertical concentrations, involving firms already considered national champions, especially in Sweden, Finland and Germany. The upper limit of this movement towards concentration seems to be reached in Sweden, where the vertical and horizontal structures are looking like those of Spain (i.e., three large vertical operators of unequal size, and a small challenger), and in Germany where two-thirds of the largest producers lost their independence during the thirty months following the initiation of the reform.

In the countries where the electricity sector was much less concentrated, local and regional authorities owned most of the electricity companies. That is why privatisation decisions there were always more decentralised, harder to control by governments, and provided greater opportunities for entry to interested foreign companies (cf. the many cases of foreign entry into Sweden, Germany, Austria, Switzerland and The Netherlands). Conversely, in the West and South of Europe where there were only nationalised domestic firms and privatisation, except in Spain, when it occurred, it was easier to control by governments if they chose to do so (choice of a 'white knight', sale by mutual agreement or stock flotation, recourse to golden shares in the newly privatised companies to keep a certain control, etc.). At this time, in the entire western and southern zone of Europe, only the Anglo-Welsh energy sector has been the object of sustained foreign entries both in generation and distribution-supply. While the Belgian private national monopoly (Tractebel–Electrabel–Distrigas) has been taken over by the French Suez group, the Belgian management has been maintained. We do observe, however, that as time advances, a European capital market is creeping into all electricity areas that are open to mergers and acquisitions including Spain, Italy (for divested generation assets), France (for SNET and CNR), Switzerland and on the East of EU, the Czech republic, Hungary and Poland. Finally, in addition to actual mergers and acquisitions, corporations are establishing many other types of links: capital alliances with minority interests, cross-shareholding, joint ventures and, even more generally, cooperation pacts (one of the keys to concentration in countries with a high degree of municipal diffusion like in Sweden and Germany).

If we now compare the terms of industrial and capital access to domestic markets in the nine largest electric countries, we notice that the conditions of access differ widely, depending upon horizontal and vertical concentration and ownership forms (cf. Table 6.9).

Capital access path in pioneering countries

First, these access paths are not the same for the three pioneering countries. In England (over 10% of the market of the European Union), where most of the electricity industry has been privatised, the two main producers serviced nearly 80% of the domestic wholesale market at the beginning of the reforms. However, due to persistent pressure from the regulator and the government, they find themselves with less than 30% of the market 10 years later. Also, during the first post-reform period their attempts at vertical integration with distributors were consistently blocked, while these last companies were allowed to establish subsidiaries to service up to 15% of their sales with the so-called 'independent production'. This independent sector does not exist in Scotland, because complete vertical integration of the two incumbent regional operators was legally remaining since 1990. Finally, since 1998, when the British government stopped protecting privatised companies from takeovers, most distribution and a substantial share of Anglo-Welsh production has passed into foreign hands. In 2003 the four biggest vertically integrated companies in England & Wales were owned by German E.ON and RWE (PowerGen, Innogy, former Eastern/TXU Europe) and the French EDF (London Electricity Group).

Conversely, in Norway where privatisation was never envisaged by the successive governments, the two principal producers (both government-owned) never attained 50% of the national market. Until 2001 they never attempted to increase the concentration of production nor to dominate distribution activities, even though production–distribution integration is allowed and extensively practised by many other Norwegian operators. However, in 2002 and 2003, the main producer has entered into some agreements and alliances with the two main distribution groups, but this move has been stopped by the competition authority in 2003. On the other hand, foreign capital was virtually banned until 1998, and continues to be excluded from production.

In Sweden, privatisations shot up in 2001 with the achievement of the take over of the second producer Sydkraft by the German E.ON, but the industrial and capital access regime does not look like neither the Norway's nor England's ones. Of course, at the beginning of the reforms, the two principal Swedish producers, the government-owned Vattenfall and Sydkraft, (at the time mixed capital owned), revealed a degree of concentration similar to the Anglo-Welsh market at that time, with nearly three-quarters of the production market. However, neither the regulator nor the government acted to reduce this share, nor to dissuade the two national leaders from further concentration in distribution. Of course, as in Norway, the prevalence of local public ownership makes hostile takeovers impossible. But the Finnish government-owned company has been remarkably successful in penetrating Swedish production and distribution networks with about 15% of the market (compared to Graningue, the EDF's Swedish affiliate till 2003, with only 5% market share), while at the same time the Swedish government's electricity company has a proportional presence in Finland. Finally, in 2001, Sydkraft passed into the hands of E.ON, while Vattenfall made a major breakthrough into the German market by scooping up the incumbent integrated operators in Hamburg, Berlin and East Germany.

Capital access in the countries of the second wave

This diversity in access paths is also found among the countries of the second wave of reforms. In Spain, industrial and capital access are limited. Outside of transmission, there were only four significant companies, all private and all vertically integrated. During the privatisation of the Spanish leader Endesa, the government refused to dismantle the national champion, wishing to maintain a situation in which the two biggest companies (Endesa and Iberdrola) control 75% of production and 80% of distribution. While the government itself blocked the merger of these two giants in early 2001, the very difficult takeover of the fourth-ranked Hidrocantabrico by the American TXU, then by the German RWE, and finally by EDF and the Portuguese EDP, has illustrated that Spain's openness to capital inflows remains precarious. When Endesa put the equivalent of 5% of the domestic market up for sale in 2002, this relative closure was again put to test: Endesa sold indeed these shares to the Italian incumbent Enel, but from which it had just received control in Italy of Elettrogen (the first divested Genco).

Conversely, The Netherlands and Germany appear much more open, despite the prominence of public ownership, owing to the incontrovertible rivalries among regional and local firms. Thus, the Dutch government's policy of concentrating production to create a national champion failed in 2000, and less than two years after the beginning of the reform, three of the four Dutch producers had already fallen under foreign control (notably the Franco-Belgian Electrabel and the German Preussen Elektra, now E.ON). Essent, the only remaining big Dutch producer, is now integrated into distribution, and has also recently penetrated into distribution in Germany. Also, the restructuring of distribution in The Netherlands has not yet started in 2004: when the Spanish Endesa tried in 2002 to get a foothold in Eindhoven and to expand further... before being blocked by the Dutch government. In Germany, on the contrary, the momentum of horizontal and vertical concentrations of production, transmission and distribution is irresistible. Five of the nine vertically integrated incumbents, the so-called Supraregional companies, have disappeared in voluntary mergers, notably to propel the two largest German groups supplying about 70% of commercialised production (RWE with VEW, Preussen Elektra with BayernWerk, renamed E.ON). The two remaining operators lost their national independence by merging or allying with foreign companies (with the Swedish Vattenfall for HWE BEWAG and VEAG, with the French EDF in BadenWerk). Finally, of the hundreds of German local firms, public or mixed, that ensure one-third of the distribution of electricity, several await a foreign 'white knight', as the recent entries of Electrabel and Essent attest.

Capital access in the countries of the third wave

The countries of the third wave have brought a further proliferation of access paths. If Italy does not bar Enel from combining ownership, but not operation, of the transmission network with ownership and operation of generation and distribution assets, it has been accompanied by partial privatisation and drastic divestments in generation and supply (approximately one-third of Enel's production facilities and one-tenth of distribution activities). These large-scale operations will draw a new map of Italy's electric industry and capital, but it will be provisional, owing to the restrictions imposed on potential buyers by the Italian government (of which the anti-EDF decree of May 2001 on the partial Edison acquisition and the restriction on Italian municipal companies entering in generation).

In Belgium, access paths are more limited. The dominant private producer Electrabel (over 90% of the domestic market) has not been dismantled. An offer has been made for its publicly owned junior partner (SPE) to ally itself with EDF but without achievement in 2004. In distribution, Electrabel has shares in 80% of distribution companies whose ownership is local mixed (the others being entirely public) and benefits from exclusive supply contracts effective until 2006. However, several Flemish cities have escaped from the tutelage of this Franco-Belgian giant, having signed with the British firm Centrica, a gas concern newly into electricity.

In France, the public producer EDF ensures about 95% of the distribution and production (see Finon, 2001, 2003). Of French production not provided by EDF, the Spanish firm Endesa has acquired the control of SNET (a power generation subsidiary of the publicly-owned Charbonnages de France), and Electrabel has signed a closed deal with the public hydro-electricity producer CNR and the railways hydro subsidiary SHEM. Furthermore, to clear the acquisition of the German firm EnBW, EDF has begun to auction off 6000 'virtual' MW, equivalent to 42 TWh (one-third of the French market of eligible clients and 10% of the total French market). In distribution none of the 170 municipal or mixed companies has been sold.

Impact of the horizontal integration

The creation of large integrated energy companies and the persistence of domestic quasi-monopolies have repercussions for the functioning of the new European electricity markets. But these repercussions will depend on the actual size that each of these new and inchoate competitive markets achieves.

- In Great Britain, for example, the physical limit of the expansion of the competitive electricity market is given by the size of the entire British market (Scotland adds 20% to the Anglo-Welsh market).
- In Spain, this limit is the Iberian Peninsula (where Portugal adds 20% to the size of the Spanish market).
- In the case of Sweden, it is the size of the entire Scandinavian market (plus 150% when Sweden's three neighbours, Norway, Finland and Denmark, are included).
- Within the continental block of 1500 TWh (the original six of the European Common Market plus Austria and Switzerland) it will be regulatory decisions and the actual development of technical and commercial infrastructures which will determine the effective dominance of the four biggest operators, the German RWE, E.ON, the Swedish–German Vattenfall Europe, the French EDF, and the Franco-Belgian Suez–Electrabel–Distrigaz).

6.2.5. *Variety of accessibility in the European markets*

The variety of the European markets decreases moderately during the period between the two 1996 and 2003 Directives. By combining the eligibility of consumers to competition and the industrial and capital accessibility of electricity industries, we obtain a first overall gauge of the effective openness of the EU markets in 2002 after the transpositions of the first Directive (96/92/EC). We find that the European electricity market was neither very open nor unified. Indeed, it consisted of at least three very different categories of competitive markets (cf. Table 6.12).

- First only a modest group of two European countries had highly open markets in regulatory and industrial terms in the year 2000. They represented 15% of the total EU electricity markets. The two countries are Norway (which is not an EU member), and England & Wales (but not Scotland). No one belongs to the continental pan-European bloc. Norway functions in a regional arrangement, while England is a physically isolated market.
- Second the largest sector of the European electricity market actually consisted of partly open countries where high eligibility is compensated by a fairly strong horizontal concentration in generation, or where a lower concentration is compensated with lower eligibility. There were eight markets in this grouping: Finland, Sweden, Scotland, Germany, Spain, Luxembourg, Austria and

Table 6.10. Industrial structures and market opening in 2002.

	High concentration in generation (C2: 0.8–1)	Medium concentration in generation (C2: 0.65–0.75)	Low integration in generation (C2: 0.35–0.55)
Large opening 90–100%	Denmark East	Finland Sweden Scotland Germany	Norway England & Wales
Medium opening 33–54%	Belgium	Spain Italy	Austria The Netherlands
Low opening 30%	France Ireland Greece Portugal		

Source: European Commision, *Second Benchmarking Report*, 2003.

The Netherlands. Joined by Italy in 2001, they accounted for 62% of the EU market.

- Third, the rest of the European electricity market was still fairly closed, notably because of high horizontal integration in generation. This grouping includes six countries without Italy, making up 27% of the EU market: East Denmark, Belgium, France, Ireland, Greece and Portugal (Table 6.10).

In 2003 when the second electricity Directive has been adopted few changes in the category have occurred. Only Austria and The Netherlands (6.7% of the EU) have joined the category of 'highly open markets' for both eligibility (100% and 63%, respectively) and generation (33% market share for the biggest three), but The Netherlands (4.4% of the EU) will actually stay 'limitedly open' because of its permanent cross-border congestions under the dedication of capacity to pre-reform import contracts. Austria and The Netherlands are the first two continental European countries entering that category where Norway and England stay.

In conclusion, not one of this nine European countries we have examined is completely open in all aspects of market access (technical or commercial, industrial or capital). The markets that are most open commercially (England, Norway and Sweden) are not the most open technically, and the countries that are more open technically (The Netherlands, Germany, Belgium and France) are not the most open commercially. Of course, there are paths of industrial and capital access that allow some of the technical and commercial difficulties to be circumvented, as we see in the case of England, The Netherlands, and Germany. But overall, on a pan-European scale, heterogeneity and fragmentation still dominate the landscape.

6.4. The Market Integration Test: Which Price Convergence?

Electricity being a highly homogenous commodity, it is well suited to wholesale trading on exchanges. Thus, making the assumption of identical conditions of delivery and service, we can compare prices between different regions in Europe. Wholesale prices appears to converge on the continental plate.

The average wholesale prices observed in 2000 varied widely (cf. Table 6.14). Among the three trailblazers (England, Norway and Sweden) the maximum gap attained 150%. In the case of Norway and Sweden this gap exceeded 22%, despite the fact that they belong to the same wholesale market (Nord Pool). Between the two countries with PX in the second wave (Spain and The Netherlands), the differential was 40%. Finally, for the two documented countries of the third wave (France, Italy) the differential was 20%. If we consider the documented countries well interconnected in the pan-European zone (Germany, The Netherlands, France and Italy), the difference reached 76% (Table 6.11).

In the following years some convergence of wholesale prices on the PX has taken place when the interconnectedness allows it. Disparities in wholesale prices appear quite substantial. Among the three trailblazers (England, Norway and Sweden) the price difference has been reduced to 40% because of the reform of the British wholesale market. Among the three countries of the second wave (Germany, The Netherlands and Spain) the disparity remains significant because of the large decrease of the wholesale price on the German market and the remaining problems of the access to the interconnections to the

Table 6.11. Wholesale electricity prices in Europe in 2000 and 2002.

	Norway	Sweden	England	Neth.	Germany	Spain	France	Italy	Belgium
2000 price	15.3	18.7	38.7	45	25	29–32.3	25–31	30	N/a
2002 price	18	19	25–28	31–40	22	30–37	22	30	22

Note: Average monthly price.
Sources: *Power in Europe*; *PLATTS 2002* for EEX, NGC, OMEL, Nord Pool, APX and PowerNext and index SWEP.
For France in 2000: Price list for the transitional system, winter peak. For Northern Italy: Swiss wholesale price SWEP. in 2000.

Table 6.12. Access price to congested interconnections in Europe in early 2001. (in €/MWh).

Interconnection	D to NL	NL to D	B to NL	NL to B	F to UK
Access price	12.6	0.07	3.54	0.01	6.77
Access price/Wholesale price*	44.7%	0	12.5%	0	24%

*Reference to Platts Pan-European Power Index in January 2001: 28.2€/MWh.
Source: Power in Europe, 2001.

Dutch market. Between the countries of the third wave (France, Italy) the difference increases to 40% because of the constraints on the interconnections, the effectiveness of the competition in France and its limitation in Italy. Finally for the members of the continental market gathering countries with large interconnections (The Netherlands, Germany, Belgium and France), the difference between PX prices has been eliminated at the exception of the Dutch price.

The remaining differences reveal restrictions in accessibility to markets between countries, in particular those by physical limitations. When the access to congested interconnections is governed by auctions, market players revealed the value of the access rights. On the good side of the bottleneck they tendered a tariff valued at the equivalent of 12%–45% of the European wholesale price (3.5 to 12.5 €/MWh) for having permanent access to those on the bad side of the bottleneck. In the opposite sense, the economic value of access rights plummets to a very low level. In 2002 the value of these rights between England, Spain, Germany and The Netherlands was in the range of 10–15€/MWh (Table 6.12).[14]

Concerning now the industrial retail market (Table 6.13) the rate of industrial switchings reveals a quite effective competition on the segment of mean and large industrial consumers in the fourth year after the directive transcription with 15%–20% of eligible customers having actually switched in Germany, France, Italy, The Netherlands, Spain and UK. But in some countries it is not with the external competition from foreigners which is obviously not working (Spain, Italy, The Netherlands). For deliveries to mean industrial consumers (24 GWh/year) the prices remain far from converging, the variance being important between the Nordic countries and the other ones, but these tend to converge. Surprisingly France, despite its limited competitive reform, has among the best prices on Continental Europe,

[14]West European Electricity Review 2002, A 'Power in Europe' Special Edition, June 2002.

Table 6.13. Industrial electricity prices and switchings in 1999 and 2002 in Europe. (Price in €/MWh, Reference Consumer: 24 GWh/year).

	Norway	Sweden	England	Neth.	Germany	Spain	France	Italy	Belgium
July 1999	25	28	49	49	63	53	49	54	55
July 2002	23	26	46	N/a	52	47	49	74	58
Switchings in 2002	N/a	10%	15%	20%	20%	20%	15%	15%	5%

Source: European Commission, *Third Benchmarking Report*, 2004.

both for large users and smaller consumers. The UK price is just a little cheaper in the industry than French price in 2002 (Table 6.13).

Among the trailblazers (England, Norway and Sweden) the gap is 100% between the Nordic countries and the UK price. In the countries of the second wave (Spain, Germany and The Netherlands), German price decreases significantly after the reform on this segment, as in Spain where the competition begins to be effective on this segment. It explains that industrial retail price is converging with the wholesale price. Besides the absence of informed industrial Dutch price, a convergence of Spanish and German prices can be observed. As to the countries of the third wave (France, Belgium and Italy), there is no convergence between Belgium and French prices but there is convergence with the German price on the Continental market. Difference is important between Italy and the other markets (Table 6.13).

Average prices in residential supply (3.5 MWh/year) vary between the nine countries, with a maximum difference of 65% for the six

Table 6.14. Residential electricity prices and switchings in July 2002 in Europe.

	Norway	Sweden	England	Neth.	Germany	Spain	France	Italy	Belgium
Residential price	65.6	69	95	98	125	86	92	142	111
Difference between residential and industrial Prices	34	43	49	(50)	73	39	43	68	43
Residential switchings in 2002	15 %	10%	15%	N/e	5%	N/aN/e	N/e	N/e	

Note: N/e : non eligible.
Source: European Commission, *Second Benchmarking Report*, April 2003.

countries in Continental Europe in 2002. In the Nordic market (Norway and Sweden) with liberalised access open to the households, this disparity is only 10%. Moreover the difference between the price of industrial and residential deliveries also shows large variety between countries, that reveals large difference in distribution costs and rents before the total opening up of the retail market. The variety between margins is mainly between Italy and Germany (68–73€/MWh) on one side and the other countries (34–43€/MWh) on the other side.

References

Bergman, L., von der Fehr, N.-H., Newbery, D. and Pollitt, M. (2000). *A European Market for Electricity?* CEPR, London.

European Commission. (2001). *Proposal for a Regulation on Conditions for Access to the Network for Cross-European Commission Border Exchanges in Electricity,* March 13th.

European Commission. (2003). *Second Benchmarking Report on the Implementation of the Internal Electricity and Gas Market,* March.

European Commission. (2004). *Third Benchmarking Report on the Implementation of the Internal Electricity and Gas Market,* March.

European Power Daily, Platts.

Finon, D. (2001). Reforms in the French power system: from weak contestability to effective competition? *Energy Policy* (29): 755–768.

Finon, D. (2003). Introducing Competition in the French electricity supply industry: institutional destabilisation of the integrated organisational model. In Glachant-Finon, Eds., *Competition in European Electricity Markets: A Cross-Country Comparison.* Edward Elgar, London, Chap. 11.

Glachant, J-M. (1998). England's wholesale electricity market: could this arrangement be transposed to the European union? *Utilities Policy* (8): 63–74.

Glachant, J-M. (2000a). *Les réformes de l'industrie électrique en Europe,* Commissariat Général du Plan, Paris, juin.

Glachant, J-M. (2003a). The making of competitive electricity markets in Europe: no single way and no 'single market'. In Glachant–Finon, Eds., *Competition in European Electricity Markets: A Cross-Country Comparison.* Edward Elgar, London, Chap. 1.

Glachant, J.-M. and Finon, D. (2000). Why do the European Union's electricity industries continue to differ? In C. Menard, Ed., *Institutions, Contracts and Organizations,* Edward Elgar, London, pp. 432–456.

Glachant, J.-M. and Finon, D., Eds. (2003). *Competition in European Electricity Markets: A Cross-Country Comparison.* Edward Elgar, London.

Glachant, J.-M. and Lévêque, F. (1999). *Points d'entrée de la régulation du Gestionnaire du Réseau de Transport de l'électricité en France,* Report for the Electricity Department of the French Ministry of Industry.

Glachant, J.-M. and Pignon, V. (2002). *Nordic electricity congestion arrangement as a model for Europe: physical constraints or operator opportunism?,* Working Paper, CEEPR series, Massachusetts Institute of Technology, Cambridge, Massachusetts, USA.

Hancher, L. (1997). Slow and not so sure: Europe's long march to electricity liberalisation. *The Electricity Journal* November, 92–101.

Hogan, W.W. (2002). Electricity market restructuring: reforms of reforms. *Journal of Regulatory Economics* (21): 103–132.

Holburn, G. and Spiller, P. (2002). Institutional or structural: lessons from international electricity sector reforms. In E. Brousseau and J.-M. Glachant, Eds., *Economics of Contracts: Theories and Applications*. Cambridge University Press, Cambridge, Chapter 25.

Hunt, S. (2002). *Making Competition Work in Electricity*. Wiley, New York.

International Energy Agency (IEA). (1996–98). *Electricity Information*. Yearbooks OECD, Paris.

Joskow, P. (2002). Electricity sector restructuring and competition: a transaction-cost perspective. In E. Brousseau and J.-M. Glachant, Eds., *Economics of Contracts: Theories and Applications*. Cambridge University Press, Cambridge, Chapter 26.

Newbery, D. (2000). *Privatisation, Restructuring, and Regulation of Network Utilities*, MIT Press, Cambridge, Mass.

Newbery, D. (2002a). Problems of liberalising the electricity industry. *European Economic Review* (46).

Newbery, D. (2002b). *Regulatory Challenges to European Electricity Liberalisation*. CMI Working Paper No. 12, University of Cambridge (UK), Department of Applied Economics, September.

OXERA (Oxford Economic Research Associates), ATOM and ECN. (2001). *Indicators of Internal Market of Electricity*. Report for the European Commission DG TREN, (downloading from the European Commission website at: http://europa.eu.int/comm/energy/library/oxera.pdf).

Smeers, Y. and Boucher, J. (2001). *Towards a Common European Electricity Market, Paths in the Right Direction ... Still Far from an Effective Design*. Working Paper, Université Catholique de Louvain, September.

Stoft, S. (2002). *Power System Economics. Designing Markets for Electricity*. IEEE/Wiley, New York.

UNIPEDE. (1999). *Electricity Outlook*. Brussels.

Williamson, O. (2000). The new institutional economics: taking stock, looking ahead. *J of Economic Literature* 38: 595–613.

Williamson, O. (2002). Contract and economic organization. In E. Brousseau and J.-M. Glachant, Eds., *Economics of Contracts: Theories and Applications*. Cambridge University Press, Chapter 3.

Chapter 7
European Gas Markets: Nascent Competition and Integration in a Diversity of Models

DOMINIQUE FINON

7.1. Introduction

The idea of building an integrated European gas market based on competitive trade, like a gas lake supplied indifferently by remote, intra-European or national sources with the help of active market places allowing arbitration and price convergence, is far from being reached.

In fact it depends upon two conditions – the deregulation of each national gas market at the national level, and at the upper level of the European market the change in contractual relations between remote gas producers and buyers, which currently make the relations rigid with long-term transactions and limit opportunities for exchange on a competitive basis. Indeed, although in marked contrast with the electricity industry there is more than 60% of cross-border gas, the reason is not market integration but an imbalance of gas sources in and around Europe; this necessitates heavy up-front investment framed by long-term transactions. This issue, together with the geopolitical issue of gas dependence, was the reason for the delay in the European legislation of market liberalisation. All the continental industries and governments were opposed to the introduction of third party access (TPA) for preserving the capability to develop new import projects and they accepted it only with great reluctance in 1998.

At the European level, four years after the adoption of the Gas Directive 98/30, the European market can still be seen as a patchwork of different markets. The United Kingdom, which owns important resources, is clearly apart as a pioneer, with total unbundling of grid and supply activities and effective competition between producers on the wholesale market and between suppliers on the retail market. Among the other countries, under the reforms decided on with reluctance under the European Directive, there are differences that leave markets more or less accessible to entries and more or less favourable to the development of competitive markets with internal and external players. How these market developments play out depends on the nature and strength of the regulatory framework.

Moreover, new market rules have introduced forms of competition that are the seeds of a more extensive competition. Some national gas markets have already evolved considerably since the introduction of the TPA provision and consumer eligibility in 1998–2000, and after new regulatory decisions that significantly weaken the incumbents' dominant position by legal unbundling of the transmission network and definition of gas release programmes in a number of countries. The implementation of the 2003 Directive has only comforted this trend in the late countries.

This chapter analyses at two levels, namely national and European, the changing shape of the European gas markets under the effects of the market reforms and their chance of integration.

- First, we characterise the former two-level European gas market marked by the key roles of long-term contracts between the producer and the bulk supplier, the legacy of which determines the constraints on competition development more strongly than in the former industrial structures in the electricity industries.
- Second, in order to characterise the potential for the development of competition, the main traits of each national gas market are identified in terms of market attractiveness and market accessibility for the incumbents' competitors. On one side, the market developments will depend on business potential in each national market, resulting at once from market size, market growth, (especially in non-mature markets and in some active market segments such as power generation), and price differences which reflect existing profits on a number of national markets. On the other side, the market developments interplay with the opportunities opened to applicants for entry by the accessibility of each market, or conversely by the possibilities offered to incumbents for preserving their dominant position and deterring entries.
- Third, dynamics of market development towards market integration are inferred at the European level from these characteristics

and from the possibility for development of new forms of gas trade among foreign producers, suppliers and users.

7.2. The Starting Point: the Two-Level European Market

The European gas market was developed on two separate levels (Stern, 1990; Estrada *et al.*, 1995; Percebois, 1999):

- The national level, with the development of national or regional transport and wholesale monopolies. These monopolies developed the existing transport networks in co-ordination with the development of national production, and later on contributed to the setting-up of the major gas importation infrastructures with the producers. Industrial organisation in distribution reflected the pre-existing structure developed for city gas distribution.
- The European level, which is characterised by a two-sided oligopoly, balanced between the major producers and the major national companies, with the exception of the United Kingdom, which has long since differed from the continental market.

On this second level, the oligopoly of sellers consisted mainly of national companies handling exports from countries outside the European Union (Sonatrach in Algeria, Statoil and the public export consortium GFU in Norway, Gazprom in Russia) and from The Netherlands (Gasunie), each of which had an export monopoly. The international oil companies that produce gas in these exporting countries were subject to this public regulation in Norway and The Netherlands. Opposite to the oligopoly of producers was the oligopoly of purchasers, which includes the national gas companies of Continental Europe that are not active in production and are in a monopoly (or quasi-monopoly) position for wholesale supply in their country – Ruhrgas in Germany, Distrigaz in Belgium, GDF in France, SNAM in Italy, Gasunie in The Netherlands and OMV in Austria, joined in the eighties by Enagas (later Gas Natural) in Spain. A high level of direct or indirect state involvement in the national transportation system allows them to control the gas dependency relationship with foreign states as in France, Italy and Austria or the national resources management policy as in The Netherlands.

Isolated from the continental market until 1997, the British market has long since been a market apart, because of its size and an availability of resources sufficient to maintain autarky. This was first achieved with a monopoly by a public gas company that was practically the only purchaser from the North Sea producers, and later through the progressive deregulation of the gas market from 1986 onwards, with the development of direct deals between producers and major purchasers including suppliers and major consumers.

7.2.1. The vertical relations between production and wholesale

The relations between national producers and purchasers take the form of long-term contracts that define a series of rights and obligations. These rights and obligations regiment the relations over a long period of time but allow price-risk and volume-risk to be shared between partners, thus allowing substantial investments to be made:

- the obligation to take-off a given quantity of gas, under Take or Pay clause;
- the price indexation clause on crude oil or oil product prices;
- the final destination clause, which obliges the purchaser to sell the gas purchased on his market alone because of country-specific price-definition clauses; this clause therefore creates de facto partitions between national markets at the resale level.

This 'two-level market' structure has not only allowed risks to be shared right along the gas chain, thus allowing substantial investments to be made in production and transport, but has also allowed a balance of market power between producers and purchasers. On one hand, the gas companies can aggregate demands, because of their exclusive right to supply on a regional or national level. Armed with their capacity for managing the quantity-risk, they are able to sign long-term purchase contracts that allow producers to develop the production and infrastructures assets necessary for the export of gas. As holders of a sales monopoly, they are also able to discriminate between various market segments according to the conditions for replacing gas with an alternative supply. For their part, the producers agree to bear the price risk – the price regulation and indexation clauses, based on the principle of 'net-back', allow gas prices to be maintained at a competitive level with rival fuels in the different gas uses.

The international oil companies involved in production (Exxon, Mobil and Shell) in exporting countries (The Netherlands, Norway) are also present further down the chain, where they obtain stock-shares in transportation and resale companies in order to capture an additional part of gas rent downstream. Exxon and Shell, who have always worked alongside each other, have preferred to concentrate on upstream,[1] in contrast to Mobil, which has always had an aggressive policy of downstream integration (Wybrew-Bond, 1999). Their upstream strategy has never involved any attempt to obtain oligopolistic control over the European market. The stock shares did not give them any industrial power in the strategy of these companies. (Stoppard, 1996; Radetski, 1998).

7.2.2. Partition between national markets

On this institutional basis, the integrated European market existed only in the bulk supply and is organised as an oligopolistic 'club'. The former European market can be described as a series of juxtaposed and entrenched national markets, supplied from outside by unidirectional stable flows coming from the same four supply sources and passing across one or two transit countries. More than half the gas consumed in the European Union crosses at least one border, but in terms of wholesale purchases, this merely involves physical interrelations between national systems. Market integration on the basis of cross-border exchanges initiated by countries with a resource surplus relates only to sales from The Netherlands inside the European Union. It will however be noted that because of the hierarchy of the two market-levels, wholesale prices on each national market follow parallel movements because of their common indexation on crude oil and oil products prices. This type of integration relies on co-operation between the gas companies grouped in a consortium to negotiate import contracts with major producers, to develop transit infrastructures together and occasionally carry out swaps between import contracts in order to reallocate flows geographically.

7.2.3. New basic conditions and feasibility of market liberalisation

This two-level market structure was well-suited to the developing gas industries, whose major transportation and distribution infrastructures still have to be set up for imports and national supplies. It no longer has the same foundations in the maturity stage, where penetration of gas into the various market segments is well advanced, as in the major national markets (Stern, 1990; Finon, 1992). The international transit network infrastructures are mostly depreciated. Increasing success in offshore gas fields and the creation of pipeline links to other countries

---

[1]Exxon and Shell play a part in the capital of Gasunie and Ruhrgas, and Exxon also plays a part in the capital of Thyssengas and ETGs in Germany.

have caused a lowering in spot and contractual gas prices, compared to older contract prices, putting pressure on old regulatory regimes. Maturity of main markets goes hand in hand with import overcapacity, overcontracting and price reductions on gas spot markets where competition is based on a logic of short-term gas-to-gas competition (Stern, 1998; Radetski, 1999). The need for import investments in the mid-term is less pressing than in the past and projects are being postponed for several years. In similar situations, the experience of the British and American gas markets deregulation demonstrates that introducing market rules at this stage can be feasible and can carry market incentives to efficiency improvement (Mestmäker, 1993; Waddam-Price, 1998).

The voting-in of the European Directive 98/30 lays down for the Member States the application of basic rules for the deregulation of national gas markets:

- the right of access to the natural gas network for direct purchases by electricity producers, eligible consumers and distributors from producers and sellers chosen by them, and vice versa,
- a minimum opening level of 20% in 2000, 28% in 2003 and 33% in 2008, which has been extended by the next 2003 directive with an minimum opening up to the commercial sector (i.e., an average rate of around 70%) in 2004 and to the residential one in 2007.

By destabilising the previous vertical relationships between foreign producers and national gas companies, the national reforms laid down by the European Directive could have two potential effects:

- removing the partitions between national markets by making trading in gas between countries easier, most notably through deals between intermediaries, gas companies, traders and multienergy companies;
- removing the vertical separation between the two market levels by allowing producers to approach intermediaries in competition, regional distributors, electricity producers and major consumers directly.

If we look at the basic conditions of national gas markets, they appear to be more vulnerable than the power markets to a European prescription that requires the Member States to implement a provision of non-discriminatory TPA, for several reasons.

First, given the storability of the commodity, technical co-ordination gives more room for organisational flexibility than for electricity markets. Second, technical interconnections are large because of the need to import to meet most needs. Third, it was not vertical integration but long-term contracts that organised relations between production and transport and between transport and distribution-supply in the previous situation. Consequently it would be easier to redefine relationships and transfer property rights to new players than in a situation of vertical integration, as it was in the electricity industry prior to the reforms. The incumbent companies do not have the same capacity to preserve the integration of their activities as in electricity. However, the resilience of the pre-reform contractual arrangements heightens the importance of maintaining long-term relations for the development of new production and infrastructures from remote sources.

However, the effects of the 1998 Directive and the reforms in transcription in national law are much more limited than is necessary for effective competition and integration of national markets for two reasons.

First, the minimum requirements for conditions of access simply do not allow transparent and non-discriminatory access. They left the choice between regulated and negotiated access to the network, without particular specifications for access to storage capacities and for transport pricing principles: they required indeed at minimum simple accountable unbundling between transportation, supply and other activities; and they did not impose to give credibility to the non-discriminatory access by the creation of a regulatory authority. It is up to the countries to go beyond. But the next 2003 directive which mandates the Member States to adopt the regulated TPA, the legal unbundling and create such an authority harmonises the national situations.

Second, the two Directives do not ask for long-term contracts to be called into question, and the effect of this will be threefold. One, the field of competition upstream is only open to a limited extent for new transactions, with very little non-contractualised gas, the so-called free gas; two, the restrictions on the final destination of contractual gas from two of the main exporters (Russia, Algeria) which are outside the EU legal jurisdiction are still maintained in many contracts, which will contribute to preserve the divisions between the national markets; three, the definition of wholesale prices remains to lay on contractual prices indexed on oil prices (crude and final products). Except in UK, the portion of wholesale supplies which does not depend on price-definition by such a netback pricing with oil price indexation is very limited. It limits the possibility of trading on a spot basis. It impedes the normal market play between gas vendors that would allow gas prices to be discovered without direct influence from other substitute prices as is currently the case on every commodity market.

7.3. The Attractiveness of National Markets

The attractiveness of a national market influences the entry of companies that compete against the incumbents. It depends mainly on the size and maturity of this market and the growth potential of its various segments, especially power generation. It also depends on the price levels in each segment of the final market, given that the relative prices in each market segment is indicative of the potential for generating profits downstream in the value chain.

In the following paragraphs the eight mature markets which correspond to about 95% of the EU demand are only considered. Occasionally for illustrative reasons we consider smaller peripheral and young markets (Denmark, Finland, Ireland) but not to the nascent ones (Sweden, Greece, Portugal) which are under the peculiar constraints of installing the first infrastructures. Some of them (Ireland, Portugal, Ireland) benefit from exemption from the 1998 Gas Directive obligations to introduce TPA provisions because of the financing constraints of installing the infrastructures of their nascent gas system.

7.3.1. Market size and growth

The most attractive markets are those that show the best combination of market growth and market size. In this respect Italy and Spain show the best potential, alongside the largest and most mature markets as Germany and UK, and the young or nascent markets in smaller economies (Denmark, Ireland, Finland, Sweden, Portugal, Greece).

Market size

In the range of size from 40 to 100 bcm/y, markets in major countries (UK, Germany, Italy, France) are the most important in terms of size, with The Netherlands (48 bcm/y) which encouraged the use of gas because of its resource endowment. The French market is of less significance (38 bcm/y) because of the importance attached to nuclear electricity; such was this importance that developments in the use of gas in electricity production and as a space heating fuel were pushed to one side. Some way behind comes the Spanish market (16 bcm/y in 2000), which has only been developing since the 1980s, but has now

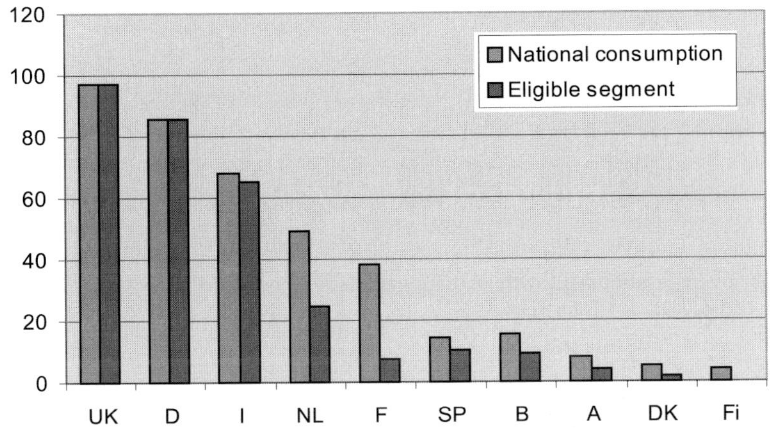

Fig. 7.1. The size of national gas markets and eligible segments in 2000 (in bcm/y). *Source*: IEA (2000). *Natural Gas* Information 2000; EC-DG TREN (2002). *First Benchmarking Report*.

equalled the Belgian market in size (15.6 bcm/y). Much further behind are the markets of other countries (Denmark, Ireland, Finland, Sweden, Portugal, Greece and Switzerland), less than 5 bcm in size; this is explained by the economic size and population of these countries and for the last ones also by the relatively young age of their gas industry. These markets will not represent major targets for potential entrants in the future. Only Austria is at an intermediate level, with a market size of 8 bcm; the reason for this is the earlier development of the gas industry (Fig. 7.1).

Market Growth

In terms of attractiveness, the advantage of size in major markets is partly compensated for by maturity, except for the new segment of the power generation which allowed upper market growths. Italy, Germany and the United Kingdom show a growth rate, excluding gas-based power generation, of 2–2.3% per year during the nineties (Fig. 7.2).

The French gas market could benefit from a catching-up effect, given that it grew at a rate of 3.5% in the nineties with some new development in space-heating and industrial uses. Countries with a 'young' gas system, such as Spain and Denmark, saw their markets grow at a faster rate, in the region of 10%, during this period.

The two groups of national gas markets (mature markets, young markets) share one common promise: their future growth will be

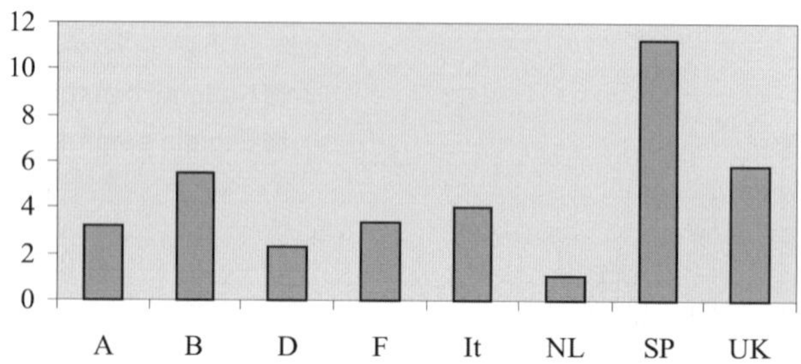

Fig. 7.2. Average annual growth of national markets between 1990 and 2000 (in %). *Source*: IEA (2001), *Gas Data Book*.

mainly based on electricity generation development. Already the commissioning of gas-fired power stations by independent producers in the mature British and Italian markets has led to renewed growth in the gas markets (5.6% per year on an average since 1990 in the first one, 5.5% since 1999 in the latter). The turn of the last century saw the commissioning of the first CCGTs in some other countries, especially Spain. This market dynamic is important for encouraging entry to other market segments. In fact, development of gas-based power generation is a lever for the electricity incumbents to enter the final gas market: they contract directly with foreign gas producers for supplying their gas turbine units and they jointly sell gas and electricity on the industrial segment. It is typically what is happening in Italy and Spain with the utilities' entry (ENEL, Edison in the former, Endesa, Iberdrola and Fenosa in the latter) to the industrial sector. Development of gas-based cogeneration on site is also a central element in multienergy and multiservice strategies (Fig. 7.3).

In future, the development potential for gas in electricity production will increase because of environmental policies and deregulation of electricity markets; this will favour the adoption of gas units, which are few, capital-intensive and cleaner than coal generation, by the electricity producers in competition and the development of decentralised production and co-generation. However, the situation of gas-based electricity production and its potential will differ from one country to another, if we refer to the portion of gas used in power generation (see graph). There are three groups of countries, each obviously with different potential for development:

- those required to respond to a very dynamic demand for electricity: Spain, Italy and the small 'catching up' countries (Ireland, Greece, Portugal) not considered here;

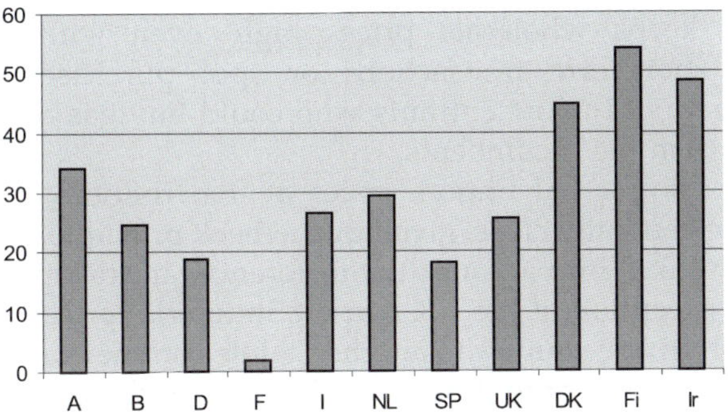

Fig. 7.3. Share of power generation segment in total gas consumption in 1999 (%). *Source*: IEA (2001), *Gas Data Book*.

- those in which growth of electricity generation is low but replacement of old equipment and progressive nuclear phase-out policies will require the installation of new units after 2010 (United Kingdom, Germany, Belgium, The Netherlands, Austria);
- those whose production capacity is based on equipment with a long life span and low variable costs, which will limit the potential for new gas outlets through centralised production by CCGTs but also decentralised cogeneration projects during the next twenty years (France, Sweden).

According to a reference forecast,[2] Spain seems to be the most promising market in terms of power generation by gas, with 17 GW of CCGT planned for 2000–2010. Next is Italy, with 15 GW to be installed by ENEL and Edison, followed by the United Kingdom, with 6 GW authorised in 1999–2000, and the small countries (Ireland, Portugal, Greece). Despite the policy of promotion of renewables (objective of 20 GW in 2010), German electricity generation remains a promising market for gas, firstly for fuelling new or refurbished cogeneration units and for centralised production, once the nuclear reactors phase out begins.

7.3.2. Differences in gas prices between European countries

The attractiveness of a national market to entrants is also partly dictated by price levels on the different market segments. High price levels can reflect low levels of competition from substitutes, market imperfections or protection of activities which are propitious to profits. For instance, gas price indexation on oil prices in long-term import contracts with

[2]Forecasts shown in *Power in Europe*, 28 April 2000.

incumbents keeps wholesale prices high, even with overcapacity, meanwhile short-term transactions or spot purchases on the gas marketplace would allow entrants who could buy gas on it at a much lower price than the incumbents.

However, contractual import prices at the frontiers of the various countries are generally close, given the netback principle of price calculation, which takes into account the differences in transportation costs. So with the exception of the UK market, industrial prices are generally quite close from one country to another, while commercial and domestic prices tend to show more differences because of the common fact that smaller the client, larger the price difference between each country.

Industrial prices

Before the implementation of reforms in 1999–2000 and the introduction of competition on the continental markets, the British gas price was clearly apart, because it was determined by gas-to-gas competition in the context of overcapacity on the British market. Between 1995 and 2000 it clearly established itself at a level 30% below the industrial prices on the continental markets. Meanwhile, some similarity of industrial gas prices on European gas markets is noticeable because of the common price indexation on oil products and oil prices, given that border prices do not differ greatly from each other.

According to Eurostat gas price collections, German industrial prices have tended to exceed the median price by 10–20%. Increases in imported gas prices under oil price moves are passed through final prices with a longer time lag and less amplitude than in other countries because German gas companies extract more profit during this period than other companies in this sector. In other countries industrial prices were more or less regulated in the opposite direction to the German prices (Fig. 7.4).

The immediate effects in 2000–2001 of connecting the British market to continental markets by the UK continent interconnector illustrate this aspect of market attractiveness. It led to a steep rise in industrial prices in UK during the period of high gas prices in import contracts on the Continent, consecutively to a rise of oil price. UK industrial prices increased because of the call for large quantities on the continental markets and the British producers' arbitrage between the two markets.[3]

[3]The British spot price was established at \$4.2/MBtu in autumn 2000 and again at \$4/MBtu in autumn 2001, compared with an average of \$2.5 between January and September 2001.

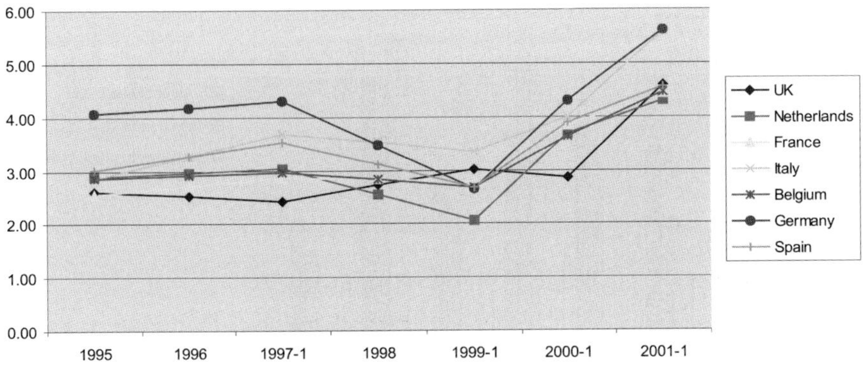

Fig. 7.4. Differences in industrial gas prices in selected countries for the period 1995–2001 (in €/GJ). Reference to the consumer category of 10 Mm3/y (or 418 000 GJ/y). *Source*: Eurostat, *Gas Prices between 1990 and 2001*.

Household prices

Domestic and commercial sale prices vary sharply from one country to another because of the difference of rents captured in unbundled retail supply and grid businesses. By referring to two different domestic tariffs in 2000, the following hierarchy in the eight mature market countries (Fig. 7.5).

- In the large domestic subscriber category, the Italian and Spanish tariffs exceed those of the other five countries by 40%. In sharp contrast, the British tariff is very much lower (50% lower) that the tariffs in the five other countries.
- In the smaller consumer category (kitchens and water heaters), the group of countries with the highest tariff includes Belgium and Spain, with Germany and Italy not far behind. The difference from the group of lower-priced countries (Austria, France and The Netherlands) is less pronounced (at about 35%) than for large subscribers. British prices are again significantly lower than the others (28% lower than the next lowest price) at present.

The general increase in wholesale prices in UK reduced the differences in 2001 somewhat, but not so much to remove the differences in profit extracted by distributors–suppliers in various regulatory environments. Because of the tighter regulation of distribution activities unbundled from the supply the United Kingdom is lower in terms of profits from distribution.

The differences between domestic and industrial prices within a country reflect the level of profits derived from gas distribution.

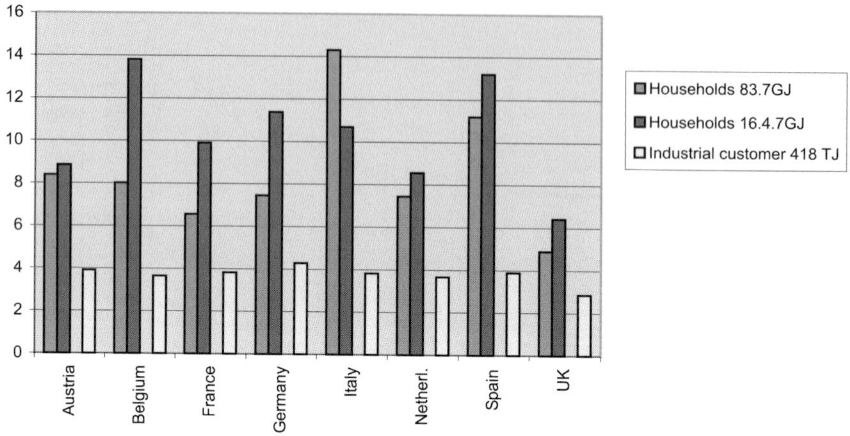

Fig. 7.5. Differences of gas prices on household market in selected countries in 2000. Reference to Eurostat categories: D3 category 'cooking, water heating, central heating' 83.7 GJ; D2 Category 'cooking, water heating' 16.74 GJ; and I4-1 category of industrial consumer: 418 000 GJ/y. *Source*: Eurostat. *Gas Prices 1990–2000*.

With reference to the industrial and domestic customer categories considered here, the differences between industrial and domestic prices are more significant in Belgium, Italy and Spain. In Spain the difference can be explained by higher costs attributable to the rapid development of gas networks, which requires finance. On the second level of price differential we find Germany, followed by France (where the average cost of distribution is a little higher than in other countries because of the lower territorial density of demand).

To sum up, some of the largest markets appear the most attractive because of market imperfections that lead to higher price levels in a number of market segments. Such are Germany and Italy, which appear to be the most promising markets for entry candidates with large size and level of prices. Spain is also a promising market because of its size and growth prospects. The UK is a more difficult one because of the small margins allowed by effective competition on the wholesale and the retail markets, and stringent regulation of transport prices.

7.4. Accessibility to National Gas Markets: National Influences on Market Rules

Competition on national gas markets will be conditioned by actual accessibility to wholesale and retail markets for national and foreign producers, suppliers and traders. Market opening by the provision of consumer eligibility must be complemented first by rules of transparent access to gas systems, second by industrial structures which facilitate entries, third by technical infrastructures (interconnection capacities) and commercial infrastructures such as gas exchanges backed onto easy access to storage facilities (IEA, 2000).

First, market rules must enforce non-discrimination for potential competitors. Even with a market opening of 100% and a relatively fragmented gas supply activity, a national market can remain weakly competitive if access to pipeline and storage facilities remains opaque with suspicion of discrimination in favour of incumbents. As theoretical works demonstrate (Baron and Meyerson, 1982; Armstrong *et al.*, 1994), if the network operator competes on deregulated stages of the value chain, the risk of deterring newcomers' entry by limiting access to transport and system services, or by overpricing, is significant. Empirical evidence is shown by British Gas's market dominance before its split in 1996 with excessive rates for transmission and balancing (Waddam-Price, 1998; Thomas, 2001). In this respect, the presence of an independent regulatory authority is crucial to guarantee fair access to essential facilities by an ex ante regulation.

Second, market dominance by the incumbent owning a large portfolio of pre-reform long-term bulk supply contracts inhibits entries, as has been said. Theoretical works show how long-term contracts create barriers to entry to the intermediate and final markets (Aghion and Bolton, 1987). The incumbent's market dominance has to be weakened by some vertical and horizontal de-integration: legal unbundling of the transmission network, and disinvestment of long-term contracts.

Third, creation of spot markets in some national hubs, where the convergence of different pipelines or LNG terminals allow competition between different gases, will increase accessibility. It will ease the development of effective competition by offering contractual diversification, price discovery, balancing opportunities and risk-management capability. It will facilitate competitive entries and bilateral transactions and give means for wholesale or industrial buyers to complete contractual purchases.

But there has been significant differences in the first wave of reforms in national gas industries, which reflect the specific features of regulatory culture, their gas resources endowment, the legacy of previous structures and the level of maturity of the market. In this sense, with all the different national situations, each market showed a different potential for the development of internal competition and entry incentives between 1999 and 2004, year of implementation of the

2003 directive which is supposed to erase some major regulatory differences. The competitive potential of each market during the period 1999–2004 will be characterised in the three dimensions of accessibility to entrants: regulatory access, technical and commercial access and industrial access with barriers to entry resulting from industrial structures.

7.4.1. Regulatory accessibility to national gas markets

The regulatory access is characterised by various elements that all together have to reach consistency: the level of retail market openness, rules that govern access to networks and storage capacities and can guarantee non-discrimination, and the presence of a regulatory authority. The first gas directive allows a wide combination of options for these key elements. In fact, the comparison shows three groups of countries:

- the pioneer group, with the United Kingdom well ahead of all the others for all the competitive regime issues;
- then a second group of countries with Italy, Spain, The Netherlands and partly Belgium, which intensified reforms in 2001 and 2002 after initially transcribing the directive only moderately; it represents 35% of the Euro-8 market;
- and then the group of laggards with Germany, France and Austria, which tend to adopt the most conservative market rules and represent 39% of the Euro-8. For political reasons France did not transcribe the 1998 Directive into its national legislation until 2003, whereas Germany very slowly defined a negotiated framework of network access in a traditional process of agreement between stakeholders.

Eligibility

In addition to the United Kingdom, where market opening has reached 100% since 1998, most Member States have gone further than the minimum market opening required. Germany opened its market up to 100% in 1998, whereas Austria, Italy and Spain have opened their market completely in 2003 and 2004. The Netherlands and Belgium have reached an opening level of 60–65% by that date. France has remained at the minimum required level till July 2004 and the transcription of the 2003 directive. The mean level of national market opening at the European level was therefore set at 80% in 2001, but this does not mean that competition will be effective and so widely extended. The effectiveness of competition depends on market rules other than eligibility, namely, rules that guarantee non-discriminatory access to available transport capacities and storage facilities. A typical example is the absence of practical access rules in Germany for medium and small consumers between 1998 and mid-2002, despite the total opening up of the market.

Third Party Access

Market accessibility differs according to the efficiency of TPA to the pipeline networks and other essential facilities: LNG terminals, storage capacity and blending (EC-DG TREN, 2000, 2002). A transparent information system on available capacities is also a prerequisite for non-discriminatory access. All this requires clear unbundling rules and access conditions that are non-discriminatory, fair and simple. Ideally the transport system would have to be totally independent of the former incumbent. By this way physical rights to capacities are 'commoditised' as are storage capacities.

At the first stage of the reforms, European countries have chosen different types of TPA and different unbundling levels, which have different impacts on the competitive pressure of entries and on the incumbent's ability to keep on preferential access to the gas system. Apart from the UK, those countries most in favour of transparent and non-discriminatory access with a clear separation of gas system operations, namely Italy and Spain, have opted for the regulated TPA. They have been caught up in 2002 by Austria, The Netherlands (which was the first country to publish its access tariffs in spring 2002) and then Belgium and Spain. The choice of negotiated TPA has been maintained by Germany[4] and to some extent by France, which has chosen a combination of negotiated access (for transport) and regulated access (for distribution).

The countries that chose regulated TPA have also overlooked, or are about to overlook, distance-related tariffs (point-to-point), which are

[4]In Germany the traditional German rule making by general agreement between the main professional associations allows to establish guidelines for the TPA. So an agreement which is being elaborated in several stages since mid-2000 is negotiated between transporters (BGW) and users (BDI, VKU, VIK) and defines the main commercial conditions. The first gas Verbändevereinbarung (VVI) has had several supplements, signed in March 2001, September 2001, July 2002, etc..., to improve transparency, access to storage, system services, congestion management, information on operational capacities.

Table 7.1. Comparison of transportation prices over 200 km in 2001 (€/MWh).

Germany	France	Belgium	Austria	Netherlands	Spain	Italy	UK
0.75–0.83	0.85	0.79	0.85	0.65	1.34	1.6	0.45–1.11

Source: European Commission (2002) – *Implementing the internal energy market*, p. 47.

penalising for trade. They adopt either post-stamp tariffs, as Spain and Belgium have done for high volumes, or more commonly nodal pricing (entry–exit) like the UK and Italy. Some tariffs are hybrid and simplified in zonal pricing like those defined in 2002 in France: the customer's transport costs are a function of the postage tariffs inside each zone; he has to pay for each transport zone which separates him from his gas supplier; and it tends to be a handicap for consumers located far from the entry points in the system.

Furthermore the level of transportation tariffs is influenced by the transparency of their calculation. Those regularly recorded by the European Commission reflect the government's will to ease gas network access in the respective group of countries. In this respect German and Austrian tariffs appear to have been more deterrent to entries than those of other countries.

Finally standard balancing obligations have been also at stake when erasing entry barriers. Most countries had preferred the more restricted obligations, i.e., hourly balancing with strict tolerance rather than the smoother rule of daily balancing adopted in Italy, Spain and more recently The Netherlands (Table 7.1).

These three groups of countries also differ in the choice of type of unbundling:

- Total ownership separation of transport and gas supply in UK, since the separation of Transco and Centrica in 1997.
- Creation of a transport subsidiary by the incumbent, with partial flotation for increasing the guarantee of independence in Italy and Spain by reducing control by the incumbent (ENI and Gas Natural respectively) over its pipeline subsidiary in 2001 and 2002.[5] Belgium and The Netherlands followed, with the splitting of Distrigaz and Fluxis (the new transmission company) respectively, and for Gasunie's gas system in 2003.

[5]ENI had to reduce its participation in the SNAM's stock to 50% in 2002 and had sold 40.24% of the newly created Gas RETE shares on the stock market in October 2001. So did Gas Natural for the Enagas stock up to 35% in 2002.

- Simple accounting and functional separation in the other countries, with internal 'Chinese Wall' or code of conduct to guarantee non-discrimination on commercial information and non-preferential treatment, as the 1998 Directive requires. Among these other countries, Germany did not go beyond the minimum requirements, while Austria implemented a legal separation in 2003. France followed in 2004 under the prescription of the second gas Directive.

An additional accessibility parameter is the clarification of technical and economic rules of distribution networks access for competitive sales to medium gas users (SMEs, commercial sector). The pre-existing vertical separation of distribution with local distributors in Belgium, Germany, Italy and The Netherlands is a factor that maintains an obstacle, whatever the regulatory will to organise the playing field at this level of eligibility and further after the opening-up had been achieved.

The requirement for storage access, which should have had a crucial role in the development of short-term transactions, is generally wholly consistent with the type of unbundling.[6] Clearly countries such as France or Germany, with remaining vertical incumbents avoid the mandatory access to storage capacities.[7]

In the same logic, countries that have chosen to give way to effective competition have sought to enforce non-discriminatory access by creating sectorial regulatory authority. However, their level of regulatory powers is not systematically correlated to the intensity of the reforms. Although the UK and Italy, at one extreme, have chosen to create an independent authority with broad effective powers (promotion and surveillance of competition, regulation of monopoly activities etc.) while Germany, at the other extreme, prefers only slight regulation with no specialist authority and ex post control by the antitrust office, the group of latecomers (Austria in 2001, France in 2002) have created authorities with effective powers, while in the intermediate group there are either consultative bodies (Belgium, Spain) or a special ministerial department in the competition directorate (The Netherlands).

---

[6]The storage capacities can be separated from the network company as it is the case in Italy.

[7]A certain range of flexibility services (balancing service, back-up) offered by the incumbents are substitutes to the access to storage capacity under normal circumstances for the efficient operation of the system. But they do not cover access to storage capacity which is independent of the system use for weekly or seasonal arbitrage.

Improving regulatory accessibility

Backed onto the European competition law and the Treaty requiring it to harmonise national laws, the European Commission has been active on a number of different levels since 1998:

- An increase in national reforms, aimed at improving access to the network and to gas companies' storage units through the regulations of the next 2003 directive,
- The limitation of technical and price-related barriers to exchanges between countries, and the creation of a dialogue between regulators and gas system operators, known as the Madrid Process. The European Commission is relying on the co-ordination of gas system operators (known as the European Gas Transporters Association or EGTA) and sectorial regulators (grouped together for 12 countries in the Council of European Energy Regulators).

The 2003 Directive introduces three main changes: the near-completeness of the opening of the final market, an increased guarantee of non-discrimination in access to each gas system, and the unbundling of distribution and supply (European Commission, 2001). First, the opening process is completed by July 2004, with the possible exception of supplies to domestic customers postponed to 2007. Of the markets in question, this move will mostly affect the French market, the opening of which would have remained limited to 28% reaches 70% in 2004. The few other markets (Belgium, Denmark and Sweden) that did not plan to open fully by 2008 see also their opening process accelerated. Second, the improvement in guarantees of non-discrimination in transmission, LNG facilities and storage access will be sought through the requirement for legal separation between network and storage,[8] regulated access by third parties with 'cost-reflective' calculation principles and transmission prices, balancing and storage service, rules of transparency on capacities available for transportation, and the enforcement of ex ante regulations guaranteeing these rules. In this respect, accessibility to the German market would be upgraded in succeeding years. Third, the unbundling of the supply and the distribution – for every distributor down to a threshold of 150 000 clients – will weaken the incumbents' dominant position and easing entries in the retail supply. All in one the regulatory accessibility is really converging between the different countries (Table 7.2).

[8]The next directive does not require the distribution network to be separated in this way.

Table 7.2. Mapping of the competitive reforms in the main gas industries before the second gas directive in 2003.

	UK	Italy	Spain	The Netherlands	Belgium	Austria	Germany	France
Eligibility	100%	100%	100%	60%	59%	100%	100%	20%
Third Party Access	Regulated TPA	Regulated TPA	Regulated TPA	Regulated TPA	Regulated TPA	Regulated TPA	Negotiated TPA	Regulated TPA
Type of transport tariffs	Entry/Exit	Entry/Exit	Stamp post tariff	Entry–exit	Distance, then Entry–exit	Distance pricing (point to point)	Distance pricing	Nodal/zonal pricing
Unbundling and TSO status	Independent (Transco)	Subsidiary with flotation	Subsidiary with flotation	Management	Subsidiary	Subsidiary	Code of conduct	Code of conducts
Standard balancing obligation	Hourly	Daily	Daily	Daily	Hourly	Hourly	Hourly	Daily
Obligation for access to storage	Yes	Yes	Yes	Yes	Yes	Yes	No	No
Authority of regulation	Yes	Yes	Consultative Body	Yes (Specialised ministerial division)	Consultative body	Consultative body	No	Yes

Source: EC-DG Tren (2003). *Second Benchmarking Report*.

7.4.2. Technical accessibility of national gas systems

Effective entry into a national market depends directly on conditions of physical access to national gas systems, the number and capacity of interconnections and LNG terminals, transportation tariffs and rules of access to interconnections in cases of border congestion (IEA, 1994). In particular access to LNG regasification terminals will provide new opportunities for LNG vendors, especially for cargo sales or mid-term direct contractual supplies. The organisation of access to LNG terminals is difficult to define and impose by regulation, because of the need for co-ordination between terminal users who are competitors.[9] The type of TPA to these facilities will have a major influence on these opportunities.

Because of the technical need for large volumes of gas transported from remote sources, the existing interconnection capacities between national markets are significant, with the exception of the three 'gas peninsulas' Spain, Southern Italy and the UK and the peripheral nascent markets (Sweden, Greece, Portugal, Ireland).

The continental markets are connected to each another with many entry points and connections because of their dependence on imports and, for Austria, Belgium, Germany, France and Switzerland, their role in the transit of gas. Generally, the import capacity significantly exceeds their import needs and is close to the total level of national needs. This regional area, with its well-connected markets and large transport capacities, could become the main battlefield in which the markets could mutually influence each other with converging prices and difference simply reflecting transportation prices. Transit restrictions for long-distance exchanges could be managed by swaps between various sources as soon as the market places are well established to ease such deals.

The British market is the least connected to other European markets because of its autonomy in terms of resources. The UK-Continent interconnector built to export gas to the continent (20 bcm/y) has a reverse flow capacity of 8.5 bcm/y, corresponding to 8% of national consumption. In 2001, however, the pipe to the exhausted Frigg deposit was reactivated to connect the British market to the Norwegian system. Spain is also a gas peninsula because of the limited connection capacities between France and Spain (4 bcm/y). Despite three connection points Italy is partly viewed as a peninsula with a kind of double gas system, one well connected to Northern and

[9]Availability of several LNG terminals will improve flexibility and economies of scale for managing off-loading of competitors.

Table 7.3. Gas import capacity into selected countries in 2002 (Pipes and LNG terminals).

	Austr.	Belg.	France	Germ.	Italy	Neth.	Spain	UK
Pipeline entries[1]	3	4	3	6	3	2	2	2
LNG terminals		1	2		1		3	
Import capacity (Bcm/y)	40.8	80.3	61.5	201.7	81.4	38.5	43.8	8.8

Source: EC-DG Tren (2002). *Second Benchmarking Report*.
[1]Export points for transit not included.

Eastern European sources and the other connected to North-African sources by dedicated pipes but loosely connected to the northern system. The Italy's and Spain's LNG terminals and interconnections with Maghreb sources via pipelines (Transmed, GEM and soon the Libyan pipe) are direct entry points for natural gas imports. More generally, LNG terminals could become a new way of importing gas on a short-term as on a longer-term basis, as it will be the case for the future imports of Egyptian gas under LNG contracts signed with GDF in France, Union Fenosa in Spain and Edison in Italy (Table 7.3).

Three other elements must be added to this view of technical accessibility about some conditions of further markets integration. First, exchanges of gas between countries can be limited by the restrictive requirements imposed in certain countries with regard to quality of gas, and the development of flows is thus hampered (Heyvaert, 2001). By comparison the American market, which refers to a single quality range, could be more easily unified. There is therefore a need for greater network interoperability and gas quality harmonisation, and this may require physical processing (in particular, the transition from H-gas to N-gas) or mini-swaps. Second, apart from the rules governing transmission tariffs, there is a great need for transparency on available transmission capacity for cross-border trade on a short-term basis and, in future, on a real time basis. It could take several years before this ideal stage is reached. Third, the legacy of long-term contracts and reserved capacities in the transit pipeline and interconnection greatly restricted opportunities for access. Part or all of the transit capacity is contracted to the incumbent companies and most TSOs have no agreement on providing information to the market on the amount of free capacity to be available.

In the context of the Madrid Process, the system operators have made some progress in defining interoperability rules since 2000, such as changes to gas quality and harmonisation of methods for calculating available cross-border capacities; they have however encountered difficulty in reaching an agreement on rules for advising the capacities available.

7.4.3. Commercial accessibility and market institutions

The development of competitive trade needs informational and commercial infrastructures to be created at different stages: the bulk and wholesale level and the retail level. The existence of such devices is a good indicator of the intensity of competitive trade beyond the legal market rules.

Existing marketplaces

Marketplaces are crucial commercial devices for supporting competitive trade by allowing gas demands and supplies to be met in an anonymous and standardised way at the intermediate stage of the wholesale exchange (De Vany and Walls, 1994). The creation of spot markets in hub places or at notional points would answer to the need to help transactions and deals by de-linking the price of gas, transmission and storage. For market participants there is a need to balance demand and supply on a monthly and daily basis and to complement supply for swing needs by short-term deals during peak periods. Development of short-term sales implies also the creation of a contract for access to the transport system (for instance a contract for a reservation of capacity for 12 months at the entry points at which the sellers will inject gas and at the exit points in their delivery zones) and the creation of a secondary market for offering a flexibility by exchanging these rights on capacities. The same applies to access to storage capacities (reservations by annual auction) and to daily balancing between injection (deliveries) and quantities removed (offtake) by clients. Hub markets provide gas supply flexibility, increase contract diversity and are natural spot-pricing reference points for regional gas prices in contracts. They are the support of futures trading for risk coverage needs (IEA, 1998).

There are however other channels of organised trade, via brokers and traders who act as intermediaries between producers, suppliers, power producers and some large consumers. This trade is developing along with the electricity trade, often in the same brokers' hands. In this respect the development of a gas price index (such as the UK's Heren index) at reference points for this grey market is a substitute for a reference spot price for helping short-term trade. The development of electronic exchanges by innovative traders such as EnronOnline before its bankruptcy in 2001, the InterContinental Exchange (ICE), Spectron or DynergyDirect has been a major step in UK since the end of 2000, first with the joint offer of transport and storage contracts, and on the continent since 2002.

Significant marketplaces are now operating in UK, with reference to a notional point, and in Belgium at the Zeebrugge hub. A few number of other hub marketplaces will emerge by 2005 on the continental markets. The creation of a marketplace needs the prospect of gas competition with several sellers present, a confluence of several pipes from different origins, an availability of 'free' non-contracted gas and the proximity of significant storage capacities, and these conditions are far from being reached in many locations and national markets. They will be helped by support from the incumbents as soon as they decide to adopt new forms of marketing and to trade gas, although initially they will still be resistant to such developments in their areas and maintain barriers (Stern, 1998).

Existing marketplaces in 2004 are different in scope and nature, as does their role in the activation of wholesale competition in respective countries. The British marketplace operates at the national level with the Bacton hub and Transco's NBP reference point with a good liquidity in view of the number of producers (about 30). Reorganised in 1999 under the auspices of the regulator to encourage trade and competition, the new British marketplace has several compartments. As well as the futures market managed by the International Petroleum Exchange (IPE) with contracts referring to the notional National Balancing Point (NBP) and the OTC market, a daily wholesale market (OCM or On-the-day Commodity Market) has been created for offering a flexibility mean for balancing and improving the flexibility mechanism managed by the system operator. Auctions for awarding of entry capacities to the transport system have been held every six months since September 1999, and are complemented by a secondary market with day-ahead auctions for trading firm entry rights in excess.

In continental Europe, the characteristics of production and of bulk supply, which is dominated by imports, does not favour the creation of marketplaces for national trade within countries. The first one was created in Zeebrugge (Belgium) by Distrigaz in 1998 at the confluence of several pipelines from Norway and the UK (Interconnector) and a LNG chain from Algeria.[10] Subsequently a standard trading contract and hub-services contract, followed by an option market, have been established. Not only does this marketplace supply the Belgium market; it is linked via dedicated pipes to Dutch buyers (Zebra pipe) and European gas companies (Ruhrgas, Wingas, Gasunie, Distrigaz,

[10]The Zeebrugge hub has a landing capacity of 40 bcm/y i.e., 15% of Western continental market.

GDF). Around 25 players are active on the marketplace, whose activity was set at one-tenth of the level of the UK NBP in 2002.

The first other marketplace candidate is the Central Europe hub located in Baumgarten in Austria, which can connect gas coming from Russia and Norway with supplies to Germany, Northern Italy, Austria, Hungary and the Czech Republic. However, this competition in gas is quite limited and it may remain mainly a reference point for OTC contracts. The second one is located in Northern Germany, in Bunde between The Netherlands and Emdem, the landfall point of several pipelines under the North Sea, and is connected to the Dutch and Russian gas flows. But here also the competition is limited. On the Italian gas market, competition could be helped by the creation of a NBP in Northern Italy; and the French and Spanish markets could be helped by a hub promoted by Total or a Spanish company.[11] Most would be dedicated to European cross-border trade and assisted by traders' arbitrage activities between markets, but they would therefore help national trade in the country of their location.

Commercial infrastructures for direct supply to final consumers

Consumers are connected to the network of the local or regional distributor with actual responsibility for delivery. Once more the British pioneering experiment shows the systems that need to be established in order to organise effective competition, especially in the domestic and commercial market sectors, beyond formal access rules. In order to allow a change of supplier, powerful data networks must be installed on this mass market for the circulation of physical and commercial information. These networks will allow a link of financial and commercial liability to be established between them and the dozens of suppliers who manage access accounts and purchases on the wholesale market. Rules for standardising addresses and account data must be defined in order to organise and facilitate the switchings. In principle, a change of supplier would also require at the outset the installation of meters allowing real-time measurements; the cost of these is a barrier to small customers. In order to limit original transaction costs, the regulator will have an opportunity to define

[11]It could be located at the future connection between Total's French south-western gas system GSO, GDF's gas system linked to Northern Europe, CEPSA's and Enagas's gas systems in Northern Spain and connections to two existing LNG terminals in Fos (Marseille) and in Barcelona, the one in construction in Bilbao (Spain) and in the next future the Total's one in project in Bordeaux (Verdon).

consumption profiles, which are referred to the average load curves for regional supplies at low pressure.

These market rules are complemented by regulatory rules on customer protection to accompany the competition between suppliers: liability of supplier and network in the guarantee of supply, surveillance of suppliers and their methods of risk cover during periods of increased prices, and surveillance of sales practices (door-to-door, telephone sales) (OFGEM, 2001). In order to limit information costs, independent comparison of various competitors' prices might be organised. In the UK the regulator publishes this information every two months in the form of 'EnergyWatch'.

Direct access by domestic consumers was introduced in the United Kingdom in 1998 following a two-year period of regional tests. It was also legally established by the laws transposing the Gas Directive in Germany in 1998. The results have been very different. In Britain, the competition had the effect of making 30% of domestic customers (six million customers) switch from the incumbent (British Gas/Centrica) to its competitors, who are mainly electricity distributors with the assets and skills associated with mass sales. In sharp contrast, in Germany, the only country that opened the sector up to competition for both electricity and gas before 2003, the opening of the domestic consumers' market to competition yielded nothing in the gas sector in 2002 and very little in electricity (switching rate of 3%).

To sum up, apart from the pioneering British market, the commercial infrastructures underlying the reforms are still in infancy in the two other groups of countries.

7.4.4. Industrial accessibility: Weakening the incumbent's dominant position

Accessibility to each national market is also conditioned by the level of barriers to entries, in particular the incumbents' position rooted in pre-existing industrial structures. The main barriers to entry to wholesale and retail markets are the vertical integration of the bulk purchase and the retail supply and the existing long-term contracts.

Vertical integration

At the starting point of the reform process the situations between countries are evidently quite diverse in terms of vertical and horizontal integration, but in fact, as pointed above, gas industries operate using a similar method of quasi-vertical integration: long-term contracts between incumbents and producers upstream, exclusivity of wholesale to distributors or complete vertical integration downstream.

France and Spain had the highest degree of integration, with two major players (GDF and Enagas-Gas Natural) covering 90% of the bulk supply with long-term contracting and around 95% of direct sales. In Italy, The Netherlands, Belgium and Austria, bulk supplies were handled chiefly by one company, respectively SNAM, Distrigaz, Gasunie and OMV. Direct sales to industry, commercial customers and households were largely dealt with by regional and local distributors. German industry appeared to be more de-integrated with a leading company (Ruhrgas) and several regional companies (BEB, VNG, Thyssengas, GVS, etc.) working on behalf of the former for negotiating their gas purchase contract from foreign producers, and a recent entrant (Wingas). These companies had little to do with retail distribution, particularly for commercial and residential customers, where hundred of local distribution companies (LDC) operate.

Beyond these structural differences, new financial links between gas companies and distributors have introduced elements of vertical integration in the most de-integrated structures. In Italy, ENI/SNAM has important stock shares in distribution companies via its subsidiary Italgas, which covers around 40% of domestic sales. Nowadays the stock of which has been completely purchased by ENI, to be integrated in its subsidiary ENI Power and Gas Trade. This is increasingly the case in Germany for Ruhrgas, especially with its acquisition of Thüga, a holding that has stock shares in 170 LDCs covering 40% of the German gas retail sale. In Belgium over 80% of distributors are indirectly linked to Distrigaz by Electrabel's stock shares in them. Another strategy adopted by the incumbents in anticipation of the reform has been to sign long-term contracts with distributors in the run-up to deregulation.

But, after the TPA introduction, change in the property regime offered some opportunities for entries by take-over or alliances. In Germany GDF purchases part of Gasag in 1999 and of TUI the Preussag group's gas supply company in 2002. ENI took with EnBW the control of the fourth regional gas distributor GVS by buying 66.4% shares from local communities.

Horizontal integration in bulk purchase

Industrial structures were vertically and horizontally integrated for bulk purchase and resale and for transport systems, thus creating a strong dominant position for the incumbents following the introduction of competition under the TPA provision. The key activity is the bulk purchase and resale, which before the reform was normally integrated with the gas system operation. The introduction of the third party access rule abolished the incumbents' monopsony, but the presence of their long-term contracts reduced opportunities for competition. Industrial organisation theory shows that long-term contracts between the upstream and intermediary activities constitute a form of vertical integration and limit de facto entries. This character is amplified if intermediaries are also vertically integrated downstream with retail activities. It limits opportunities for competition-based business and leads instead to a horizontal integration of distribution and of supply in one region or one country.

At the time of the reforms, the existing long-term contracts represent the largest part of the supply in national markets. The additional supply to be developed from imported sources over these incumbents' contractualised supplies is less than 10% of the total gas demand in the European Union in 2010 and less than 25% in 2020, according to Eurogas, the association European gas companies (see Fig. 7.6). This is indicative of these contracts' limiting effects on entries. Without a contract release program or the cancellation of current contracts, the situation will only change gradually, as and when the current contracts lapse and if entrants can find new direct purchasers (distributors, power producers, large consumers) by offering them much better conditions.

Here again the progressive British reform is instructive. It has opened the way for lowering barriers to entries. In 1993–1994 the incumbent still benefited from a quasi-monopsony position despite the common carrier provision introduced in 1986. It was therefore compelled to release some of its long-term contracts in order to reach a market share of 40% on the market segment. Then, after development of entries, it was split into two companies in 1996: Centrica for supply and some integrated production (16% of British production in 1998) and Transco for transport and distribution. By 2000 Centrica

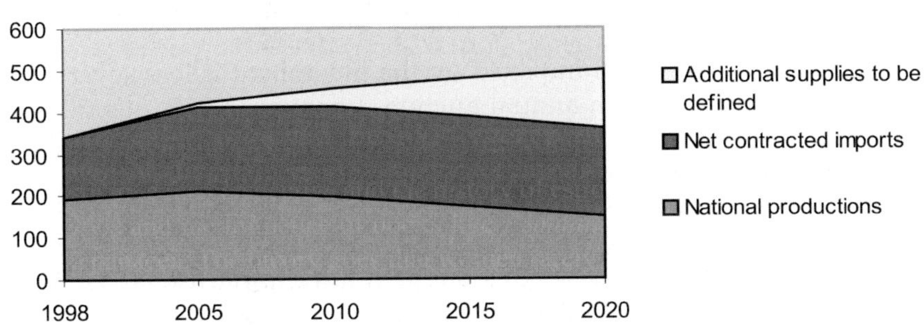

Fig. 7.6. Situation of existing long-term import contracts in future European Union supplies (in bcm/y). *Source*: *Eurogas, 2000.*

had kept an estimated market share of 20% in industry and 70% in domestic. Some of the former incumbent's long-term contracts were cancelled in 1997, some were transferred to entrants and the pre-existing ones covered less than 70% of the gas supply, the rest being new contracts and spot sales. There were many entries in the field of gas supply: around 25 active entries in the industry and commercial segments and around 15 in the household sector, completely opened to competition since 1998. The British experience therefore showed that the transfer of some of the contracts to new purchasers and suppliers, and the legal separation of supply and network, must be managed jointly in order to trigger the competitive forces.

The European Union cannot legally require reform of the industrial structures and the property regime to help the development of competitive forces through entries or privatisation, unless there has been abuse of dominant position or merger that creates a dominant position. We can therefore observe very different situations, because changes depend upon governments, legislators or regulators and their will to extend the field to include competition. This creates different structural situations for encouraging candidates to enter the wholesale and retail markets (Table 7.4).

Gas release programmes are the main means of creating internal competition. In four countries, Italy, Spain and more recently Germany and France, the regulator has required some release of bulk purchase by long-term contracts, under a system similar to the British programme. This method of entry has favoured national players, especially the national electric utilities.

- In Italy, this means a reduction to 75% by 2003 and to 61% by 2009. In this way, part of the amount covered by a SNAM contract signed with Gazprom has been transferred to ENEL and Edison (for 2 bcm/y each), and part of a 6-bcm contract signed with Libya will have to be transferred partly to Edison (4 bcm/y) and partly to GDF (2 bcm/y).
- In Spain, the restriction imposed on the incumbent is less severe: in 2001 Gas Natural, in an annual auction, transferred 25% of a 7-bcm contract signed with Algeria to different foreign companies. The first winners were Iberdrola, Union Fenosa and Endesa and, of the foreign candidates, BP.
- In Germany under the remedies imposed to E.ON for the clearing of its take over on Ruhrgas, this one will have to release at last 15% of its contractual gas in 2003–2004.
- In France for creating more competition in the south western area where Total owns the gas system but where all the bulk purchases

Table 7.4. Industrial accessibility of gas markets in 2002.

	Concentration of bulk purchases	Unbundling transmission and supply	Integration transmission-distribution-supply	Concentration in the retail supply (industry, commercial households)
UK	c. 50% (C1)	Independent (private)	No	50% (C1)
Italy	85% (C1), 95% (C2)	Subsidiary (with 65% floating)	Partial	67% (C1)[1]
Spain	90% (C1); 95% (C2)	Subsidiary (with 40% floating)	Yes	70% (C1)
The Netherlands	83% (C1)	Independent in 2003 (public)	No	46% (C1), 49% (C2)
Belgium	97% (C1)	Subsidiary (with floating)	Yes (Indirect)	95%(C1)
Germany	54% (C1), 65%(C2)	Integration (private)	Partial	35%(C1)[2], 54% (C2)[3]
Austria	80% (C1)	Subsidiary	Partial	
France	90% (C1)	Integration (public)	Yes	88%(C1), 94% (C2)

[1]SNAM direct sales to industry and power generators (51%) and Italgas' sales in retail (16%).
[2]E.ON group control on distribution.
[3]Ruhrgas and E.ON group sales.
Source: DRI-WEFA – *Results from opening the gas market* (Report to the European Commission) – August 2001; EC-DG TREN (2002b).

are still concentrated by GDF, the regulator organised a small gas release programme in 2004.

By taking these steps, governments have offered opportunities for developing entrants' ambitions in the field of gas supplies. In Italy Edison's ambition is to cater for 14% of demand for gas by 2008, with 12.5 bcm/y from the contracts taken from ENI/SNAM and its own purchases in Algeria (4 bcm/y) and Qatar.[12] ENEL's ambition is to reach a market share of 15% by 2010 covering every market segment, including the domestic one; it is continuing its own purchase development from Algeria and from Egypt and Nigeria. In Spain Endesa, Iberdrola and Union Fenosa have each developed a strategy

[12]*Gas Matters*, May 2001.

in order to reach a market share of 15–25% by 2010, by selling mainly to industrial users and power generators: they are buying from ENI, Algeria (Sonatrach, In Salah consortium) and Egypt.

Another method of de-concentration depends upon disjunction of equity gas sales in countries with large production. In The Netherlands, dramatic changes in industrial structures occur in 2003–2004 under the influence of the oil majors' new strategy of directly selling their equity gas. Previously in transmission, bulk purchase, wholesale supply and export businesses, the Dutch gas industry was organised under the semi-public monopoly Gasunie, which was also partly owned by Exxon and Shell at the same level of equities (25% each). Production was also handled mainly by a semi-public company NAM which owns and operates the giant Groningen field, with a respective two oil majors' ownership of 25% of equity shares. Exxon and Shell withdrew from the Gasunie stock share at the time of its legal unbundling. At the same time as they keep their share in NAM they market their equity gas by inheriting their part in the Gasunie purchase contracts and resell on the Dutch and export markets.

A third method for de-concentrating the upstream is to encourage the build-up of new infrastructures independent of the incumbents' gas system: imports pipelines, LNG terminals. These methods were used in two cases before the introduction of a TPA provision. This was in Germany the entry of Wintershall through the building of an important gas system (MEGAL and STEGAL) and the catching of around 15% of the wholesale market in the nineties in The Netherlands, where a number of distributors (PNEM/MEGA and Delta) and industrial consumers installed their new own pipe, the Zebra pipe, and purchased via the UK Continent Interconnector. (They bought 2.5 bcm/y over eight years from Centrica in 1998; this represents 15% of industrial gas consumption in that country). This method of entry, using special pipes, has yet to prove its worth in the new competitive environment. The construction of LNG terminal installations appears to be the preferred way for entrants to growing markets, possibly in joint ventures between entrants. Spain, Italy, and in the near future France and the UK are concerned. Spain is the best example: as well as the new LNG terminal of Gas Natural in Bilbao, the three power utilities and the international oil companies are doing this in order to enter into supplies in the industrial segment and supplying their gas CCGT plants. A terminal (8 bcm/y) is being developed by Iberdrola and Fenosa near Valencia with possible association with Endesa and BP in future and another terminal is planned in North-West Spain, at Ferreol, between Fenosa, Endesa and

Sonatrach. CEPSA, the second oil company, is also planning a gas pipeline (Medgaz) from Algeria together with Sonatrach.

Vertical integration between wholesale and retail supply

The separation and fragmentation of the distribution activity in the pre-reform structure is far from being a condition for development of competition in supplies, even if the prior presence of several suppliers could be thought to be more conducive to it. In fact we observe some barriers coming from past vertical relations and from some new strategic moves by incumbents into downstream activities.

A barrier to competition in supplies to the industrial and commercial segment is the quasi-vertical relation between major gas companies, regional distributors and LDCs. The former negotiated purchase contracts for the largest regional distributors or supply them by wholesale contract; they can have also control of the stock of some of them. The German situation is typical of this (DRI-WEFA, 2001). Ruhrgas, which supplied 10 regional grid operators in 2001, did not make a direct offer to their industrial clients by bypassing them. Conversely, as Ruhrgas directly or indirectly supplied different regions, the other regional operators, including Thyssengas (RWE group) and E.ON's gas subsidiaries, did not take the risk by offering gas directly to industrial customers in other regions. *A fortiori*, the large LDCs did not look to compete outside their area. The many financial links held by Ruhrgas, Thyssengas and other large regional distributors further lessen the incentive to compete. A similar problem occurs in Austria, in Belgium[13] and to a lesser extent in France and Spain, where the few regional and local distributors had no direct purchase contracts with foreign sellers and are supplied by the national incumbent. But, when these companies have no financial links with the incumbent, they are able to establish a partnership with entrants to supplies, as in Belgium the so-called pure Intercommunales (which represent 15% of the supply outside industry) have done with Centrica in Belgium.

These situations are not rigid and could change in opposite directions. Economic factors tend to increase vertical and horizontal concentrations after the reform meanwhile new regulatory measures play occasionally in the other direction. The German gas industry is affected by mergers and acquisitions. The number of gas distributors has dropped from 670 in 1995 to less than 400 in 2002. Some regional

[13]The majority of the 23 intercommunal electricity and gas distribution companies are financially controlled by Electrabel, which is part of Suez as Distrigaz.

distributors have merged (creation of Avacon). Cities have privatised their municipal gas distributions by selling them to major transporters (Ruhrgas, Thyssengas and GSG). These transporters control part of the regional and local distribution while at the same time supplying most industrial consumers in their area. After its acquisition of Ruhrgas in 2002 E.ON has kept more than half of its anterior stock shares in gas supply at regional level (Contigas and Thüga) and at municipal level which gave more than 30% of the retail market, after the divestures imposed for clearing the acquisitions. RWE, the other electricity major, also controls several regional gas transporters, including Thyssengas and several *stadtwerkes*. In Italy, ENI Gas and Power has increased its presence in distribution through its subsidiary Italgas, which controls a number of local distributors (27% of the clients), while ENI also makes direct sales to major industrial consumers.

But gas release programs and corporate unbundling which are imposed on gas companies play in the other direction by allowing entries of vertical companies. In Italy, after the announcement of the gas release programme, ENEL, one of the main entrants, expressed interest in entering into distribution activity by taking over Camezzi, a gas distributor supplying 15% of domestic clients. In Spain, Germany and France gas release programmes allow oil and gas companies to develop their positions in the supply to industrial clients. In the future the second gas directive will require corporate unbundling between supply and distribution network (up to 150,000 customers) in latecomer countries that definitively open the game if competitive forces exist upstream. In the United Kingdom, with the help of the incumbent's position by ownership unbundling and gas release, it has been possible to enter the retail supply with up to 60% of the market share on the industrial market and 30% on the retail market.

Property regimes and capital accessibility

The national gas industries have been exposed to entries by foreign players but to a lesser extent than have the electricity industries. Opportunities of takeover or alliances are mostly restricted by vertical (or quasi-vertical) integration of transport, distribution and supply businesses as in UK, France, Spain, Belgium and Italy. So it is by the ownership of the gas companies, either because it remains public as GDF in France or because it is already controlled by an oil company or by a large multiservice group such as Distrigaz by Suez. National mergers also limit opportunities of foreign entries, as in Germany.

The presence of foreign interests are marked in some countries and gas activities:

- In gas production GDF and Ruhrgas have penetrated in British and Dutch North Sea offshore in the recent years.
- In retail supply activities oil and gas companies have developed business in the British market (Statoil, Total with Agas, GDF with Volunteers Energy).
- In distribution and supply companies foreign capital entries are determined by the initial fragmentation of industrial structures. Entries would depend upon the decision of local and regional communities to privatise their distributors. So German, Austrian and Dutch gas industries are most open to cross-border takeovers. The foreign presences will however keep a small share. In Germany, ENI took over the regional distributor GVS with the electric company EnBW in 2002; GDF took 50% of GASAG, the Berlin distributor, bought TUI (Preussag group) and has a stock share (5%) in the East German distributor VNG and in Austria a stock share of 25% in ETSAG. Centrica has developed alliances with the Belgium independent municipalities.

7.4.5. Ranking national markets accessibility

If we now compare the different national gas markets in terms of general accessibility before the second 2003 directive, a comparison of rules, technical openness and industrial structures shows differences in individual situations and in predisposition to 'gas-to-gas' competition at both intermediate and final level. Accessibility is not correlated to the degree of openness of the final market because of the importance of other institutional and industrial conditions. Neither it is fully related to the degree of transparency in access, if the technical openness allows entries by bypassing the incumbent, for instance.

In order to rank countries according to accessibility levels and competition potential, a synthetic accessibility index is assessed on the basis of the average of eight quoted properties:

- regulatory accessibility, which cover six properties (openness of the final market, four characteristics of network access transparency – type of unbundling, type of TPA, transport tariff levels, balancing and storage access rules – and enforcement of non-discrimination guarantees by regulatory authorities), each with quoted properties rated from one to five.

- technical accessibility levels to reflect number of entry points and import capacities, with a rating from one to ten because of the importance of this character for the entries of competitors.
- commercial and industrial accessibility, which covers the existing industrial structures and the presence of gas marketplace in the country, and is rated from one to ten.

The synthetic accessibility indicator (see annexe for the calculation) makes appear a clear dissociation between the same three groups of countries which have already been identified.

First, the market that is the most accessible and open to competition is clearly the United Kingdom, which is clearly distinguishable (see Fig. 7.7). Its overall index totals 43 out of a maximum of 50. The number of competitors in the wholesale market and the retail supply market reflects a high degree of market accessibility. Only the limited capacity for interconnection with the continental market and the complexity of the balancing rules are restricting factors, but the level of gas resources in Britain compared with demand does not allow this to limit competition, especially because financial instruments allow for swaps between the British and continental markets. In the future, new interconnectors which are planned with Norway and The Netherlands and a number of LNG terminals will ease in foreign gas entries when British production declines.

In the second group of gas markets with significant reforms before the second gas directive, Italy, with an index of 33, is clearly conducive to competition as the moves with a corporate separation of

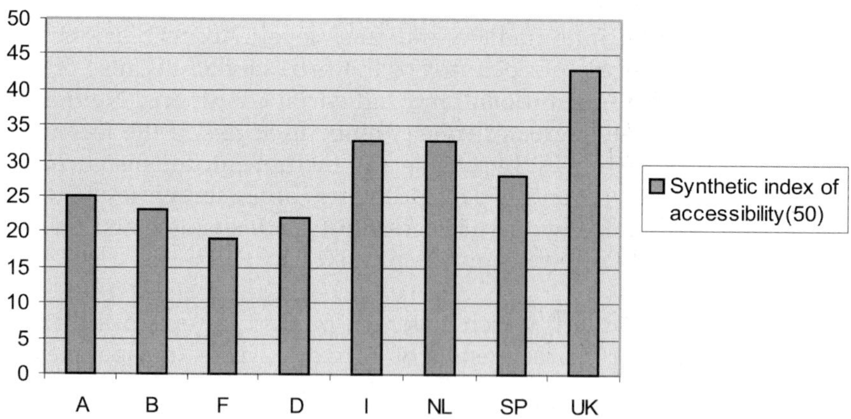

Fig. 7.7. Ranking of overall accessibility of national gas markets in 2002. (Composite index).

network and its progressive release of import contracts. Internal competition is developing under the effect of strong regulation of the transmission assets backed by splitting of the transmission network and the gas release programme. The Italian market could however be hampered by its geographical location. Spain, with a score of 27, is on the same track, but competition is restricted by the limited capacity of the pipe connections to the continental market. In these two countries reinforcement of connection capacity with other gas systems, development of new LNG terminals and new pipe connections with North Africa could open the competitive game more widely in future. The Netherlands, with a score of 33, is in a better position because of the scale of the interconnections and the closeness of two gas hubs. With potentially clear unbundling in the future, Gasunie's role in bulk purchase is being completely replaced by a flexible co-ordination by Gröningen Maatschap and the nascent competition between producing companies. The vertical separation of distribution and the closeness of the Zeebrugge hub has already had the effect of activating competition, as the direct contracts with the British producers demonstrate. In addition, the next separation of the oil majors' marketing of their Dutch equity gas, the total unbundling of Gasunie activities and the emergence of another hub on the Dutch–German frontier, will boost competition .

Of the third group of latecomers, Belgium and Austria are the most open to the development of competitive forces, with a score of 25. Originally marked by the dominance of a national incumbent, Distrigaz and OMV, they are now undertaking major changes with the adoption of the regulated TPA, the separation of the network company from supply activities (especially with the floating of the Belgian network company) and the closeness of the gas marketplaces, thanks to major transit infrastructure connections. The German market could appear a priori favourable to competition, because of its scattered industrial structures and its central position within Europe, with numerous points of entries allowing a gas exchange to emerge before 2005, but its score is only 22. This means that it is hampered by a low level of transparency and non-discrimination guarantee in the TPA rules in the context of the professional agreement (the so-called VVI), the absence of a specialised regulator, the persistence of vertical links between the national and regional grid operators and the distributors, and the increasing concentration within the sector, all of which bring its accessibility index down. France's score is 19, with a market distinguished by the presence of a major gas company covering all or most of the market, a delay in the transcription of the gas directive up to 2003 and the continued public status of the

Table 7.5. Industrial switchings estimates in 2002.

	UK	NL	IT	SP	A	B	D	F
Accessibility index	43	33	33	28	25	23	22	19
Industrial switching[1]	16%	15%	10%	n.a.	6%	n.a.	5%	20%

[1]This refers to clients consuming more than 0.1 million m^3 per year.
Sources: EC-DG TREN (2002b) – *Third Benchmarking Report*.

incumbent which does not give competitors confidence to enter the market. However, the French market has been made more accessible by rules and regulatory institutions that favour access to the network, as soon as the formal reform has been implemented; in 2002, because of informal rules implemented by GDF before the vote of the reform it was considered just as easy to enter the eligible segment in France as in some other countries, especially Germany, Belgium and Spain (EC-DG TREN, 2002).

The impact of market rules, technical access and industrial structures on the activation of competition could be evidenced by the development of industrial switchings in 2002 in countries where incumbents were still in place. There is a correlation with accessibility indexes, at the exception of France where the number of switchings are those registered since the opening up of the access (Table 7.5).

To sum up, the European market has been fragmented for a long time, with national markets that show differences from each other in the process of reforms after the first directive. Some of these markets have moved quickly towards a set of rules more in keeping with the competitive standard. These differences are barriers to the establishment of a competition playing field at the European level. Difference in regulatory barriers are being erased by the Madrid process and the implementation of the second 2003 Directive. But the main barriers remain from the characters of the European production with remote sources of which the development necessitates large investment which have to be secured by long-term contracts.

7.5. Increasing Competition on the Two-Level European Gas Market

The importance of outside sources in supplies to national markets has conditioned the organisation of long-term relations between remote state-owned producers and European gas companies. It will condition it in the future because the gas import share will increase; but it will be in a new way, which will ease competition in bulk supplies and stop the separation between the two levels in the European market by allowing producers to sell direct to intermediaries and large users and vice versa (Golombek *et al.*, 1998; Egenhofer and Labory, 1998; Oostvoorn and Boots, 1999; Ellis *et al.*, 2000). In particular, market growth makes entry with LNG transactions easier.

We will examine the evolution of the remaining barriers to the development of competition on the upstream European gas market, which are of three types: first, the continuing barriers to trade posed by technical and economic conditions of transit and transportation; second, the legal restrictions of competition in production and exports in producing countries; and third, the de facto restrictions in the bulk supply market which result from existing long-term contracts.

Indeed the challenge for new market players is the lack of available capacity at entry points and on the network, combined with the transport costs for delivering supplies from origination to customer. In addition, more harmonised and stable tariffs for access and balancing services are a prerequisite for a functioning competitive market. The legislative action of the European Union and various countries has significantly affected some of the trade barriers (European Commission, 2000, 2002). However, market relations are also subject to internal changes: new trading methods will spread, the most notable of these being the creation and consolidation of the gas exchanges. The competition dynamic that will thus be developed at the edge of the set of vertical relations will render the European gas market more flexible, through the negotiation of shorter-term contracts and the renegotiation of clauses in long-term contracts, especially those relating to prices.

7.5.1. New trajectories of the European gas bulk supplies

In future, upstream competition in gas supplies will be restricted by the need to increase imports from remote sources and to preserve long term contractual framework. British offshore production has reached a plateau of 90–100 bcm/y, which will stay steady for about fifteen years and then could fall sharply. Supplies will then rely progressively on import entries. In The Netherlands offshore production will not compensate for the decline of the Gröningen deposit (the resources of which are estimated at 1250 bcm and account for half of all gas sold), and the government is anxious to control the decline of reserves by presently limiting production to 80 bcm/y and envisaging a decrease up to the exhaustion of the production around 2025. Norway is the only European Union country (or assimilated) with potential for growth in production and exportation, from 60 bcm/y to 85 bcm/y between now and 2010 for its reserves South of the 62nd parallel. The resources North of this parallel look promising but expensive, with the development of the Snohvit deposit and its transportation in the form of LNG.

Imports from countries outside Western Europe will increase steadily in years to come, indicatively from 170 bcm/y now to 420 bcm/y in 2020 in a central scenario (Eurogas, 2000); Concentration will probably be in two countries, namely Algeria and Russia. Algeria's aim is to increase exports from 50 bcm/y in 1998 to 65–70 bcm/y in 2010 and 85 bcm/y beyond that. It could meet up to 17% of the Western European demand and 30% of Europe's import requirements. Russia, meanwhile, is likely to increase its exports to the European Union from 70 bcm/y in 1999 to 100–150 bcm/y in 2010 following the commissioning of a number of new major pipelines. This will meet about 35% of all Western Europe's demand for gas and 60% of its import requirements.

However, the LNG projects in various areas appear most promising, especially for the rapidly developing Southern markets. In the mid-run LNG imports will offer the best opportunities for the development of a flexible competitive market. World LNG trade will expand significantly in the future, as the number of new ship orders bears witness. The cost of LNG chains has dropped significantly since 1990, and since the installation of the first liquefaction trains it has benefited from scale economies. The break-even between LNG chains and pipeline projects in terms of distance is decreasing, thanks to technical progress and lower costs. A spot cargo market could also be established in the Atlantic Basin after 2005, and its function is anticipated to go beyond the current function of flexibility and adjustment of the present LNG cargo trade (Bower, 2001). To do this, it is essential to have a transparent access to gasification terminals, as the second 2003 Directive is supposed to guarantee for the future. LNG projects flourish around Europe, most notably in the regions between 4000 and 5000 km from Europe, because of the oil companies' increasing interest in gas production and their will to 'monetise' their gas discoveries quickly. As well as the Snohvit LNG project being studied in Norway, several LNG trains with Southern European destinations have been commissioned in Nigeria and Trinidad and others are being installed or studied in Egypt, Venezuela and Angola. Qatar and Abu Dhabi which have developed LNG trains for the Asian market have overcapacities and sell LNG in the framework of short term contracts to Spain. Against this dynamic background, the offer of LNG could find new opportunities for expansion to current importers (Italy, Spain, France and Belgium) and expansion into other European countries (Northern Europe, Portugal, Greece and Turkey).

7.5.2. The reduction of technical and price-related barriers to the transit access

The export and international transit infrastructures set up from the 1970s onwards are relatively dense (Cedigaz, 2002). The transport capacities were created as and when long-term purchase contracts were signed and are reserved for the associated contractual deliveries. In the medium term (2010–2015), the development of new imports via pipelines can mostly be achieved through existing land and submarine infrastructures. The remaining transport requirements will mostly be met by strengthening compressor capacities and installing new pipes across shorter distances, as it will occur with the export pipes coming from Algeria, Norway and Russia for this decade (Stern, 1998). The national networks will also require selective additions in order to meet transit requirements (as was recently the case with the pipeline built in 2000–2001 in North-East France and running into the Transitgaz pipeline in Switzerland for a contract between Norway and Italy).

New interconnections will be developed in accordance with companies' strategic choices, on the model of the UK-Continent interconnector developed by British Gas and its associates in 1996 by decoupling the fuel contracts taken on by their promoters and the capacity contracts. The project announced in 2002 by Gasunie and Centrica for the installation of a new interconnector between The Netherlands and the UK, or the project of Marathon Oil's Eimdal-Bacton line between Norway and the UK could be similar.

Moreover, with significant improvements in network access rules for transit and transport, communications between markets will make it easier to set up 'swaps' to optimise the transport of contract-based flows. Profits may be realised by swapping flows and choosing the most effective combinations of supply source and market, especially in contracts concluded between sources and markets a long way from each other (such as Norway and Italy or Spain and Russia). In this context the long-term TOP contracts, which still have considerable life left, will generate needs for swapping and trading operations.

Clarification of transit access rules

By abolishing the 1992 Transit Directive, the second 2003 Directive also requires alignment of the conditions for access to gas transit, laid down in the long-term contracts relating to standard conditions for

TPA to the network. In keeping with this principle, the capacity reservations imposed in the long-term contracts would have to be abolished, a step that would have the effect of greatly increasing competition for the allocation of network access rights. In this logic, the acceding countries that are important to international gas transit, i.e., Czech Republic, Poland and Slovakia, should have to apply these rules. However, in the 2003 Directive, the access for transit flows would not be regulated but still negotiated, and among transit flows allocation of capacities on the dedicated transit pipes will still depend on the discretionary choice of the owners. Progressive evolution will come from the fact that after 2010–2015, part of the reserved capacities in the transit pipelines and interconnectors will disappear when the contracts terminate. Most of the quantities covered by future contracts will borrow existing transportation capacities. The use of dedicated pipelines in international transit (like Stegal, Midal, TENP in Germany and Belgium) should be shared between gas purchasers once the contracts have been completed.

Transport and interconnection tariffs

Exchanges may also be hindered twofold by internal transport tariff calculation methods and by cross-border tariffs. Directive 98/30 says nothing about any principles that might facilitate exchanges. By allowing freedom of choice on negotiated access by third parties, it had created greater possibilities for differences between transport tariff calculation methods than does regulated access. In addition, it had not specified any principles for cross-border tariffs. For exchanges between countries, an 'entry–exit' tariff system in each gas system which is supposed to be the most efficient tariff hindered exchanges between countries because of the need to conclude separate contracts for the successive transits involved in a cross-border transaction and the superimposition of 'entry–exit' tariffs[14] (Brattle Group, 2000). This leads to the creation of a distance-based tariff system and causes pancaking, which discourages transactions. It greatly reduces opportunities for arbitration between national markets. It appears that defining a specific entry–exit tariff applicable to cross-border transaction only, across several different countries, would be the only means of avoiding the requirement to pay cross-border charges without distorting the cost-reflectiveness.

[14]Brattle Group (2000), *Third Party Access to Natural Gas Networks in the EU*. London, Brattle Group.

As the harmonisation of principles of transportation tariffs depends on agreement between gas system operators and regulators, it is a long process beset with conflict. The European Union has finally voted in December 2003 a direct regulation to conditions for access to the gas transmission networks with internal transmission and cross-border tariff guidelines likely to facilitate exchanges.[15]

7.5.3. The extension of market rules to the foreign producers-exporters

The policy of integration of European markets needs also to act with producers located outside the European Union's legal framework. The three major external sellers (Algeria, Norway and Russia), and The Netherlands, had organised their gas trade with big gas companies under the regime of state monopoly even when foreign companies are involved in the exploration and production of their gas fields. However, the legal framework of the European Union concerning competition in production and exports is now being applied in The Netherlands and extended to Norway, while those in Algeria and Russia are changing under pressure from the international lending institutions (IMF, World Bank). It affects also existing long-term contracts of which some clauses do not comply with European competition law.

Market rules in production and exports

In the context of the association agreement of Norway with the European Union,[16] market rules were reluctantly adopted by Norway in 2002 for production, exports and transportation (Claes, 1998; Sunnevag, 2000). In its exploration and production Norway has had to follow up the Licensing Directive of 1994 with non-discriminatory criteria. For the exports, the monopoly of gas marketing by the public consortium GFU has been discontinued. Each North Sea producer could sell its equity gas independently, for the new developments as well as for existing GFU contracts. The joint selling scheme of the two Norwegian producers Statoil and Norsk Hydro was officially abolished in 2002, and their new sales must now be negotiated with new contracting parties up to five years. Recent contracts (Statoil-BP, Statoil-Centrica agreements, different Snohvit LNG agreements) have therefore been negotiated and established outside the GFU and in a

[15]Document COM (203) 741/12.10.2003.

[16]The so-called European Economic Area agreement.

separate way for each company. A complementary reform is the introduction of the TPA for access to the North Sea gas transmission infrastructure. A special state-owned company has been created for this purpose with regulated tariffs. Despite the major nationalistic stakes of controlling gas profits, these changes are considered as acceptable as soon as there are symmetric change in the opening of the gas markets in the European national markets downstream.

Algeria and Russia, the two other major external gas sellers, are not under the framework of the European Union legislation for organising their gas activities. However, their progressive integration into the international economy is encouraging them to reform their public sectors as well as the legal rules (Aïssaoui, 2001; Stern, 2002). Their oil and gas industries, which are major providers of hard currency through exports, are also subject or about to be subject to some reforms, which would open some competition around exports in the future. The former legislation of the two countries allows shareholding in producing companies. In Algeria, a number of partnerships with foreign companies were established in 2001 and 2002 for the development of new gas and oil fields. Joint ventures are allowed to sell their gas directly, as should have been the case with the In-Salah consortium composed by BP and Sonatrach, with the limitation that the national company still participates indirectly in export sales. A new act voted in 2002 also creates provision for negotiated TPA for access to transport infrastructures and the possibility of independent players building their own pipes.

In Russia, reforms are slower. In production, foreign entries are still hazardous and have no efficient legal protection. New players in the gas production are exclusively national, namely the Russian oil companies (Lukoil, Yukos, etc), but they are exerting pressure to export directly to Western Europe. However, reforms in gas network access are still limited: the provision of negotiated TPA to the Gazprom pipelines, voted in 1997, has as yet gone unheeded (Moe, 1998; Locatelli, 2001). The dismantling of Gazprom into a transport company and several producers allowed to export gas to Europe has been contemplated on numerous occasions in recent years. But the separation of the transport network and the abolition of Gazprom's export monopoly will not be voted in the near future because of the inconsistency between selling conditions on the internal market and on the exports market in Western Europe (IEA, 2002). Some entries into production in joint ventures with Gazprom will succeed from 2003 onwards, but for only supplying a newly deregulated part of the Russian internal market. This however is not a stepping stone to a more radical reform.

The abolition of the contractual clause of final destination

The final destination clause in existing contracts for Algerian and Russian exports limits changes in trade between Member States because it restricts the possibility of buyers' reselling gas outside their respective territories.[17] The European Commission is therefore asking for this clause to be deleted from the present and future contracts. Such a change would weaken the quasi-vertical integration link between foreign producers and major gas pipeline companies. If this happens, the companies could then compete directly against foreign producers as well as against other gas companies and traders on other national markets with the gas that they buy from them, by seizing the opportunity to differentiate between spot prices and contractual prices. Remember that the latters are defined in *net back* with reference to the buyer's location and to peculiar segmentation of the buyer's resale market.[18]

The exporting countries concerned, i.e., Algeria and Russia, oppose this for two reasons. First, they want to prevent their purchasers from competing against them by looking to draw off profits downstream through the resale of gas on other European markets or in North America for LNG transactions. Second, their vertical relations with the major gas companies will be forcibly unbalanced. The exporters will bear the price-risk asymmetrically by taking it upon themselves when oil prices are down but profiting from it considerably less when they are raised.

There is no guarantee that Russia and Algeria will agree to abolish the final destination clause if the European Union imposes the abolition unilaterally and without compensation via a temporary provision guaranteeing sharing of profits with purchasers (Hached, 2001). However, they are in an unfavourable position because of the possibility that the European Commission will exert pressure on the buying parties. At the most, some contracts could be adapted by including a profit-sharing clause if the buyer resells gas on another, more profitable market. A 'claw-back' clause could also be integrated, allowing the sellers to interrupt supply during periods of price-peak on spot markets in order to sell directly on these markets, as is the case in the North Sea gas contracts that use the UK Continent

---

[17]Article 81 of the Treaty of Rome.

[18]In fact, it would allow first-tier purchasers to enter different national markets by playing on the flexibility margins of off-take clauses in order to resell on the short-term market during favourable seasonal price fluctuations. They would thus be encouraged to reach an interseasonal balance between gas stored after long purchase at an indexed price based on oil prices and opportunities for income on the short-term markets.

interconnector to supply continental buyers (*International Gas Report*, April 2002).

7.5.4. Changes in contractual bulk supply

Existing long-term contracts have long since maintained a quasi-vertical integration between the four main exporters and incumbents for the major part of physical flows. They limit the field of competition to a minor portion of wholesale purchases for the next fifteen years. They impose a price formation method that is different from a gas-to-gas competition, with the oil price indexation clause. As they limit de facto the development of a liquid market able to generate a reference price on the continental market for a certain time, it is not possible to link price definition in the contracts to the spot price, as is the case on some electricity markets. In the recent past the European Commission, as the promoter of the competition, opposed the preservation of existing long-term contracts in their present form, but it does not have the legal means to impose this change in the European Union without the consent of Member States and associates. So alongside the pre-reform contracts, short-term transactions will see a slow development. New long-term contracts that remain necessary for the development and finance of the new projects will reconcile flexibility with securisation needs on shorter duration.

The development of short-term transactions

If we refer to the British market, spot sales are likely to develop at the wholesale level as and when *free gas* is made available, existing contracts terminate and trading hubs are established. There will be a need for available transmission capacity and opportunities for arbitration between national markets. LNG trade will be partly affected by this trade, under conditions of available capacity in LNG terminals. The short-term purchases on the market will cover large purchasers' and intermediaries' 'swing gas' needs, alongside their contractual 'base load' purchases. This flexibility of the demand addressed to the spot market is ensured by the portfolio of suppliers' interruptible contracts, which rely on the capacity of purchasers to respond to a rise in prices by reducing gas demand through 'fuel switching', by the substitution of processes and by interruption of production as occurs in the electricity generation on a competitive market.

Some major external producers are contemplating a new opportunity for extracting value not only by selling gas on a regular basis, but also by offering swing services, which are much more profitable. The remaining integrated operators (Distrigaz, Gas Natural, GDF, Ruhrgas, etc), even after losing monopoly of access to their storage capacities, will become active in this short-term market by making use of the flexibility of the off-take clause in their long-term contracts. The entry of traders into the various national markets depends on opportunities for competition, which are in turn increased by their action; when the liquidity of the spot market increases indeed with the help of entrants' activities, so does the frequency of transactions, while the total quantity exchanged under each contract is reduced.

Changes in pre-reform contracts

The long-term contracts signed before 1998 reduced the scope for competition in the market for swing gas in the near future. It could however be able to some extent to destabilise contractual relations drawn up at an earlier date, most notably the off-take and price clauses. The example of the cancellation or renegotiation of the rigid TOP contracts following the deregulation of access to the network in Great Britain and USA is an indication of this risk (De Vany and Walls, 1994, 1995; Wybrew-Bond, 1999). The cancellations and renegotiations were made at a price, namely the protracted legal disputes and the huge amounts of compensation paid to the producers, price which is inferior to the advantage to purchase gas in another way (IEA, 1998). In Europe, however, along the lines of the transactions cost theory (Williamson, 1985), the large flexibility and incompleteness of the contracts, brought about by the *Take or Pay* flexibility (between 85% and 115% of the nominal annual off-take) and the *re-opener* clauses aimed at adjusting the price indexation formulae, guarantee that they will last; the rigidity of the American contracts did not allow this flexibility and instead caused a break-up.

For the off-take clauses, the possible circumventing of the major purchasers (gas pipeline companies) by their major suppliers on the respective national markets, and the possible resale of contract-based gas by gas companies on other markets, may lead to disputes between contractors, to say nothing of the arrival of new competitors. For the incumbents, little can be expected from the possibility, laid down in the 1998 Directive, of a government that applies the national interest clause to protect the gas company's TOP commitments by decreeing a reduction in the scope of competition in the name of supply security. Application of the clause would be closely monitored by the European Union, which considers that in any competition environment, market shares lost by an incumbent in the area served by him can be recovered on other markets.

The price indexation clause in many contracts should also be renegotiated without compensation, for three reasons:

- The influence of seasonal changes in spot prices, with summer prices set at a lower level than long-term contract prices. The indexation formulae for oil prices are themselves a source of destabilisation when a short-term market emerges, favouring new entrants over the incumbents who are tied to a price indexed on oil prices by their long-term contracts. This advantage increases as oil prices rise. In addition, because of the time lag of the price formula, the movements of oil prices are recovered on the contractual gas price after a gap of between 6 and 12 months. This gap is no longer compatible with a competition market in which substitutions may be made between competing types of gas and oil or other substitutes.
- Endemic overcapacity, which will lead to the establishment of a spot price with an annual average set at a level below that of contract prices indexed on oil or oil product prices.
- Changes in the value of gas on the market downstream, with increased importance for gas outlets in the electricity production industry for the first-tier purchasers.

In the near future, therefore, the indexation of gas prices in a certain number of contracts could take account of seasonal fluctuations in gas prices and the valuation of gas kWh on the electricity markets.

The new contracts

The appearance of market places on the continental market, and the progressive affirmation of a short-term reference price, will affect the forms of long-term trade used. They will however remain dominant because of the need to guarantee the development of new import operations via this type of transaction; but they will be adapted to suit the new trading environment, as has already been the case with the new contracts. The European Commission has admitted the need for them with the reserve of a shorter contractual period than before (10 years).

In the next ten years none of the infrastructural developments made in connection with additional supplies will require long-term contracts to guarantee their financing or limit their risks. The major infrastructures – the pipelines running from Norway to the Continent,

Algeria to Italy (Transmed), Algeria to Spain (GEM) and Russia to Slovakia, or certain LNG terminals such as Fos (France), Zeebrugge (Belgium) or Cartagena (Spain), could see their capacity increased with just modest investments. The density of the European large-scale transport network will reduce the specific magnitude of the infrastructural investments required for connection to high-density markets, limited some foreseeable investment in compressors. It will also open up the possibility of long-term swaps.[19] During the 1980s and 1990s, several investment-intensive operations were successfully developed without a previously signed long-term contract (Interconnector, Jagal/Megal, Zebra) and with reduced guarantees. Sellers also signed five-year contracts with the aim of disposing of additional production. This is the case with Norwegian contracts signed since 1995, after the major contracts used as a base for developing Troll and Sleipner and associated pipelines (Europipe, Norfrapipe and Zeepipe), as they relate to marginal developments and rely on the flexibility of the new infrastructures (Bartsch, 1999).

The stake of development in import gas projects after 2010 would most probably require a longer-term commitment. On the British market, with its high level of liquidity allowing limited volatility, many contracts were drawn up for a period of 3–5 years and with indexation on spot prices after a period where new contracts were signed for a period of 5–10 years. On the continental market, with higher capital restrictions, the long-term contracts to be signed will last for 10–15 years or less, compared to 20–25 years in the recent past. The oil and gas companies are planning new developments based on contracts with a period of 7–10 years, whereas the LNG contracts (Egypt to France and Egypt to Spain) signed in 2001–2002 were established for a period of 15 years. Some LNG contracts on the new international market are signed for a duration of four years for limited quantities of 1 bcm/y around. Interestingly in the prospects of development of gas imports in UK, Centrica signed a 10-year 5-bcm/y contract with Statoil in mid-2002.[20]

Concerning the price indexation, there will be a much wider range of indexation formulae, especially for electricity producers and first-tier purchasers whose resale base will expand to include electricity generators. Some price indexation should partly include electricity

[19]An example of the effects of this increased density is the contract signed between Statoil and BP in June 2001, which assumes delivery at a National Balancing Point whatever import route is followed (Heimdal-Frigg pipeline or routing to the continent via the interconnector, etc).

[20]For information on the new gas contracts we refer to *Gas Briefing International* issues.

prices, as it is the case for contracts between the Egyptian exporter and the newly diversified Spanish companies. At a later stage, spot gas prices will be introduced as an element of indexation for new contract prices, and may become the sole indexation element for new contracts when markets are sufficiently developed and liquid to allow dissociation of price movement into a seasonal component and a long-term trend, as has been the case in Great Britain since 1998. The Centrica-Statoil contract refers to the NBP price as a solid reference price for sales on the British market. Therefore, for new contracts, the contract price of gas would thus move progressively closer to the price of a competition-based market.

To conclude on the increasing competition, the European gas industry and its institutions are shaped, and will remain to be, by the basic conditions of the production. Upstream from the national wholesale markets, gas supply sources are directly international in most European countries (5–6 major supplier countries, only two of which are European Union Member States), while almost all electricity supply sources are still local or national (hundreds of power stations scattered all across Europe). Besides the geopolitical reality of the gas supplies which makes government more reluctant to question the 'pre-reform' contracts these existing contracts do not let sufficient field for 'free gas' which is not 'contractualised' and allow a sufficient liquidity on the continental marketplaces. In addition, after the end of the existing contracts, there is a general expectation that, on the continental markets, the total value of spot transactions will not exceed a modest 8–10% in comparison to total consumption, because of the need to set the development of new import operations against a framework of new long-term contracts. The single but important difference will be the more flexible way of contracting helped by the gas buyers' new portofolio approach.

Annexe of Chapter 7

Calculation of European market accessibility indices

Table A1 shows the rating indices, for which each aspect of regulatory accessibility has been assessed on an empirical basis on a scale from 1 to 5 in relation to the best inclination for allowing entries and gas exchanges, which is rated 5.

In Table A2, the total regulatory accessibility index is rated up to 30 maximum, technical opening up to 10, and industrial accessibility up to 10.

Table A1. Level of regulatory accessibility in the eight selected gas markets in 2003.

	Austria	Belgium	France	Germany	Italy	The Netherlands	Spain	United Kingdom
Market opening	4	2	2	5	4	3	4	5
Unbundling	2	3	1	1	3.5	4	3	5
TPA	2	2	1	1	3	3	3	5
Transport tariffs	3	2	2	1	3	4	3	5
Balancing and storage	2	2	2	1	3.5	3	3	4
Regulatory institutions	2	2	3	1	5	2	2	5
Total	15	13	11	10	22	19	18	29

Table A2. Synthetic index of regulatory, technical and commercial accessibility of national gas markets in 2003.

	Austria	Belgium	France	Germany	Italy	The Netherlands	Spain	United Kingdom
Regulatory accessibility (Max.: 30)	15	13	11	10	22	19	18	29
Technical opening (Max.: 10)	6	8	6	8	5	8	4	4
Industrial and commercial accessibility (Max.: 10)	4	4	2	4	6	6	6	10
Total (Max.: 50)	25	23	19	22	33	33	28	43

References

Aghion, P. and Bolton, P. (1987). Contracts as a barrier to entry. *American Economic Review* 77(3): 388–401.

Aïssaoui, Ali, (2001). *Algeria and the Political Economy of Oil and Gas*. Oxford University Press, Oxford, (OEIS series).

Armstrong, M., Cowan, S. and Vickers, J. (1994). *Regulatory Reform: Economic Analysis and the British Experiment*. MIT Press, Cambridge (Mass.).

Baron, D.P. and Meyerson, R. (1982). Regulating a monopolist with unknown costs. *Econometrica* 50(4): 911–930.

Bartsch, Ulrich (1999). Norwegian gas: the struggle between government control and market developments. In Mabro Robert and Wybrew-Bond Jan, Eds., *Gas to Europe: The Strategies of Four Major Suppliers*. Oxford University Press, Oxford, pp. 201–254.

Bower, David (2001). Liquefied natural gas: moving markets and technology advances. *Petroleum Economist*, April, 3–10.

Brattle Group (2000). *Methodologies for Establishing National and Cross-border Systems of Pricing of Access to the Gas System in Europe.* Brattle Group, London.

Cedigaz (2001). Natural gas in the world — 2001 Survey. Cedigaz, Paris.

Claes, Dag Harald (1998). *The Process of Europeanisation: the Case of Norway and the Internal Energy market.* Arena working papers WP 02/12.

De Vany, Arthur and Walls, W. David (1994). Natural gas industry transformation, competitive institutions and the role of regulation. *Energy Policy* 22: 755–763.

DRI-WEFA (2001). *Report for the European Commissions: Results From Opening the Gas Market*, August 2001.

EC-DG TREN (2000). *State of Implementation of the EU Gas Directive (98/30/EC): An Overview.* European Commission, Brussels.

EC-DG TREN (2002). *Implementing the Internal Energy Market (First Benchmarking Report).* European Commission, Brussels.

EC-DG TREN (2002b). (Commission Staff Working Paper), *Second Benchmarking Report on the Implementation of the Internal Electricity and Gas Market.* European Commission, Brussels.

Egenhofer, Christian and Labory, Sandrine (1998). *The Development of Competition in European Gas Markets*, Centre for European Policy Studies, Brussels.

Ellis Andrew, Bowitz Einar and Roland Kjell (2000). Structural change in Europe's gas markets: three scenarios for the development of the European gas market to 2020. *Energy Policy* May, 28(5): 297–310.

Estrada Javier, Moe Arild, and Dahl Martinsen Kare (1995). *The Development of European Gas Markets: Environmental, Economic and Political Perspectives.* John Wiley Publisher, Chichester.

Eurogas (2000). *Annual report 2000.* Eurogas, Brussels.

European Commission (1999). *The Security of Gas Supply in the European Union* (Communication to the Council of Ministers). COM (1999) 571 final, Brussels.

European Commission (2000). *Opening up to Choice, Launching the Single European Market.* European Commission, Brussels.

European Union (1998). Directive 98/30/EC, *Directive on Common Rules for the Internal Market of Natural Gas.*

Finon, Dominique (1992). Maturité des industries gazières et viabilité du régime concurrentiel. *Economies et Sociétés* Series EN No. 5, January–February, 189–221.

Golombek, R., Gjelsvik, E. and Rosendhal, K.E. (1998). Increased competition on the supply side of the Western European natural gas market. *The Energy Journal* 19: 1–18.

Heyvaert, Grieb (2001). The hub services: the Zeebrugge case. *Economies et Sociétés* (Series EN), No. 8, pp. 85–99

IEA (1998). *Natural Gas Pricing in Competitive Markets.* OECD/IEA, Paris.

IEA (2000). *Regulatory Reform: European Gas.* OECD/IEA, Paris.

Locatelli Catherine (2001). *The Dynamics of Industrial Organisation in Economies in Transition: the Example of the Russian Gas Industry.* Russian-European Centre for Economic Policy, Working Paper Series, Moscow, July, 19 p.

Mestmäcker, Ernst, J. Ed. (1993). *Natural Gas in the Internal Market. A Review of Energy Policy.* Graham and Trotman Publishers, London.

Moe, Arild (1998). Joint management of oil and gas resources in Russia. *Post-Soviet Geography and Economics.* (7): 588–605.

OFGEM (2001). *Improving the Customer Transfers, The Way Forward.* OFGEM Report, June.

Percebois, Jacques (1999). The gas deregulation process in Europe: economic and political approach. *Energy Policy* January 27(1): 9–16.

Radetski, Marian (1999). European natural gas: market forces will bring about competition in any case. *Energy Policy* January 27(1): 9–16.

Stern, Jonathan (1990). *European Gas Markets: Challenge and Opportunity in the 1990s*. Dartmouth Publishing Company, London.

Stern, Jonathan (1998). *Competition and Liberalisation in European Gas Markets: A Diversity of Models*. The Royal Institute of International Affairs, London.

Stern, Jonathan (2001). *Traditionalists versus the New Economy: Competing Agendas for European Gas Markets Over the Next Two Decades*. Briefing Paper, RIIA, London.

Stoppard, Michael (1996). *A New Order for Gas in Europe?*. Oxford Institute for Energy Studies (Monographs), Oxford.

Sunnevag, Kjell (2000). How sustainable is the framework for Norwegian gas sales? *Energy Policy* May 28(5): 311–320.

Thomas, Steve (2001). Gas as a commodity: the UK gas market. In Künneke Ralph and Arensten Maarten, Eds., *Gas Markets Transition in Europe*. Elsevier (to be published).

Waddam-Price, Catherine (1998). The UK gas industry. In Helm Dieter and Jenkinson Tim, Eds., *Competition in Regulated Industries*. Oxford University Press, Oxford.

Wybrew-Bond Jan (1999). What does the future hold for the European gas business? In Mabro Robert and Wybrew-Bond Jan, Eds., *Gas to Europe: the Strategies of Four Major Suppliers*. Oxford University Press, Oxford, 2p.

Williamson, Oliver (1985). *The Economic Institutions of the Capitalism*. The Free Press, New York.

Chapter 8

Competition and Market Integration in Europe: Towards a Multienergy and Multidomestic Oligopoly

DOMINIQUE FINON AND JEAN-MICHEL GLACHANT

The reforms mean that the cards have been and continue to be redistributed. Competitive forces in the sense of Michael Porter (1985) can develop with the help of changes in regulations and industrial structures as a result of the reforms. In each country rivalry of direct competitors – historic incumbents, local entrants from the other energy industries, foreign operators, traders and new suppliers – on the wholesale and the retail markets depends upon the attractiveness and the accessibility of the different markets. Threat of entries which depends upon these market characters and the protection of the incumbents takes the form of the so-called electricity and gas convergence among different ones, in particular the multienergy competition. Reforms have removed the barriers that once existed between the electricity and the gas markets.

Development of competitive forces also depends upon new types of relations between national markets. What possibilities do exist for unification on a European scale between the electricity and gas markets following the analysis of the national markets and the new regulatory set-up between countries? But as shown in the categorisation of electricity markets, in the recent years, there have been two opposed moves in the concentration of national markets: a group of countries where de-integration of the industrial organisation has been mandated and another where the industry retains or evolves towards a high level of vertical and horizontal integration round the former incumbents and under the auspices of the government. Moreover these incumbents seized opportunities open by the privatisation to extent in the other markets by acquisitions. The strong move towards the oligopolisation of the European electricity and gas markets is paradoxically the pendant of the market integration. Indeed the main argument for maintaining intact the incumbents or encouraging horizontal integration even after deregulation has been that the playing field will be soon pan-European and the incumbents will be competitively exposed in a larger European market.

8.1. New Competitive Forces: The Electricity and Gas Convergence

At what points do the electricity and gas markets converge at the current stage in the deregulation process? There can be no doubt that the gas and electricity markets are influencing each other more than previously and will influence each other still more in future. Several influences play at different levels: influence between short-term prices via direct and indirect arbitrages between gas and electricity markets, determination of the electricity price with reference to new gas power plant generation units, influence upstream especially from the gas markets to the electricity markets and influence downstream via multi-energy competition. The whole set of influences is carried out by diverse strategies of the so-called gas–electricity convergence adopted by companies coming from different initial core energy business (electricity utilities, gas companies, oil companies) and which diversify their initial activities along four particular tracks under internal and external growths: portfolio convergence which consists to be present in the two energies without developing major synergies; industrial convergence which is based on the value extraction of the gas resource by electricity generation (mainly the oil companies); industrial convergence also for electricity companies which manage fuel risk by upstream integration in the gas pipeline or LNG imports; and commercial convergence by energy trading activities and dual fuel offers to clients.

Influence between wholesale electricity and gas prices

The fact that the gas is becoming a major input in the electricity generation in the majority of European countries creates some interferences between wholesale gas and electricity markets on two scales of time (Table 8.1).

- This influence can be felt in the short or very short term through the reference of gas prices as the major elements of variable costs

Table 8.1. Share of power generation in gas consumption in mature and young gas markets in 1999 (%).

Major mature markets								Young markets				
Au	B	D	F	It	NL	Sp	UK	DK	Fin	Ir	P	CH
34.1	24.6	18.7	2.0	26.4	29.4	18.1	25.5	44.6	54	48.4	62.5	7.6

of production by units for the producers' bidding on the daily markets. If we refer to the British markets with mature competition and spot markets in the two energies, the price of the gas used for the electricity generation influences the price on the electricity market as the price of electricity could influence the attractiveness of using gas plants and so for the level of gas price. Traders as producers realise recurrent arbitrages between the two energies, the spread of relative prices determining the choice of produce or not by the electricity producer, the priority going to the higher price. So real time shutdown decisions could be observed in the centralised system, as for the production on site (CHP, etc.) by multiservice companies. Certain large gas consumers could ask for the partial indexation of their contractual price on the electricity price. And the spot prices are partly correlated as soon as a certain liquidity exists. However the degree of correlation between the price moves on the two markets could not be so high: in Britain correlation coefficient reached 0.3 between the day ahead gas price and the peak EFA spot (Chauvet, 2000).

- Finally to reach a certain convergence on the continental markets in the next future two conditions are necessary: to reach a certain level of liquidity on the two markets and to reach good conditions of transparency and flexibility of access to the gas and electricity networks and interconnections, that could only be the case in some years (see Chapters 6 and 7).
- For the medium and long-term the extension of electricity production is based mainly on natural gas, and the consequent role of the long run marginal cost of the 'gas' kWh as the reference long-term price, has placed gas prices at the centre of the price-making on the electricity wholesale markets which are under anticipation of under-capacity, or will place them in the future. The influence is felt through the raising of prices to a level that makes entries profitable through investment in combined cycle gas turbines (CCGT). It will however be noticed that on the 'continental plate', the wholesale electricity market prices (€20–25 per MWh) were

below the complete costs of a new gas power station, the so-called long run marginal cost (€35–40 per MWh) until mid-2003 and then climb towards this level in an oligopolistic collusion revealed by the fact that no under-capacity was foreseeable by that time. Meanwhile gas price has only exerted a limited direct influence on the spot price, given that the first CCGTs have been commissioned in 2001–2002 after a situation of quite high prices on the concerned markets with invesment in CCGTs, i.e., Italy and Spain. Moreover gas power production is exposed to the competition of the existing coal plants, in particular during the periods of high oil price on which gas price is partly indexed. Even in Great Britain, we have seen a temporary increase in coal-based electricity production in disfavour of gas in 2002, in a period of increased fuel cost.

The opportunities of cross-entries between industries

Renewing or increasing producers' power generation fleets on the basis of a new fuel such as gas will increase the contestability of the gas production in the upstream of the electricity generation. It favours the appearance of new entrants, either through the movement of some independent electricity producers upwards gas production as in Britain in the nineties, or by the movement of oil and gas producers downwards in electricity production as it has occurred in UK with some merchant plants or oil companies' joint ventures with RECs, in Scandinavia (Statoil), The Netherlands (Shell), Italy (ENI), Spain (BP, Repsol, Sonatrach) and could occur soon in Germany (Exxon, Shell) and France (GDF) by developing production units. This is, of course, provided that access to gas is not pre-empted by the historical electricity operators. We are currently noticing in Spain and Italy that the need for investment in production against a background of saturation of existing interconnections determines high wholesale electricity prices (with reference to the complete cost of the CCGT kWh) indeed attracts entries, that intensifies competition between electrical operators and oil and gas companies.

Conversely it has also been noticed that the main entrants into gas sales are the electricity producers, who are seen to be purchasing international gas contracts for their own needs in power generation for securing their supplies and for the extension of their businesses. Endesa, Iberdrola and Fenosa in Spain are entering in the gas supply business as ENEL and Edison are doing in Italy. In Germany this extension is made by electricity utilities' takeovers on gas transporters and distributors: E.ON extends its activities in gas with the takeover of Ruhrgas, RWE with the control of Thyssengas, EnBW (associated with ENI) with the control of GVS.

The multienergy competition

In addition, businesses selling gas to final customers have the potential to compete seriously with the electricity industry by offering dual fuels if the final electricity markets are opened to competition and vice versa for the electricity suppliers. The potential of multi-energy competition is mainly concentrated on the SMEs (industry, commercial clients) and residential clients, the large clients being few receptive to this type of offer. The multienergy competition is supposed to create some advantages in terms of costs for the electricity or gas companies which remained integrated in distribution: optimisation of grid and network O&M economies, integrated metering, customer service savings, with the opportunities of cross-selling. But given the provisions of legal unbundling of activities, multienergy competition is increasingly based on the search of marketing advantages. Observation of final markets allows indeed to stress two principles of success: the advantage owned by initial mono-energy incumbents in this type of competition, and the inducement of a competition-based development of electricity markets by the conditions under which the gas markets function.

- First if we consider the mass market of the residential and commercial clients, intensive activities have been observed since 1998 with numerous initiatives in the countries where the commercial and residential market segments have been opened to competition (UK, Scandinavia, Germany): creation of brand, sales on Internet, multiservice sales, alliances. Barriers for the competition in the supply are high, with important cost of entries and prospects of profitability by clients limited. The threshold above which this activity is reckoned to become profitable is two million customers.[1] They are partially overcome by gas or electricity incumbents in the other fuel markets with the help of dual fuel offers. They have much lower overhead costs for that by comparison to another entrant and they benefit from scale economies and brand image. In this way the conquest of segments of the gas market by an electricity incumbent is targeted mainly within its territory, in order to cultivate customer loyalty with a mark-up on the second product much lower than the mark-up on the first. This electricity supplier will typically offer a lower price for gas than the gas business, in order to attract that business' traditional clients, and vice versa. The goal is that customers are more loyal if

[1]European Utility Retail Survey, 2001, reported in *Energy Economist*, September 2001.

they receive two products and are getting a good price for one of them.

- The greatest influence on the electricity retail market is coming from gas sellers or distributors who were single-energy before the reforms. It has been noticed in Britain that since its creation in 1999 Centrica (British Gas) attracted several million customers on the households segment in the electricity supply (37% of customers). In Italy ENEL has great expectations of competing with ENI Power and Gas (ex-SNAM) in the gas supply to industrial and residential clients. In Spain the greatest influence on the industrial electricity market segment was exerted by Repsol and Gas Natural. Conversely on the gas market the real competition in retail sales is coming also from the former electricity incumbents. So it is in UK where the most successful competitors of Centrica on the gas supply have been the supply business of the former regional electricity distributors.
- Observation of countries in which distribution was assured by distributors of both electricity and gas as Germany or Austria shows that the opening of the retail market has had but little effect in terms of entries because of the cost of entering the market without an initial sales network and the tougher conditions of competition in dual fuel. The German model which retains bundled transportation and supply and in which the major power utilities take over the major gas companies (Ruhrgas, Thyssengas, GVS, etc.) and develop integration downstream by taking over municipal multiutility distributors is much less receptive to competitive forces. It corresponds to a multiutility model, and not to the multi-energy model.
- Second, there is a 'Double Competition' principle whereby the way one of the two markets (gas for instance) function may induce competition-based development of the other market (electricity for instance). In fact, in Great Britain during the first half of the 1990s, the formation of a competition-based wholesale gas market and the new possibility of direct transactions between gas producers and users allowed new electricity producers to enter. By the end of that decade, it was the historical existence of distinct mono-energy incumbents that allowed cross-competition entries into the retail markets for both forms of energy at the moment the total unbundling of supply and networks was decided.
- Certainly the British situation of interaction between wholesale and retail markets in gas and electricity is still a rare situation in other European countries, with its total opening of the market, the presence of twelve regional distribution companies (now seven

suppliers) of large size and a complete unbundling in electricity and gas distribution. However, one can clearly see the development of a significant strategic tension between gas and electricity in several other countries, especially Spain, Italy, and The Netherlands, where unbundling is effective, or about to be so. In the future France and Germany will be also the field of such an evolution when the 2003 Directive will impose clear unbundling in transport between network activities and supplies to mean and large consumers. In France the competition will mainly develop between the respective gas and electricity incumbents.

So electricity and gas convergence is a multiform reality, but not so clear cut than the 'common wisdom' anticipation was at the beginning of the second wave of reforms. Gas and electricity markets interact but without synchronicity and clear direct effects. In this new environment the major players' learning process in this new businesses has *in fine* selected a number of strategies that will be analysed in the later chapters. First, electricity companies develop external growth for expanding in the gas business and play mainly on their integrated structures and size. Second, in a logic of internal growth and with the strength given by their gas resource and large finance surface, oil companies are integrating their gas supply business, developing multienergy trading activities and electricity production assets. Third, the former gas companies are the most vulnerable: either they become integrated in multienergy companies or bet on their market knowledge to preserve their integrity.

8.2. The Potential for European Integration of Electricity Markets

In electricity, European integration of national markets is, and will remain, limited at the level of short-term transactions. With the exception of Scandinavia, the real basis for the constitution and operation of most electricity short-term markets is mostly confined within national borders. And this basis will remain either national or regional by regrouping three blocks of countries (Nordic countries on one side, France–Germany–Belgium–Austria on another side, and finally Iberic countries) for a long time.

As the developments on electricity markets demonstrate (Chapter 6), the effective accessibility of national markets in terms of technical, regulatory and commercial opening may improve significantly, especially through increased interconnection capacity and harmonisation of market and network access rules following action by the coordination of national regulators and the European Commission.

In the immediate future, price levels and fluctuations on certain wholesale voluntary markets (the *Power Exchanges* or PXs) may move closer. In Continental Europe the movement will be around a French–German block widened to include Belgium, Switzerland, Austria and, at the end of its pre-reform imports contracts, The Netherlands. The creation of a continental PX, if decided, would be a major step towards European electrical unification, and would undoubtedly be preceded by the quotation of a harmonised price for a reference marketplace. Without anticipating the level of spot exchanges between countries alongside national spot exchanges which could remain quite low, arbitrage activities will allow such integration as soon as the major barriers to trading will be suppressed.

To reach this situation, only intensive harmonisation of the network access rules (transport tariff rules, rules for allocating and pricing interconnection capacities, rules for managing losses and congestion and balancing rules) could speed the rate at which the pan-European wholesale market matures in the long term (10 years). Under these conditions, it could show a level of homogeneity similar to that currently found in the Nordic power market. However, one should not underestimate the current barriers to the formation of more unified European markets, for several reasons.

- *Slow evolution towards technical coordination.* Movements towards such conditions under the so-called Florence process initially suggest the difficult setting-up of intensive co-ordination of technical and informational resources between system operators.
- *Gap in connections capacities.* Technical accessibility which is characterised above (Chapter 6) is not completely relevant. First, in the past connections have been designed for security reasons by coordinated monopolies and new congestions are becoming apparent because of developments in market flows. First, all the existing interconnections have been designed as links between national systems and not as infrastructures essential for the unification of the market at the European level. Second, institutional restrictions on capacity reservation coming from pre-reform contracts increase these gaps. Even significant interconnections, corresponding to over 15% of peak demand, may be insufficient when cross-border flows reach very high levels by combining short-term purchases and restrictions by long-term pre-reform contracts, as for flows towards Italy or The Netherlands.
- *Limitation of opportunities of efficient inter-countries exchanges.* A priori the weight of history in each individual electrical industry will be felt for a long time to come wherever change is involved, as

production fleets have all been built on national bases using an autarkic logic, in view of the technical requirement to permanently balance supply and demand. If this situation could generate incentives to exchange because of the differentiation of hourly production cost between countries, the main potential of exchanges is in fact exploited by the pre-reform long-term contracts. In addition, the presence of production overcapacity on the 'continental plate' is reflecting in common wholesale prices between the complete variable costs of fossil fuel generation (the cash cost) and the complete long run marginal cost. In a first stage (2000–2003) the wholesale prices arising from this overcapacity on the continental plate (€20–25 per MWh) were clearly below the long-term price aligned to entry costs by the construction of a combined-cycle gas power station (€35–40 per MWh). In a second stage (since mid-2003) it could have been risen to an average level of 30€/MWh by the oligopoly of the four majors without attracting entries.

It is putting back major investments until after 2010 and possibly much later than that, having the effect of freezing the European market around the pattern of national generation structures inherited from the past. In contrast, in two out of the European peninsulas (Iberian, Italian), separated from the continental market by insufficient interconnections, demand/supply tensions tend to maintain a local dynamics of high wholesale prices and own development of investments which consolidates the logic of national market.

Moreover on the longer run entrants and incumbents' probable choices of investment in the mainstream generation technology (CCGT) could homogenise the cost structures of competitors and contribute to the limitation of exchange between countries.

- *The differences in industrial structures and property regimes.* As pointed out in our analysis of the industrial and commercial accessibility of markets (Chapter 6) and by the EC's benchmarking reports, a factor in the persistence of fragmentation of the national electricity markets is the continuing differentiation between industrial structures in the various countries. The most notable aspects of this are the horizontal and vertical concentration of generation and sales in a number of countries (Belgium, France, Germany, Italy, Spain among the major markets), which reduces the opportunity of direct foreign exchanges. Moreover as purchase of assets becomes one of the privileged means of penetrating new markets at the European level, opportunities are limited in countries with low fragmentation of the electricity industry or persisting public property regime as in France.

- *The vertical reintegration.* In addition, the competition-based model with vertical de-integration and development of production businesses not integrated with final supply businesses (the so-called 'merchant plants' model) appears to be declining sharply in the countries of the first-wave reforms. The difficulties encountered by operators in this area in the British market following the transition to a decentralised market in 2001 by the NETA in a situation of over-capacity, reveal the risk of investing because of the eventuality of sharply decreasing wholesale prices: several producers short or absent in supply activities indeed have made bankruptcy in 2002. The banks have become very wary of financing this type of entry into production and appreciate the bankability of projects in relation to vertical integration or long-term contracts backed to captive markets. Strong vertical integration between electricity production and final supply is therefore likely to persist, or develop itself, as in the United Kingdom where the sector had been originally de-integrated and the regulator had decided in 1997 to remove restriction on the vertical integration. It will maintain the national fragmentation of the European electricity markets.

In these circumstances, beyond the restricted integration of wholesale markets, what are the possibilities of unification of final markets? The main effect of the future development of a reference marketplace on the 'continental plate' would be to harmonise wholesale market prices up to 2005–2008 between the main countries in this area, especially France, Germany and Belgium, Austria and Switzerland. For the final supply the sales forces and client access channels remain distinct and still be partitioned between the various countries. Some companies envisage to unify its sales forces and trade names in the countries covered by a common unified wholesale market, as Fortum does in Scandinavia in 2002, while others prefer to keep the brand name of the local companies that they took over for marketing consideration as does Vattenfall in Germany.

On the large customers segment (10 GWh/y and above) traders, brokers and marketers' arbitrage activities help to unify markets, in particular on the 'continental plate'. But it has not yet been fully integrated at the European level: in 2003, the major European businesses were still looking in vain for a supplier capable of contracting with them for the whole of their sites in the European Union. For mean-size industrial and commercial clients final prices would converge slowly with the help of the most active traders and marketers entered in the most open markets. For small commercial or residential clients it will partly or not at all converge because of the disparities in regulation of the distribution price between countries (see Chapter 6).

8.3. The Potential for European Integration in the Gas Markets

Overall, the gas markets would appear easier to open technically on a Europe-wide scale than the electricity markets, more accessible commercially and thus more open potentially to competition from new entrants in each national market. But developments of Chapter 7 invite to conclude as well that the emergence of a unified multihub gas market is as difficult as in the electricity sector because of the lack of distributed sources and the contractual commitments. We present successively these two sides of the coin.

1. First there are three reasons for which the gas markets could be more open technically and more accessible commercially:

- Because of the huge international supply, capacities for interconnection for the transportation of gas supplies across national boundaries reach between 50% and 100% of national consumption almost everywhere (at the exception of UK), while it averages only 10% in the electricity industry.
- Gas commodity is more flexible and more easily exchangeable and routable than electricity. Gas can also be stored and its volumes altered within the transport network without endangering the function of that network, unlike electricity flows in transport networks. The physical realisation of commercial transactions in gas on a European level will always be easier and less haphazard than the conclusion of electricity transactions.
- Finally, at the moment of the reforms, the structure of the gas industry was less integrated vertically and horizontally than that of the electricity industry. The requirement for technical co-ordination between upstream and downstream is less demanding; and it is the long-term vertical contracts that have ensured stability of relations between producers, transporters and suppliers to final customers. Distribution was separate from transportation and wholesale purchase-resale in most countries. As a result, vertical integration of production and sales was rarer in the gas industry, and it is therefore easier to enter the various stages in the value chain. Moreover, it is easier to the regulator to impose some horizontal de-integration by gas contract release programs than by divesture of incumbents' equipment in the electricity sector. So that

the entry barriers by the purchase contracts monopolised earlier by the incumbents are more erasable.

Do these differences from the electricity market therefore show that an integrated European gas market could be established more easily despite the delay initially encountered? After 2000 the institutional conditions are developing favourably. At the access rules level, we are moving towards a regulated TPA consolidated by the legal separation of transport networks and storage capacities from the supply in all countries; this separation is already effective in six countries and will take effect in France and Germany in 2004. The transcription of the 2003 directive on electricity and gas markets consolidate these evolutions. The harmonisation of the various tariffs for cross-border transportation with the definition of a specific 'nodal pricing' on the European level, and the laying down of rules for communicating information on available capacities is slower than in the electricity industry. But it may be consolidated by the European 'réglement' on access to the gas transmission networks and the cross-border carried in December 2003.

In the gas sector, therefore, a regulatory situation where the respective TPAs are transparent and non-discriminatory at the national level and well harmonised at the European level might exert a much more significant effect on the advent of competition-based markets at the European level than in the electricity sector. Because of the existing technical capacities, there is no reason why a short-term European market should not emerge once the total of non-contracted or free gas has increased sufficiently after the normal phasing of some present long-term contracts. That will feed the liquidity of the emerging spot markets, in particular because the major purchasers are adopting a portfolio strategy. The level at which gas exchanges are carried out at the wholesale level is naturally the European level, given the location of most productions outside the European Union. The emerging gas exchanges are multinational in nature, as is the case in Zeebrugge and Baumgarter. A plausible scenario would be the emergence and the co-existence of four to five exchanges in a number of hubs, with price differences that reflect transportation cost differentials. The short-term prices themselves would be constantly affected by numerous arbitration procedures. Confident in the price discovery by these markets producers and major bulk suppliers should progressively agree to index the contractual price on a reference spot price.

2. However the emergence of a unified multihub market with a European reference price does not appear to be envisaged up to 2010.

In order for this multihub market to establish itself, the gas profession considers that at least two continental hubs have to reach a certain level of liquidity alongside the British NBP spot market. This is far to be the case for the Zeebrugge hub in 2004 after five years. In the meantime, at no reference point on the continental markets can the price be used as a base for fixing the prices of the bilateral producer/ purchaser contracts, in contrast to the British NBP, which processes over 20% of the physical flows coming to the British consumers.

- *The pre-reforms contracts.* The previous existence of vertical relations between major producers and major suppliers via the 'pre-reform' contracts will put the brakes on the development of these competition-based exchange. The pre-reform contracts do not let sufficient field for 'free gas' and do not allow a sufficient liquidity on the continental marketplaces. Internal competition in each country will remain limited, with a wholesale price based on the contractual reference to oil prices for most of the major suppliers' purchases. The 'gas release' programmes in Spain and Italy open the national game to two or three new actors, mainly national ones, but whose contractual purchases will be made at a price that remains indexed to oil prices.
- *The persisiting needs of long-term contracts.* After the phasing-out of the existing contracts, the evolution of the forms of trade will be restricted by the characters of every new bulk supply projects. New gas supply sources will be mostly international in most European countries (5–6 major supplier countries, only two of which are European Union Member States), while almost all electricity supply sources are still local or national (hundreds of power stations scattered all across Europe). The development of a regional or worldwide LNG market, with current projects in Nigeria, Trinidad, Egypt, Norway, Qatar, Abu Dhabi, Angola, Venezuela in the Atlantic and Middle East zone, is an emerging trend which could not influence significantly the competition in the European supplies in this decade at the exception of some markets (Spain in particular). Consequently there is a general expectation that on the continental markets, the total value of spot transactions will not exceed a modest 10% of the total supply, because of the need to set the development of new import infrastructure projects against a framework of long-term contracts (IEA, 1998). Away from the need to secure investments by this type of contract, gas remains a commodity that is imported from a long distance and therefore carries significant geopolitical stakes from some sources, which are part of the risk in the transactions and generate the need of

securisation of investment by contracts. So it would be much longer than in UK to obtain a reference price for the contracts by the gas exchange prices.

- *Weak effects of transparent access.* In this sense, a quite moderate freeing of internal competition will only result from the yet significant mandatory measures such as implementation of a credible regulated TPA and disappearance of some incumbents via the legal separation of the network and the integration of their gas supply activities in a general energy supply division of an oil and gas or multienergy companies (ENI with SNAM, E.ON with Ruhrgas, Suez-Tractebel with Distrigas, Shell and Exxon with Gasunie). Moreover, the strategy of penetration by asset acquisition cannot have the same effect as in the electricity industry, given that national gas productions have been excluded from the acquisition game.

In these circumstances, competition development on the respective markets will come mostly from two sides, first from outside suppliers, the former gas companies that enjoy flexibility of purchase contracts, oil company entrants, and second from the search of cheaper gas from the outside markets by gas suppliers, especially via some purchases on the gas exchanges by means of 'swing' supplies. Among these, the most important driver of competition will probably be the oil companies' entries on the gas markets by focusing only on large buyers segments: electricity producers, large industrial consumers and major distributors.

As for electricity, the integration of national gas markets will always be hampered by obstacles on the unification of markets for retail sales, in particular to commercial clients, small and medium enterprises and a fortiori to residential customers. Only the unification of major suppliers' sales forces to major industrial clients will be realised. It will be mainly by oil companies, as is already the case in their trading department.

Under these conditions, the emergence of a European market will show itself first of all in the increasing influence between the various national wholesale spot markets as it is already the case between the British NBP and the Zeebrugge spot price, and second in the mutual influence between final prices by the intermediation of arbitrages on the big purchases.

8.4. The Oligopolisation of the European Electricity and Gas Industries

As shown in our categorisation of electricity and gas markets, in the recent years, there have been two opposed moves in the concentration of national markets: on one hand a minority of countries where the concentration has decreased – mainly the United Kingdom and to a lesser extent Italy – and on the other hand those where the electricity and gas industries either retain a high level of integration (France, Belgium) or they went from a quite fragmented structure to a vertical and horizontal integration, in particular Germany and Spain, and to a lesser extent Austria and in Scandinavia with moves in vertical integration. In markets which are characterised by former centralised public or private monopolies (France, Belgium, Italy, Portugal) the concentration remains high, at the exception of Italy.

So E.ON, RWE, EDF and Suez-Electrabel appear to be dominant on their home markets and from there on the main integrated area of the new European electricity scene. Meanwhile Vattenfall-Europe in the North and ENEL in the South are candidates to join them in the club. Other companies – the Iberian majors, and the other Nordic firms (Fortum, Statkraft) – are confined to their national and regional markets where they could eventually exercise market power.

The process of concentration is nearly achieved on national markets. The merger of EDF and GDF, occasionally contemplated in France, would probably never be concluded. The spectacular E.ON-Ruhrgas merger in Germany in 2002 and some 'electricity–gas convergence' acquisitions in Iberia and in Italy in 2003 are the last major national moves. At the inter-country level the lately penetration of the German groups and EDF on the most accessible British markets in 2001–2002, the Vattenfall Europe's creation stradding on the German and Nordic markets in 2002, the GDF's extension in Germany and Italy in 2003, the ENI's extension in Iberia (Galpenergia) and Germany (GVS) in 2001–2002 have been major moves. It could be followed by the latecomer ENEL's Europeanisation. But opportunities are nowadays limited even if a number of smaller companies are left quite vulnerable to takeover by the dominant groups.

This oligopolisation at the European level is a major element of the market integration which could have negative effects. Dominant position of large companies at the national level is already an obstacle to effective competition which is the only way to have positive price effects for the consumers. The extension of competition by international trade could then be a way to alleviate the risk of dominant positions, but it will be restricted by the cross-frontier development of companies.

8.4.1. Impact of market integration on dominant positions

Indication on dominant positions and risk of market power exercise is currently given by indexes of concentration (concentration ratio CR,

Table 8.2. Index of industrial concentration of national electricity and gas markets in selected countries.

	Austria	Belgium	France	Germany	Italy	NL	Spain	Sweden	UK
Electricity generation in 2000	45%	96% (C2)	92%	64%	69%	59%	83%	90%	36%
Gas final supply in 2000	67%	53%	90%	50%	72% (C2)	48%	94%	47%	42%
Gas bulk supply in 2002	80 %	97%	90%	50%	95% (C2)	83%	95% (C2)		50%

Source: European Commission. *Second Benchmarking Report*, October 2002 and chapter on Gas markets (Gas bulk supply), Table 2.25.

Herfindhal-Hirschman Index HHI) used by the antitrust authorities to discriminate between these two groups of countries. Referring to these indexes, all the markets are well above the level of concentration ratio which presumes market dominance (CR1 > 33%; CR3 > 50%) for the electricity production, the gas bulk supply and the gas final supply, at the exception of UK and The Netherlands (Table 8.2).

But there is an effective mitigation of the risk of abuse of market power by regionalisation of some markets such as the Nordic ones and since 2000 the continental ones in electricity. So larger relevant markets than the national ones by grouping national markets have to be considered by reference to the interconnection capacities (Table 8.3). Referring to the HHI prognosis – unconcentrated if below 1000, moderately concentrated between 1000 and 1800 and highly concentrated if above 1800[2] – the Nordic markets appear to be as unconcentrated, as the UK one is supposed to be. And as far as the oligopolisation of the continental market is concerned, the HHI index shows a moderate concentration with a level of 1500, that means a dilution of the dominant position of the four majors.

However the definition of relevant markets to areas larger than national markets has to be carefully discussed for two reasons: first the temporary congestion on interconnections during the peak load episodes in the year; second the remaining regulatory and institutional barriers to trade when the technical capacities exist.

[2]Utilisation of the HHI index by the U.S. Department of Justice (http://www.udsdoj.gov/atr/public/testimony/hhi.htm).

Table 8.3. Comparison of HHI index of concentration in electricity generation in selected regional and national markets in 2002 (in round figures).

UK	Nordic market	Germany–Austria–Switzerland	Continental market	Portugal–Spain	Germany	France–Belgium–NL	France or Belgium
500	500	1000	1500	2000	1400	5000	8200

Source: Oklo Institute (2002).

8.4.2. The debate about market concentration

The economic arguments in favour of market concentration and large companies will point the economies of scale and scope. Meanwhile market integration could afford some competitive pressures. Critics against market concentration point the risk that even without abusing their dominant position the majors could orient the price making on the wholesale markets and deter entries. They point also the risk of the undue political influence of them.

The defence of dominant companies by the market integration

From the opposite point of view, competition is preserved as soon as two conditions are respected: a transparent and non-discriminatory access to the grids and sufficient interconnection capacities to allow large communications of national markets. The integration of a national market with the neighbours is the main argument for preserving incumbent and defending concentration: it is ample to ensure that competitive pressures are effective and consumers will get real choice either in the country or outside. This rationale is theorised in terms of market contestability: credible entries thread creates sufficient incentives to productive and allocative efficiency. So, in this view, this problem is not the number of competitors and their relative size in each national market because the 'relevant market' to be considered would be the pan-European one, or at the least the continental one.

From this view, big-sized companies present a number of economic advantages, in particular to exploit scale and scope economies for investing, to compete in an equitable way with the other European giants on their respective markets and to rival on the European capital market of company acquisitions. Tenants of this view argue also in terms of political advantage: big sizes and vertical integration help to pursue governmental goals in matter of energy security or environmental policies. It could be a way to realise objectives of energy security based on the development of large, indivisible and capital-intensive equipment in the electricity as in the gas industry. So it is for the development of import and transit infrastructures and the signature of large long-term contracts with remote producers. Reaching big size allows to assume more important risk of investment on large operations of investment. So was one of the rationale of the acquisition of Ruhrgas by E.ON in 2002 to prepare the future by the consolidation of gas purchases to the Russian players. So it was of the defence of the horizontal and vertical integration of electricity industry in France for maintaining the nuclear option.

An extreme position is to defend the advantage of the oligopolistic control of the market price by tacit collusion between the electricity majors in order to avoid a too large price volatility, in particular when the equilibrium between supply capacity and demand begins to be unstable when the reserve ratio is diminishing (See Bergougnoux, 2000 for instance). So tacit collusion to maintain price higher than the short-term marginal cost would create good conditions of investment by avoiding price spikes and risk on supply stability in the future.

The critics of market concentration

In an inter-market area in which there will be an actual integration of markets, the risk would be that the few geographically extended champions control the market price by tacit collusion on the inter-related wholesale markets in this area, as points the Oklo Institute (2002). Being present in adjacent countries by controlling production and supply assets, these companies would have the possibilities to create artificial congestion on the interconnections.

Moreover dominant position of large companies at the national level is an obstacle to effective competition which is the only way to have positive price effects for the consumers, even if the markets are communicating. Price set under oligopoly conditions could be far harder to correct than in pre-reforms period. At the political level indeed the largest companies are in a position to influence the definition and the way of implementation of market rules and to influence policy choices beyond the normal representation of their industrial interests.

It infers that differences of industrial structures between countries give an undue advantage on the European playing level to those enthroned as national champions by their government. The critics consider that there is a reprehensible tolerance from the national and European authorities for highly concentrated markets that paved the way for the emergence of a narrow oligopoly able to control the

European market and made excessive profit to the detriment of the consumer (Oklo Institute, 2002; Thomas, 2003). They regret the absence of the European Union's legal power to mandate divestures in the energy industries.

To conclude, as market integration is accompanied by a strong move towards the oligopolisation of the national and European markets, this raises the issue of bounding this move. The objectives are economic and political: limiting market power and balancing the political influence of the largest electricity and gas players on energy and environmental policies. These issues will be addressed in the following chapters.

References

Bergman, L. and van der Fehr, N.-H., Newbery, D. and Pollitt, M. (2000). *A European Market for Electricity?* CEPR, Londres (Monitoring European Deregulation Series).

Bergougnoux, J. (2000). *Les Services Publics en Réseau*. Rapport au Commissariat Général du Plan, La Documentation française, Paris.

Chauvet, N. (2000). Convergence gaz-électicité, *Revue de l'Energie*, November, 521, 557–560.

European Commission (2000). *State of Implementation of the EU Electricity Directive 96/92 EC. Country by Country Overview.* State of play by the end of May 2000, European Commission, Brussels.

European Commission (2002). *Implementing the Internal Energy Market. First Benchmarking Report*. European Commission, Brussels.

IEA (1998). *Natural Gas Pricing in Competitive Markets*. OECD/IEA, Paris.

Oklo Institute (2002). *Power Generation Market Concentration in Europe 1996–2000. An Empirical Analysis*. Report to the European Parliament, May 2002.

Porter, M. (1985). *Competitive Advantage: Creating and Sustaining Superior Performance*, Free Plan, New York.

Thomas, S. (2003). The Seven Brothers. *Energy Policy* 31: 393–403.

Section III
Strategy

Chapter 9
Strategic Configuration: Options and Perspectives

ATLE MIDTTUN

9.1. Introduction

This section presents the emerging patterns of strategic configuration for major European energy companies. Following the general theme of this book, we are particularly concerned with the extent to which the EU deregulation actually leads to the Europeanisation of the energy industry, beyond the boundaries of the individual nation-state, and to what extent energy industry business strategies move beyond the traditional electricity and petroleum configurations towards broader multienergy integration.

The discussion in this section integrates these questions in a broader analysis of strategic configuration of large European energy firms. The first part presents some of the strategic options open to energy industry and subsequently reviews some of the theoretical perspectives on drivers of strategic reconfiguration. The second part presents a statistical analysis of patterns of strategic reconfiguration of the fifty largest European energy firms. The third part explores some of these patterns more in detail through a qualitative analysis of selected company cases.

9.2. Options for Strategic Configuration

Deregulation of infrastructure opens up numerous alternative modes/patterns of strategic configuration, leaving it up to the firms to optimise their positions. This chapter explores some of the main options available and some of the theoretical perspectives on drivers of strategic configuration.

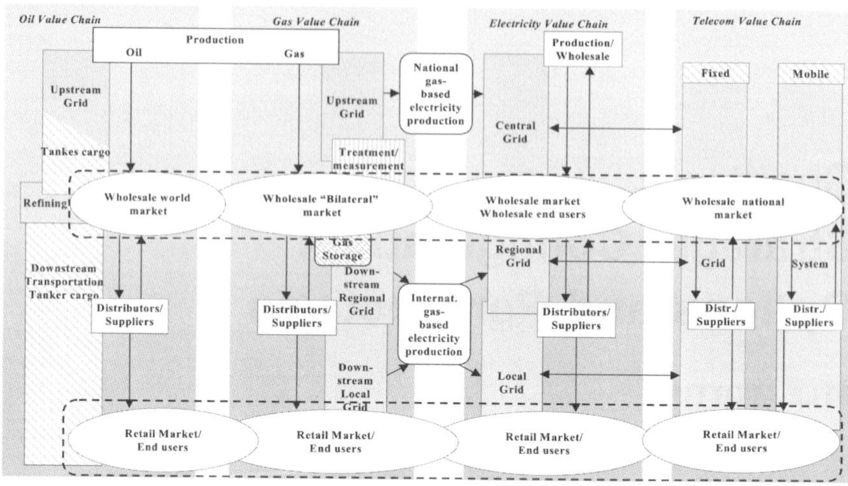

Fig. 9.1. Infrastructure value chains.

Infrastructure industry, such as electricity, gas, telecommunication and water supply, are complex as they span a number of highly integrated functions. Strategic reconfiguration in such industries, therefore, implies careful evaluation, not only of elements, but also of their actual and potential relations to other parts of the value chain(s). As a point of departure, we should therefore analyse reconfiguration of energy industry as involving a complex set of activities in several value chains (Fig. 9.1), each of which includes of a set of production, transmission, wholesale, and retailing activities, which the firm may choose to engage in or leave out.

The principal options for strategic configuration open to European energy companies are:

- decoupling and specialisation.
- horizontal integration between two or more chains.
- vertical integration within one or more chains.
- various forms of diagonal integration between upstream and downstream sides of different value chains.

9.3. Five Ideal Types

In the numerous combinations possible, we can distinguish between a few ideal types (Fig. 9.2), each representing combinations of sectoral and functional combinations such as vertically integrated

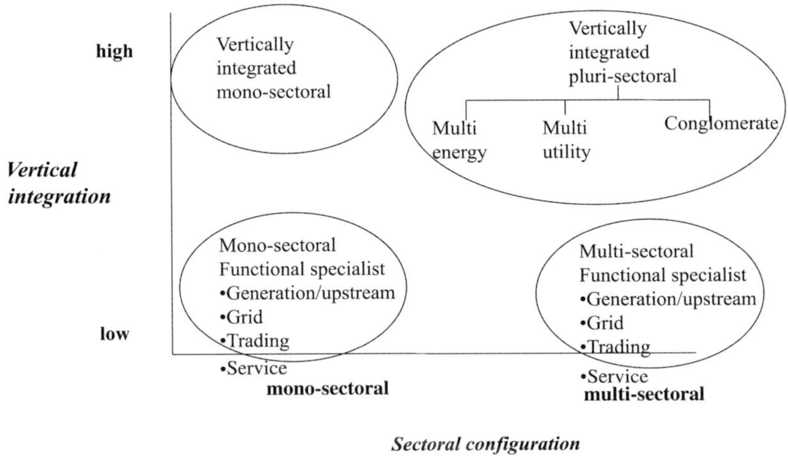

Fig. 9.2. Ideal types of configurations.

mono-sectoral, functionally specialised mono-sectoral, vertically integrated pluri-sectoral and multi-sectoral functional specialists. To simplify, we have integrated two ideal types; the vertically integrated pluri-sectoral and the conglomerates into one wide category.

9.3.1. *Vertically integrated mono-sectoral configuration*

The vertically integrated company is obviously one of the dominant ideal types in the energy industry. In the petroleum industry, the major companies have traditionally been engaged broadly throughout the whole value chain from exploration to customer supply in gas stations and heating services.

In the electricity industry the vertically integrated company is also one of the basic types, where the company typically generates electricity from one or several nationally available resources and transmits it through distribution outlets to its customer base. While the transmission grid may be shared with other generators, the distribution and supply business would typically be owned and exclusively supplied by the generation unit of a vertically integrated company.

Gas companies have traditionally been less integrated, in the sense that they have often left the gas extraction to the petroleum companies, and have been more focused on the transmission, distribution and supply side. Nevertheless, their engagement in customer supply, distribution and transmission, and also in long term contracting of gas from the large petroleum suppliers, has traditionally put them at least in a semi-vertically integrated position.

9.3.2. Vertically integrated pluri-sectoral configuration

The deregulation of both gas and electricity facilitates a move from mono-sectoral vertically integrated companies to broader multienergy companies, which are engaged in two or more energy sectors, and typically with considerable assets in each of them, although not necessarily fully vertically integrated.

An even broader configuration, the multiutility company, transcends the energy sector and includes engagements in one or more infrastructure sectors. Similar to the multienergy company, the multiutility company, as we define it, includes assets, but does not necessarily involve full vertical integration in each value chain. Given the accessibility of other deregulated industries like telecom and transport, broader multiutility configurations are also facilitated by deregulation.

9.3.3. Full conglomerate

The broadest configuration in the broad multifunctional configuration category is the conglomerate, which is a company with extensive multi-sectoral engagements also outside infrastructure sectors, and large asset bases in several of them, although not necessarily fully vertically engaged. Its activities typically span wider than the multiutility into non-utility industrial and commercial sectors. The conglomerate therefore, when big enough, may appear as an industrial group more than a single company.

9.3.4. Mono-sectoral functional specialist

Besides the broad multifunctional engagements, the energy sector also features more focused and specialised configurations. The natural starting point in this category is the mono-sectoral functional specialist that is diametrically opposite to the conglomerate. It is characterised by a focus on one part of the value chain, upstream, midstream or downstream and develops a specialised managerial competency in this limited area. In its most typical form, the specialist is also limited to operation in one sector (mono-sectoral specialist). However, deregulation facilitates broader multi-sectoral refocusing within the same function. This allows a move towards multi-sectoral functional specialist roles.

9.3.5. Multi-sectoral functional specialist

One of the emerging functionally specialised configurations in the energy market is the multi-sectoral service company. Like the multi-utility company, the multiservice company transcends the energy sector and may branch into a wide set of sectoral engagements. As implied in the name, the multiservice company is service focused. That is, it focuses on activities in the interface with customers rather than upstream activities. Typically the multiservice company operates with little or no assets.

Other functionally specialised multi-sectoral configurations are, multisectoral wholesale trading specialists and generation/production specialists across the energy and potentially also other sectors. Similar to the multiservice company, they single out one segment of the value chain and specialise in this segment across sectors.

Among functionally specialised configurations, grid management stands out as a particular case. Because of its natural monopoly character, this function is subsumed under specialised regulation, which, in some institutional contexts makes it mandatory for this function to be separated out.

9.3.6 Summing up

Given that deregulation of energy industry should, in principle, allow freer and more competitively exposed choice of configuration, the variety of configurations present in the European market invite the question: why, and under which circumstances does a company choose one or the other? The question of how and why various functions of energy industry's and other related industries' value chains are or should be commercially organised may be rephrased in terms of the more general question of determining the boundaries of the firm. In principle, one might select any position in the matrix in Fig. 9.1 as a point of departure and examine the commercial advantage of adding on any other position or value function from any other value chain. The firm's boundaries should then be set around the bundle that gives the largest sustainable net commercial benefit (Afuah, 2000). However, this begs the question of by what criteria the commercial advantage of the bundles of value creation may be judged?

9.4. Drivers of Strategic Configuration

The literature on drivers of strategic configuration is extensive and spans across economics, organisation theory and institutional analysis;

each perspective focusing on specific explanatory variables. To the core economics repertoire belongs the focus on drivers such as production cost and transaction cost as well as strategic positioning, inherent in oligopoly theory. From an organisation theory perspective, drivers for strategic configuration are to be sought inside the firm, among other things in the form of firm capabilities and learning. A broad institutionalist literature stresses national institutional contexts and business styles as major shapers of strategic configuration. This line of reasoning often suggests considerable path-dependency inherent in culturally established practices and national institutions. Finally, a pragmatically motivated industry perspective argues for industry-specific factors driving strategic configuration.

9.4.1. Production costs

The production cost perspective evaluates the commercial advantage of industrial organisation, in this case functional clustering in deregulated energy/infrastructure industries, in terms of how the bundling of a given set of activities might affect the productivity of the firm. The focus under this perspective will typically be on the advantages of scope/synergy and scale, that allows the firm to increase output relative to input. With new options open to the firm following deregulation, the firm should be in a position to select from a much larger range of possible combinations of products.

Production cost analysis has traditionally been based on strong simplifying assumptions. In price theory, knowledge about alternative production possibilities is traditionally seen as explicit, freely transmissible and easily encapsulated in what Joan Robinson (1956) calls 'blueprints'. It is nevertheless recognised that technological progress has a dynamic influence on this calculus, which may also affect the organisational boundaries of the firm. In a more dynamic analysis of productivity, the firm might therefore be assumed to move its product function along the boundary of technical possibilities. However, defining those possibilities involves an uncertainty that undermine calculability and moves production cost analysis towards a more qualitative approach.

9.4.2. Transaction costs

Transaction cost theory focuses on costs of contracting associated with imperfect knowledge and asymmetric information as major determinants of organisational boundaries. Following Williamson's modern codification, the central concept in transaction cost analysis is asset specificity. Assets are highly specific when they have value within the context of a particular transaction, but have relatively little value outside the transaction, which opens the door to opportunism (Williamson, 1975). Once the contract is signed and the asset deployed, one of the parties may threaten to pull out of the arrangements – thereby reducing the value of the specific assets. Other things equal, the firm will thus be motivated to select forms of strategic configuration that include asset-specific activities into the enterprise, whereas the firm may prefer to contract other non-asset-specific activities in the market.

The joint evaluation of production and transaction costs in mainstream organisational economics literature has traditionally entailed treating production costs and transaction costs as separate and independent variables that jointly determine the bundling of economic functions in one firm (Williamson, 1975; Oster, 1999). Firm boundaries would thereby be extended to include new value chain activities under high internal productivity gains and high transaction costs. Inversely, value chain activities would be externalised to market transactions under low internal and productivity gains and low transaction costs.

However, even with a basically neoclassical analytical point of departure, the literature is divided on configurational forms. On the one hand, the literature lists a long set of arguments in favour of complex bundled positions such as the multiutilities concept, which allows multiservice providers important savings in key business processes such as customer acquisition, customer service, billing and eventually branding. All these savings are seen to derive from economies of scale and scope, following from horizontal integration which allows sharing of information and systems, and access to customers (Robinson, 1997; Briones *et al.*, 2002).

The broader multi-sector organisation like the multiutility is also seen to have competitive advantage in web-based customer interfaces, where genuinely bundled offers, rather than the simple co-location of products, can allow enhanced cross-selling by the creation of real value to the customer in terms of cross-product discounts. This literature also argues that the multiutilities' use of bundled products can increase customer-loyalty by creating multiple/simultaneous points of contact and by making switching decisions more difficult for customers. Loyalty schemes also become more relevant as the share of wallet captured increases in size. It is argued that the key success factors will revolve around the attractiveness of the product combination and the ability to develop a brand that can represent products and services that traditionally have sold independently.

On the other hand, another part of the literature takes a critical view on the multiutility configuration. The reasons are twofold.

First, a common statistical finding from the many empirical studies is that increasing business diversification is accompanied by declining firm performance, particularly as diversification becomes unrelated (Hitt and Ireland, 1987; Buhner, 1987). Even stronger critique has been raised against conglomerates. Several recent academic papers and the business press claim that conglomerate firms destroy value and do a poor job of investing across business segments (Maximovic and Phillips, 2002).

Second, for most firms, this literature argues that organisational talent has an industry-specific component. The conglomerates have a discount in the stock market relative to single-segment firms.

Following the critique of broad business configurations, a group of scholars argue strongly for international specialisation (Mascharenhas, 1999). International specialists are firms that focus on one line of business but with an international scope. They compose the growing middle ground between diversified multinational companies and local firms. The argument in support of large international specialists is that they concentrate their efforts in one industry, resulting in a major push internationally and along upstream and downstream stages of their industry. If these firms were to spread their resources across industries, they would lose the power that comes with concentration. Large international specialists are therefore also able to advertise extensively within their product-segment and develop strong brand recognition with end-users.

Furthermore, this literature argues that their large size and volume encourage international specialists to perform multiple functions in-house. They have ample resources to conduct research. Because of their growing size and increasing vertical integration, large international specialists are able to exert increasing bargaining power with respect to their suppliers and distributors.

This may be part of the reason why a number of researchers (Rummelt, 1974; Berger and Ofek, 1995) have found that firms pursuing strategies of related constrained diversification are on average more profitable than other firms. Business diversification is often accompanied by over-investment in and cross subsidisation of poorly performing segments. In contrast, international diversification has been found to be positively related to firm performance (Geringer *et al.*, 1989; Hitt and Ireland, 1994; Tallman and Li, 1996).

9.4.3. Strategic positioning

While the previous perspectives implicitly argue for strategic configuration on the basis of internal commercial development of the firm, the strategic positioning argument views company integration as a function of potential market power and the rents accruing from it. The cost structure in terms of production and transaction costs, or in terms of the more widely defined co-ordination costs assumed by the capabilities tradition, is obviously a basic concern. However, the focus in the strategic positioning argument is rather on how acquisitions of new commercial activities will increase the firms' ability to acquire and maintain extra profits by limiting competition in its commercial domain (Porter, 1980; Hax and Majluf, 1991).

Competitive advantage from strategic positioning may accumulate through a number of mechanisms. Firms might wish to expand their boundaries to secure property rights to scarce resources, they might seek to acquire information advantages or they might want to create various forms of friction to impede imitative competition. Firms might acquire positions that are difficult to imitate, derived from size advantages, preferred access to resources or customers, and/or restrictions on competitors' options.

Related to the strategic positioning perspective is also portfolio-diversification and risk management. According to this perspective (Mintzberg *et al.*, 1988), firm organisation would be motivated by the need to spread risk across different markets and to secure a balanced/stable cash flow. Having positions in mature industries with stable revenue allows the firm to support ventures into expected future high-growth areas, which are not yet generating sufficient revenue to be self-sustaining.

Control over strategic positions in the value chain may therefore, in this perspective, be more critical to success than simple cost efficiency. Ability to influence decisions in monopoly parts of the energy- and telecommunication value chains may be seen as providing interesting strategic advantages even after deregulation. New specialised regulatory regimes for grid management may, after all, have deficiencies that give the incumbent strategic advantage. In particular, the so-called negotiated transit access will obviously have both information and potential commercial advantages for the incumbent company.

However, the costs of strategic positioning will obviously have to be offset against co-ordination costs as well as against regulatory risk. Furthermore, the portfolio diversification and risk management arguments are also dependent on the perfection of capital markets.

The greater the market imperfections, the larger are naturally the incentives to internalise risk management within the firm.

9.4.4. Capabilities and learning

Taking an organisation theory point of view, the so-called capabilities perspective has questioned both the production cost and transaction cost analysis as a basis for industrial configuration, and argues for breaking down the production cost–transaction cost partition. The capabilities literature argues that the problem of imperfect and asymmetric information is not only present in the realm of transactions, but also in the realm of production.

According to Langlois and Foss (1999) much knowledge – including, importantly, much knowledge about production – is tacit and can be acquired only through a time-consuming process of learning by doing. They further argue that the knowledge about production is often essentially distributed knowledge that is only mobilised in the context of carrying out a multiperson productive task, and not a single agent possesses that. This implies, they argue, that some sort of qualitative co-ordination – for example, through direction and command – is needed for its efficient use.

Taken to its logical consequence, the capabilities perspective leads to firm-specific competitiveness and thereby to firm-specific economic organisation and boundary setting. In a world of tacit and distributed knowledge and differential capabilities, having the same blueprints as one's competitors is unlikely to translate into having the same costs of production. In such a world, they therefore argue, firms will not confront the same production costs for the same type of productive activity.

Organisational capabilities may, however, also build bridges across the seemingly contrary positions on complex versus simplistic/focused strategic configuration. The downsides of the multiutility, pointed out by some of the economics literature, may be modified by organisational measures, e.g., development of a holding structure, where managers may improve cost control and quality performance by modularising the firm into focused business units with performance accountability at the service provider level.

9.4.5. Path dependency and national styles

Competitive advantage may, however, derive not only from firm level factors – including the firm's ability to take dominant market positions – but also from the national political economy of which the firm is a part. A national styles literature focuses on how differences in national, regional and sector institutions generate significant variations in firm and market structure and operation. This general argument is developed under several labels: business systems (Whitley, 1992), social systems of production (Campbell *et al.*, 1991) and modes of capitalist organisation (Orru, 1994). The essence of this literature is that industrial configuration may differ extensively between nations, as national industrial 'milieus' draw on specific traditions and competence in their national surroundings.

In the context of deregulated infrastructure industries, it seems natural to stress the interaction between the regulatory context and business strategy formation. National regulation may, for example, affect strategic configuration through the impact of the depth of the reforms in terms of business restrictions and profitability limitations, through direct mandatory changes of horizontal and vertical integration, through national champion policies, public ownership and financial benefits allowed by regulatory provisions.

Implicitly, and sometimes also explicitly, the national styles literature draws on a path dependency argument, asserting that industrial systems cannot develop independently of previous events and that local positive loops propagate traditional patterns into future strategic decisions (David, 1993). This interpretation implies that small events may play an important role in determining the future course of long-term development. Thus, the path dependency and national styles literature foresee that institutional, social and organisational factors will continue to produce national differences in strategic orientations, even under international competitive conditions.

The national style argument may be phrased both in efficiency- and non-efficiency terms. The efficiency argument is built on assumptions that differences in resource bases, competencies, organisational practices and institutional structures are based on different and equally competitive commercial strategies. A non-efficiency-based argument for national diversity might be that various organisational and institutional mechanisms specific to each national business style, function as filters to cushion international market forces, and shelter national business strategies from their cruel selection.

Either way, the national style perspective implies that nation-specific factors such as ownership patterns and nation-specific institutional arrangements will override the general factors highlighted in the production cost, transaction cost, capability and strategic positioning perspectives. The national style perspective therefore predicts nation-specific patterns of industrial organisation also in the energy industry.

9.4.6. Industry-specific factors

Besides the national economy, the industrial sector has also been defined in the literature as a locus of competitive differentiation. Masson and Bain have argued that market structures are specific from industry to industry (Bain, Masson and Qualls, 1976; Bain, 1968), with industry-specific variation in concentration and variation in entry barriers.

Taking a resource-based view of the firm, Rummelt (1984) and Mahoney and Pandian (1992) have pointed out that there are unequal returns in industries and that impediments block the flow of investments across industries so as to maintain sustainable differences.

The industry perspective is theoretically eclectic and relies heavily on empirical findings. Industry explains 20% of variance in competitive performance in the statistical analysis of the US firms. Powell (1990) argues that industry factors explain a significant proportion of the overall financial performance variance.

9.5. Towards an Empirical Analysis

The limited extent of deregulation at the present stage obviously limits the analytical relevance of perspectives based on strategic configuration under full competitive exposure. The analysis of industrial configuration in this section must therefore obviously also relate to the institutional specificities embedded in the regulatory design and practice and in the market construction and operation, which have been analysed in the previous two sections (see Sections I and II of this book).

The limited integration of a European competitive market for energy may, for example, obviously bias configurational focus more to national and regional domains with more homogeneous competitive pressure. However, the national style, path dependency and capability perspectives suggest maintenance of configurational diversity even under a level competitive playing field. These positions contrast with a competition and configurational convergence hypothesis, that may be constructed on the basis of the production cost and transaction cost perspectives.

The empirical analysis of industrial configuration is complicated by the availability of several vehicles of economic integration. As we have argued in the preceding section, some of the specialist types may be fairly assetless and could be integrated through trading/service, etc. The Enron trading model, for example, was considered to be one of the interesting prototypes, which implied value creation based on arbitrage between different energy markets with trading competencies and trading systems as the most important asset. Other types of integration might be stronger asset based and involve integrating broader value chains in electricity and petroleum as in the Finnish Fortum case, where the two state companies Imatran Voima (electricity) and Neste Petroleum were merged.

Taking the above analytical perspectives as a point of departure, the following two chapters explore the strategic configuration in European energy industry in a statistical and more casuistic-qualitative analysis, respectively.

References

Afuah, Alan (2000). Redefining firm boundaries in the face of the internet: Are firms really shrinking? *Paper Presented at the ICOS Seminar, University of Michigan 12/8.*

Bain, J.S. (1968). *Industrial Organization*. John Wiley, New York.

Bain, J.S., Masson, R.T. and Qualls, P.D. (1976). *Essays on Industrial Organization in Honor of Joe S. Bain*. Ballinger, Cambridge, MA

Berger, Phillip G. and Ofek, Eli (1995). Diversification's effect on firm value. *Journal of Financial Economics* Amsterdam Jan. 37(1): 39, 27.

Briones, Alex, McKenna, Matthew and McArthur, Neil (2002). Europe's retail markets: Building a better mousetrap? *Public Utilities Fortnightly* Arlington Feb. 1, 140(3): 34, 6.

Buhner, Rolf (1987). Assessing international diversification of West German corporations. *Strategic Management Journal* Chichester Jan/Feb. 8(1): 25, 13.

Campbell, John L., Hollingsworth, Roger and Lindberg, Leon N. (1991). *Governance of the American Economy*. Cambridge University Press, New York.

David, P.A. (1993). Path dependence and predictability in dynamic systems with local network externalities: A paradigm for historical economics. In Dominique Foray and Christopher Freeman, Eds., *Technology and the Wealth of Nations*. Pinter Publishers, London.

Geringer, J.M., Beamish, P.W. and da Costa, R.C. (1989). Diversification strategy and internationalization: Implications for MNE performance. *Strategic Management Journal* 10(2): 109–119.

Hax, Arnoldo C. and Majluf, Nicolas S. (1991). *The Strategy Concept and Process*. Prentice Hall, New Jersey.

Hitt, Michael A., Ireland, R. Duane (1987). Building competitive strength in international markets. *Long Range Planning* London Feb. 20(1): 115, 8.

Langlois, Richard N. and Foss, Nicolai J. (1999). Capabilities and governance: The rebirth of production in the theory of economic organization. *Kyklos* 52(2).

Mahoney, J. and Pandian, J.R. (1992). The resource-based view within the conversation of strategic management. *Strategic Management Journal* 13(5): 363.

Maksimovic, Vojislav and Phillips, Gordon (2002). Do conglomerate firms allocate resources inefficiently across industries? Theory and evidence. *The Journal of Finance* 57(2): 721–767 (47 pages) Libraries Worldwide.

Mintzberg, Henry James, Robert, M. and Quinn, James Brian (1988). *The Strategy Process: Concepts, Contexts and Cases*. Prentice Hall, Englewood Cliffs, N.J.

Orru, Marco (1994). *The Faces of Capitalism*. Paper presented at the Sixth Annual International Conference on Socio-Economics. HEC, Paris.

Oster, S. (1999). *Modern Competitive Analysis*. Oxford University Press, Oxford.

Porter, Michael (1980). *Competitive Strategy*. The Free Press, Macmillan, London.

Powell, Walter, W. (1990). Neither market nor hierarchy: network forms of organization. *Research in Organizational Behavior* 12: 295–336.

Robinson, Joan (1956). *The Accumulation of Capital*. Macmillan, London.

Rummelt, R.P. (1974). *Strategy, Structure and Economic Performance*. Division of Research, Harvard Business School, Boston.

Tallman, Stephen and Li, Jiatao (1996). Effects of international diversity and product diversity on the performance of multinational firms. *Academy of Management Journal* Briarcliff Manor Feb. 39(1): 179 (18 pages).

Whitley, Richard (1992). *European Business Systems: Firms and Markets in their National Contexts*. Sage, London.

Williamson, O.E. (1975). *Markets and Hierarchies*. Free Press, New York.

Chapter 10

Configuration and Performance of Large European Energy Companies: A Statistical Analysis

ATLE MIDTTUN AND TERJE OMLAND

The review of the literature on strategic configuration in Chapter 9 has pointed out several drivers with both complementary and partially contradictory implications. This includes a transaction- and productivity-cost argument, which determines firm boundaries according to transactions and production cost characteristics; a strategic positioning argument which sees company integration as a function of potential market power, and the rent accruing therefrom a capability argument which brings in knowledge and firm specific competencies as a basis for configuration; a path dependency and national style argument which sees national-specific factors such as ownership patterns and institutional arrangements as determinants of configuration; and finally an industry argument that sees industry specific variation in factors such as concentration and entry barriers as major drivers of configuration.

Taking this literature as a reference, this chapter explores patterns of strategic configuration and performance in European energy companies based on a statistical analysis of 50 firms taking the strategic options and theoretical perspectives on drivers of strategic configuration as a point of departure.[1]

[1]Based on Annual Reports from 1996 to 2001.

The analysis includes the largest electricity and gas companies in selected North and West European countries with significant presence in their national market.[2]

While many variables are more or less directly related to annual accounts published in annual reports from the companies, some of the configuration-variables are more complex and need further explanation. Vertical integration for electricity firms is calculated as the ratio of generation to supply or of supply to generation. For gas firms, the vertical integration ratio includes the ratio for generation to supply and control over transport and wholesale. Further definitions are given in an appendix. Given the dynamic evolution of the newly deregulated energy markets, we have designed the analysis as a panel study with samples from the yearly accounts of the three years 1996, 1999 and 2000.[3] A list of companies is given as an appendix.

The limited scope of our empirical study implies that we cannot explore all the above perspectives in depth.[4] They must rather be seen as heuristic frameworks against which we position our interpretations.

10.1. Exploring Patterns of Structural Configuration

One of the remarkable features of the structural configuration of European Energy industry is the pluralism of forms and engagements, which seems to contradict at least a strongly formulated universal production and transaction cost hypothesis. A variety of structural configurations of energy companies exists side by side along such core

[2]Given the explanatory approaches at three levels: the firm level, the national level and the industrial level, we have, with the limited resources available, made a compromise between the selection of large European firms and the need for a nationally diversified sample. In some countries there are only one or two major electricity or gas companies, so the number of companies represented can differ from country to country. We have sampled across both the electricity and gas/petroleum sectors, with a European focus, including both large members of Eurelectric and Eurogas.

[3]Mergers and acquisitions from the 1996, 1999 and 2001 sample implies that the 1999 and 2001 samples contain some companies not present in the 1996 list.

[4]Data are especially limited with regard to production cost, transaction cost and capability perspectives that demand extensive access to internal data and observation of the firm. We have, therefore, designed simplistic indicators for structural analysis, which only includes production and transaction costs indirectly, represented through their effect on strategic configuration/organisational structure, including such indicators as size, internationalisation, diversification and vertical integration. The national style perspective and the industrial arguments are, however, represented through selection of companies from nine West-European countries and across two sectors, electricity and gas.

structural dimesions as vertical integration, diversification, size and internationalisation. It is, for example, possible to be both specialist in one function, but also vertically integrated; to be diversified, but also sector focused. It is possible to be international but also nationally confined, and small as well as large.

Based on their positioning along the vertical integration and diversification axes, we have in the following analysis split our sample into four basic structural configurations: I functional specialists, II diversified specialists, III vertically integrated specialists and IV conglomerates.

A majority of companies can be characterised as belonging to a *mono-sectoral, vertically integrated group* (the III position in Fig. 10.1). The build-up of strong vertical integration is here presumably facilitated by a limited sectoral focus. Inherited planned economy positions, where sectoral focus sometimes was embedded in the regulatory conditions may, in some cases, explain this. However, there are obviously also functional production-cost and transaction-cost arguments that could be brought to bear here. This group includes companies like Endesa, EDP, EDF, Wintershall and GDF, all electricity and gas companies that are both generating and selling energy. Endesa, EDP and EDF are mainly focusing on the electricity market,

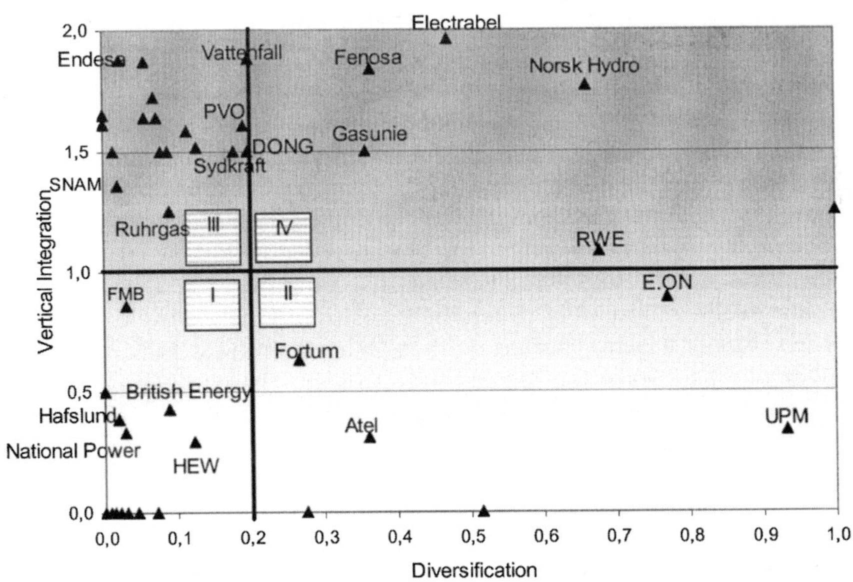

Fig. 10.1. Patterns of diversification and vertical integration in 2001.

while Wintershall and GDF have their main market in the gas/petroleum sector.

A few of the larger European energy companies, however, retain a *functional specialist position* (I in Fig. 10.1) indicating perhaps the virtue of focused specialisation, given a specialist position inherited from a pre-deregulation past. In this category we find companies like HEW, National Power, British Energy, Hafslund and Statkraft.

A number of companies can be defined within a group characterised by *extensive diversification and limited vertical integration* (the II position in Fig. 10.1). Resource needs and industrial focus arguments may obviously be brought in to explain why it may be rational to maintain limited vertical engagements when spreading out across multiple sectors. Examples here can be found in large firms like E.ON, UPM, Fortum and Atel. These companies are all into more than one sector. E.ON is a multiutility conglomerate with energy, chemicals and real estate division. The energy division is a multi-utility division, with engagements in water, natural gas and electricity. E.ON is however in the process of refocusing and has recently retreated from engagements in metal and telecom. However, E.ON is still expanding geographically through building/buying up daughter companies in Scandinavia, USA and Eastern Europe. Another diversified company, also belonging to this group is the Finnish company UPM, one of the world largest paper companies, which is also engaged in generating electricity both for internal use and to be traded on the open market. UPM here follows a strong Finnish tradition of integrating power supply into energy intensive industrial activity. The Finnish state company Fortum belongs to this category through their multienergy engagements in oil, gas and electricity.

Some companies also manage to combine *advanced vertical positions with considerable diversification* (the square IV position in Fig. 10.1). It should be noted, however, that many companies in this category remain in the moderately diversified position. One example is Norsk Hydro. Norsk Hydro started with a calcium nitrate factory, and a power plant to serve the factory with power. The company has later also come to include an oil and gas division, an agricultural division and minor divisions in seafood and petrochemicals. The seafood division has later been sold out. Another company that is represented in this position is RWE. They are diversified through their engagement in gas, electricity, oil, water, waste and some service divisions. RWE's position as a vertically integrated company is explained by their activity both in generation and supply.

10.2. Size and Internationalisation

Even in a sample of the largest European energy companies, one may observe large variation in size and internationalisation. As a general pattern, the smallest companies are mainly focused on their home market, the medium sized companies fall in the medium range when it comes to internationalisation and the largest companies are the most international.

The German companies E.ON and RWE are both large companies with some degree of internationalisation. E.ON gets almost half of their income of about 80 000 million euros outside Germany. RWE and EDF follow E.ON in size. These two companies have between 30 and 40% of their income from abroad. On the other hand, smaller companies like Norsk Hydro and UPM have around 90% of their income outside their home market. The main group of companies in our material, fall into the category below 6000 million euros in revenues and under 10% of their income outside their national market.

Relating size and internationalisation in Fig. 10.2 to the four classes of combinations of vertical and horizontal integration in Fig. 10.1, we can observe that the largest companies are found in the IV position (vertically integrated and diversified) which is obviously the most resource-demanding position. These are also the most international

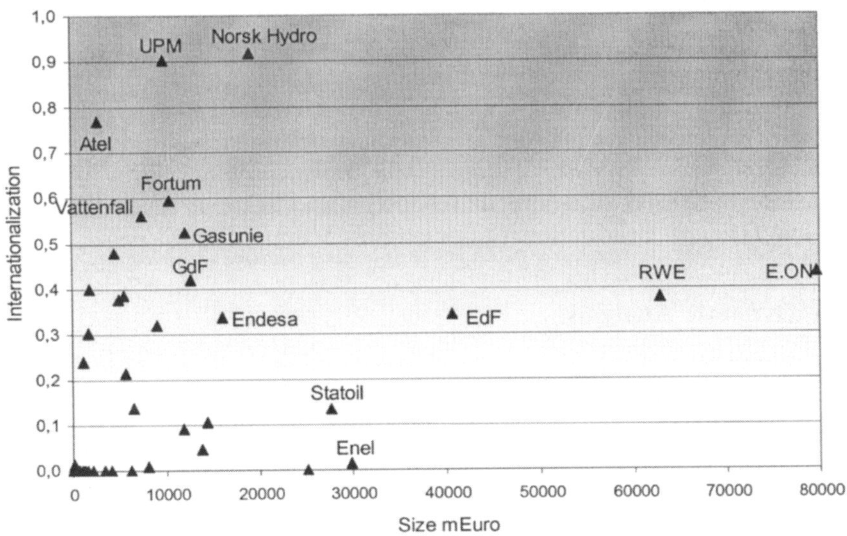

Fig. 10.2. Size and internationalisation 2001.

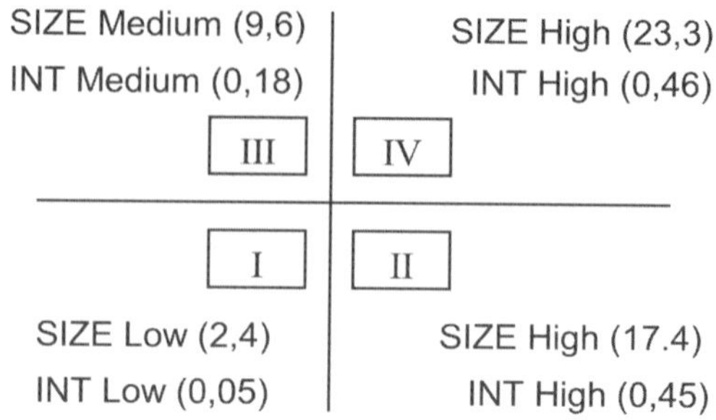

Fig. 10.3. Diversification, vertical integration, size and internationalisation 2001. Size in 100 million Euro and internationalisation = earnings in foreign markets as a share of earnings in national markets.

companies, diversified, but with limited vertical engagements. The smallest among the large European companies are to be found in position I (specialised both sectorally and in terms of vertical integration). These are also the least internationalised. See Fig. 10.3.

From a theoretical point of view, this observed configurational pluralism can be interpreted in several ways. First, it may be taken as evidence for strong path dependency of organisational capabilities and contexts under the pre-deregulation national and regional monopoly regimes, which still linger on through implicit trade barriers. Second, it could also be taken to reflect diversity of resource bases that motivate build-up of different systems with different organisational implications. In the first case, varieties of configurational forms would reflect the limited competitive pressure under monopolistic planned economy and its continuation in less explicit but still efficient trade barriers, which allows efficient and non-efficient forms to exist side by side. In the second case, a variety in configurational forms would rather be based on diversity of resources and competencies that can achieve efficient production under common market exposure.

10.3. Dynamic Analysis

The dynamic picture of structural development in our observation periods – the late 1990s and the beginning of the new century – is complex, but with several discernable patterns (Fig. 10.4) indicating again that a simple efficiency driven functional hypothesis, based on production and transaction costs is hard to justify.

Figure 10.4 divides configurational development into first dynamic period of 1996–1999 (with solid lines) and a second period of 1999–2001 (with broken lines). The first period exhibits three dominant

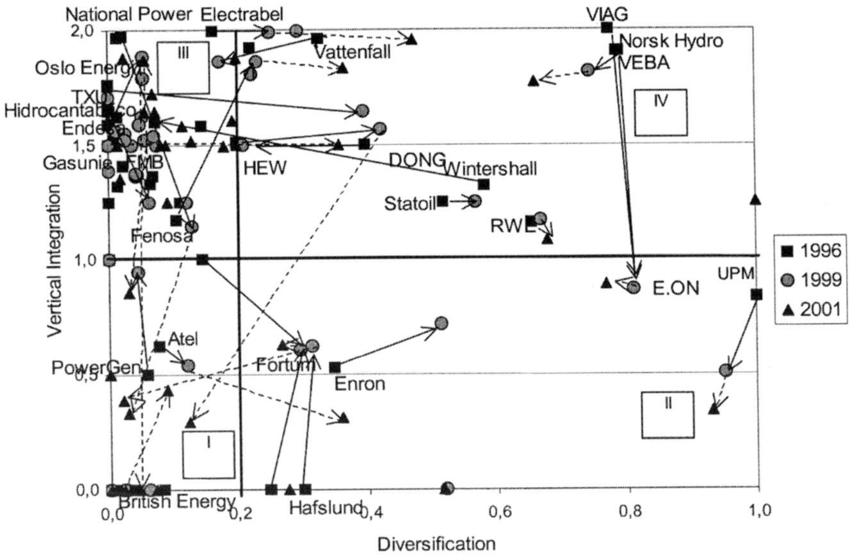

Fig. 10.4. Vertical integration and diversification in a dynamic picture.

patterns as far as diversification and vertical integration is concerned: the first pattern, which characterises above all the square III configuration – the vertically integrated, mono sectoral firms – is one of stability. Many of these firms retain their pre-deregulation configuration throughout the first period studied. Examples in this category are Enel, EDP and Birka.

A second pattern is one of diversification and moderate deverticalisation, which characterises a number of firms moving out of the square III position and into the IV quadrant. Examples are Fenosa, TXU, Electrabel and E.ON. E.ON's deverticalisation relates to the fact that it has bought up companies and thereby increased its generation of electricity, while its supply of electricity to end user has remained fairly steady.

The third pattern is one of vertical integration, typically characterising companies moving up from the lower end off the I and II quadrants, but also, in one case from a quadrant III position. The move towards stronger vertical integration is often also accompanied by moderate move towards diversification.

Finally, in the first period, there are a few examples of companies moving out of diversified or vertically integrated positions. The most dramatic move back to sectoral specialisation from a diversified position was made by Wintershall, which took a large step from a diversified position to a more specialised, vertically integrated position as the company split out its petrochemical industry division, which stood for more than 50% of Wintershall's total revenue. The result was that the oil and gas division, in 1999, stood for more than 95% of the revenue, compared to around 40% in 1996.

Also the second period (1999–2001) exhibits a large variety of strategic configurations. Some of the patterns that were seen in the first period are continued, but some new developments have also emerged.

Companies like FMB, HEW, National Power, Powergen and Oslo Energi all specialised functionally by reducing their vertical integrated position from 1999 to 2001. In the case of HEW, their supply activities in 2001 grew to five times their 1999 position. This increase was essentially due to the inclusion of the newly acquired VEAG's electricity sales as well as to the strong increase in the volumes in the HEW power business. The company's generation capacity remained fairly stable in this time period. Oslo Energi is another example of functional specialisation in this period. The company sold out its distribution division in 2001 and retained an exclusive focus on generation. During this process the name Oslo Energi followed the distribution division, and the generation division took the name E-CO Energi.

Some companies continued diversifying throughout the 1999–2001 period, but they were fewer than in the pervious period. Electrabel and Fenosa are two of the corporations that have followed this pattern and moved towards a more diversified position, from the late 1990s into the first years of the new century. Increased influence of the natural gas division explains the move by Electrabel. For Fenosa, diversification is explained through its international investments, mainly in Latin America.

The functional and sectoral specialists that focus on one sector and in one part on the value chain, have generally retained their position throughout the second period.

As far as size and internationalisation are concerned, the main trend has been to grow in both directions throughout both periods, although scaling up dominates over internationalisation in the first period (Fig. 10.5). VIAG and VEBA (merged to E.ON) and Enron stand out as the prime examples of scaling up on both dimensions in the first period. Enron tripled its size in three years and also became more international. RWE, Statoil and EDF also expanded in size in the 1996–1999 period. On the other hand GDF, together with Gasunie have decreased their international activity. E.ON has increased its international engagement and now has half of its income from outside Germany. E.ON also illustrates the strong European movement towards consolidation into larger energy companies.

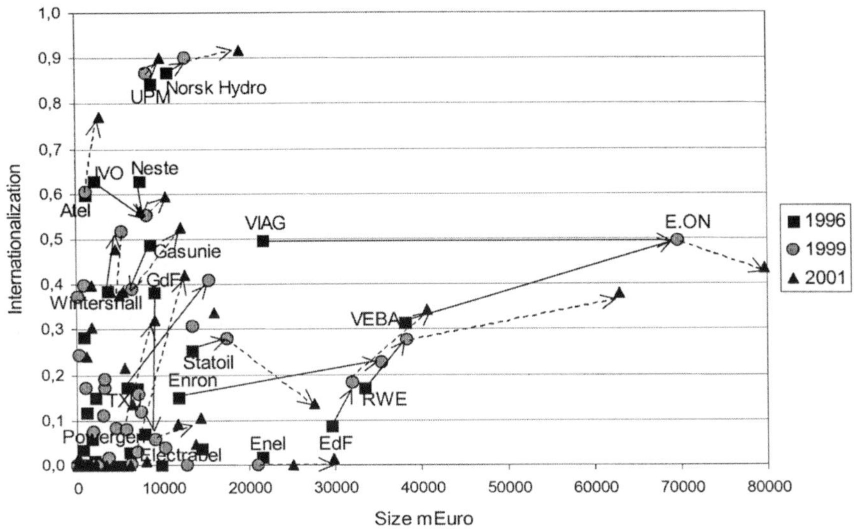

Fig. 10.5. Size and internationalisation in a dynamic picture.

The second period follows the first when it comes to structural configuration. Several companies have continued to expand in size and also to increase their international activities. The large German companies, E.ON and RWE, have both continued to scale up, but have taken different directions as far as internationalisation is concerned. E.ON saw a decrease from 1999 to 2001, while RWE increased its international engagement from approximately 30 to 40% in this period. Mid-size companies have internationalised even more. Gasunie, Atel, Electrabel and Powergen have all expanded extensively outside their national home base. In the case of Electrabel it is mainly its trading activities that have led to increased international engagement. Powergen has increased its position in the US and more than 30% of the company's revenue was related to this market in 2001 compared to approximately 8% in 1999. On the other hand there are also exceptions such as Hafslund, Statoil and Wintershall, which have all reduced their international engagement, but for different reasons. Hafslund's generation plants in USA did not give the same profitability in 1999–2001 as in 1996–1999 due to low electricity prices. This had also an impact on the overall result for Hafslund, which saw their total income fall from 1999 to 2001. In the same period Statoil and Wintershall had an increase in total revenue, but the growth was mainly created in their home markets.

10.4. Exploring Patterns of Economic Performance

Like structural configuration, economic performance also varies extensively between the large companies, indicating large variations in functional market pressure. Presumably this leaves considerable scope for other factors to play a major role in determining commercial outcomes.

Economic performance is a complex multidimensional concept and may refer to different aspects and time-periods. While return on assets remains a bedrock measurement of profitability and financial performance, other analytical techniques have been developed to evaluate the increasingly complex and diverse nature of business strategies and operations. Return on equity is one obvious and important additional measure. But there are other important performance and productivity measures as well, including operating earnings per full time equivalent employee, efficiency ratio, pre-tax operating margin etc.

We shall in the following confine ourselves to traditional balance-sheet variables such as return on assets (ROA) and return on equity (ROE), but we have also calculated indicators for labour productivity (LAP). For the energy companies that are listed on the stock exchange we have also calculated price earnings ratios (P/E ratio).

10.5. The General Pattern

On average, our sample of large energy sector companies in Europe generally performs at a medium level (Killion 2003), measured by standard indicators such as return on assets and return on equity. The average return on assets (ROA) is 5.8% and the average return on equity (ROE) 18.4%.[4]

Within the energy/utility industry, there are considerable differences between the gas- and electricity sectors in terms of return on assets and return on equity, but the differences are not excessive. Generally speaking, the gas industry outperforms electricity industry on the average by 10.6% versus 3.7% ROA and by 28.5% versus 14.1% for ROE (Fig. 10.6 and Table 10.1). The higher ROA and ROE figures for gas and electricity may reflect less regulatory pressure on parts of the gas industry, particularly the interface with the upstream side, or a different composition of activities.

Due to Wintershall's extraordinary earnings in this period, we have used the median to show the differences between the electricity and

[4]Both the return on assets and the return on equity ratio are measured by using return before tax (Eklund and Knutsen, 1997).

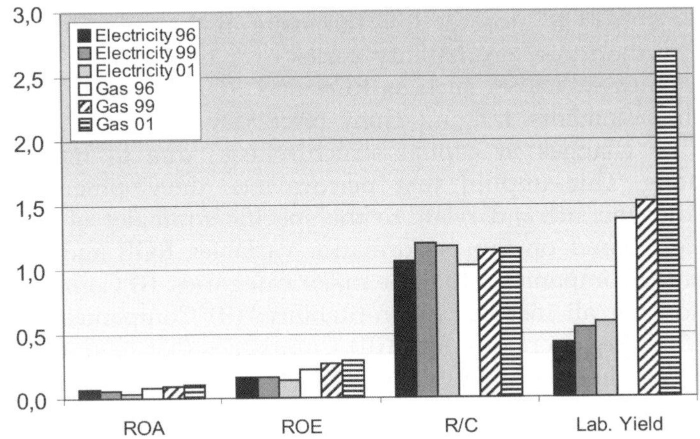

Fig. 10.6. Economic performance measures for electricity and gas industry 1996 and 1999, medians.

Table 10.1. Median profitability over time in our material.

	1996		1999		2001	
	Electricity	Gas	Electricity	Gas	Electricity	Gas
ROA	0.051	0.072	0.049	0.062	0.030	0.077
ROE	0.142	0.185	0.132	0.202	0.115	0.217
Rev./Cost	1.186	1.102	1.143	1.102	1.101	1.072
Lab.Yield	0.409	0.627	0.466	0.900	0.531	1.659

gas sector. Even if we compare the two industries without Wintershall, however, gas will still be ranked higher. The most striking differences, however, come in the field of labour yield per employee, where the gas-sector outperforms electricity by a factor of 3 to 1. The relatively stronger and more labour intensive downstream engagement in retail distribution in the electricity industry, compared to gas, may be one explanation. However, the lower degree of politicisation and regulation in the gas industry, may also possibly account for a stronger commercial focus in personnel policy.

10.6. Dynamic Analysis

The general energy industry patterns exhibit extensive performance variations between firms. The ROAs in our sample varies from almost 0 to more than 17%; and the variation in ROE is from 0 to more than

50%. The spread is more or less the same in the electricity and gas sectors. Furthermore, profitability varies over time.

Performance variables such as ROE and ROA mirror a number of underlying elements ranging from price developments in product markets to changes in capital structure e.g., due to mergers and acquisitions. This implies that performance development may be highly idiosyncratic and relate to the specific strategies of individual companies. Based on the performance variables ROE and ROA, we may sort the companies into three major categories: (I) Companies that have had over all increase in profitability; (II) Companies that have had profitability decreases, and (III) Companies that have maintained relatively stable profitability. For both electricity and gas industries, we shall, in the following, comment on selected cases from each of these groups to indicate the complexity of the underlying patterns. The analysis of individual companies is undertaken against the backdrop of general price development in product markets, characterised by dramatic changes for both sectors (Fig. 10.7).

First, among companies in the electricity sector with a positive profitability development, we find companies like Statkraft, FMB and Enel. In the case of Statkraft, their positive development is partly due to a one-time compensation from E.ON for the termination of an agreement to build a cable between Germany and Norway. This extraordinary payment implied a strong positive development of financial ratios for the period 1999 to 2001.

Like Statkraft, both FMB and Enel also increased their profitability during the 1990s but mainly through better return figures, as compared to a fairly stable asset and equity-base mainly in the second

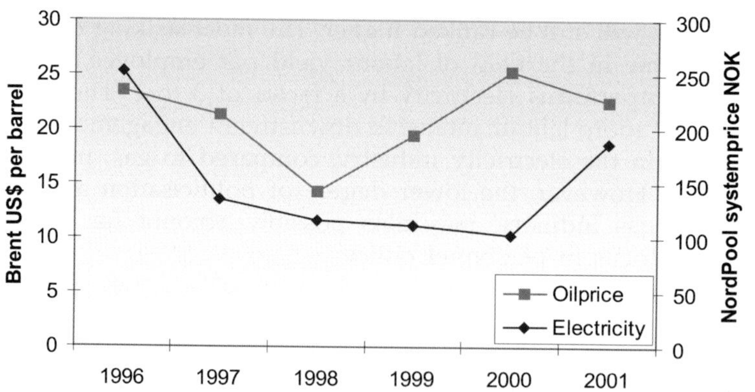

Fig. 10.7. Oil- and electricity prices. Gas prices have traditionally been linked to the oil price.

period. There are also companies that have seen an increase in their ROE, but not in their ROA. In this category HEW and EPZ stands out as the two most typical cases. As far as the German company HEW is concerned, they have managed to maintain a high level and even increased profitability throughout both periods. The large increase from 1999 to 2001 came mainly because of a tripling of results from 2000 to 2001.

The extensive profitability increase in EPZ comes out of the split of the company in May 2001. The power stations in the provinces of Noord-Brabant and Limburg were transferred to the acquiring company, Essent Energie Productie B.V. (including the stations staff). Hence, 2001 is EPZ's first year of operation as the new company.

Second, among companies that had a negative profitability development, we find companies like Powergen, National Power and British Energy. All these companies do, however, display different patterns of development. National Power saw a stable result in 1999 compared to 1996. From 1999 to 2001 they reduced their ROA, but at the same time their ROE increased. The decrease in ROA is due to a lower result in 2001, given a stable asset base. In the same period the company's equity base was reduced to only one third of the 1999 volume. One reason for these changes was the split of National Power split into Innogy and International Power.

Powergen started out with a positive development between 1996 and 1999, but then saw a dramatic profitability setback between 1999 and 2001. The decrease in ROA from 1996 to 1999 is related to an increase in debt, which explains why the changes were larger in ROA than ROE. In the 1999–2001 period, Powergen did not manage to show normal results, mainly because of some extraordinary costs incurred through their business reorganisation and restructuring programme.

The French state company EDF also came out with negative results. In the period between 1996 and 1999 their ROE dropped, but at the same time the company saw an increase in ROA. Some of the bacground for this development was a restructuring of EDF's balance sheet according to a plan signed with Government in 1997. Through this plan EDF acquired the general transmission system in France and the state interest in concessionary facilities of FRF 18 346 was reclassified as a capital contribution of FRF 14 119 million, which gave the company a revaluation surplus in equity of FRF 2 522 million. This increase in equity resulted in decrease in the ROE throughout the first period. In the second period, the ROA was also reduced and the company scored relatively low on both indicators. These transactions as well as acquisitions financed through their debt and equity could explain EDF's reduced results.

Third, in the middle range in terms of ROA/ROE we find companies like Electrabel, Iberdrola, Fenosa and Sydkraft. Fenosa made a positive move in 1999, but this was mainly due to extraordinary incomes through sales of their telecommunication industry and shares in Unión Fenosa Generactión. After these extraordinary incomes in 1999, Fenosa saw a reduction back to 1996 level in 2001. Electrabel, Iberdrola and Sydkraft maintained fairly stable profitability in all three years. They have all increased their assets base, but managed to keep the return grow at the same pace (Fig. 10.8).

Gas companies also show great variation as far as profitability is concerned. Wintershall, Statoil and BG Group all score high on profitability increase from 1996 to 2001. For Wintershall the result was partly due to structural moves, which resulted in major extraordinary incomes in 1999. As mentioned before, they sold out some of their activities and narrowed their focus. These extraordinary incomes in 1999 explain some of the profitability reductions from 1999 to 2001, but even so, the 2001 results are much better than in 1996.

Statoil's profitability has also grown over the 1996–2001 period, but with varying direction in the 1996–1999 and 1999–2001 period. As seen in Fig. 10.7, there was a dramatic drop in oil prices that later led to restructuring of petroleum companies. This restructuring cost Statoil extensively and resulted in a higher growth rate in assets than in return. From 1999 to 2001, the restructuring paid off and coincided with an increase in oil price. This combination produced very good financial results.

The BG Group, which also displayed over all positive results, made several strategic decisions since 1996. In 1997 British Gas went for a demerger of Centrica and was renamed BG Plc. Late in 1999 BG Plc. completed a financial restructuring which resulted in the creation of a new parent company the BG Group. Even this company ended up with a demerger where Lattice Group was split out as a separate company in 2000. The financial outcome for the BG group can to some extend be related to these structural changes but also to the growth in product prices.

In the group of companies that decreased profitability from 1996 to 2001 we find among others the Danish petroleum company DONG and the German company Distrigas. In the period 1996 to 1999 DONG invested heavily into engagements that did not pay off right away. In the second period the investments also produced higher return but not to the 1996-level. Distrigas built up an asset-base through investments in pipelines connected to Zeebrugge in 1997 and 1998 and this

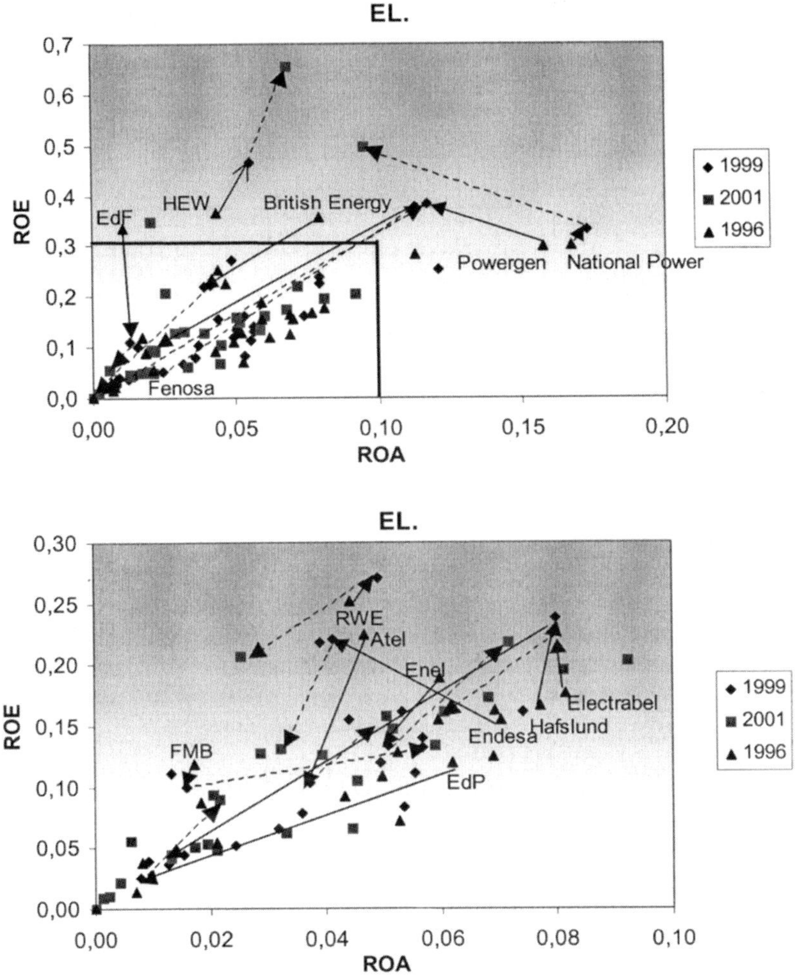

Fig. 10.8. A. ROA and ROE in electricity industry, high performance companies. B. ROA and ROE in electricity industry, low performance companies.

resulted in a good year for Distrigas in 1999. In 2001 Distrigas was demerged into a natural gas transport company (called Fluxys) and a company focusing on natural gas supply and sales and on international capacity trade (called Distrigas). This changed the asset-base and income figures for Distrigas resulted in a negative trend compared to the figures they operated with in 1999.

Similar to electricity companies, some gas companies have also maintained stable profitability from 1996 to 2001. However, the long-term stability may also be decomposed into short-term variation. An example here is the Norwegian conglomerate Norsk Hydro which reduced their profitability between 1996 and 1999. This was partly due to low oil prices in 1998 and partly due to a costly restructuring. Norsk Hydro had to live with low prices both on the fertilizer and the petrochemical market in 1999. By 2001, however, Norsk Hydro managed to get back to the 1996 profitability level. However, the company has clearly embarked on a round of further focus and demerger (Fig. 10.9).

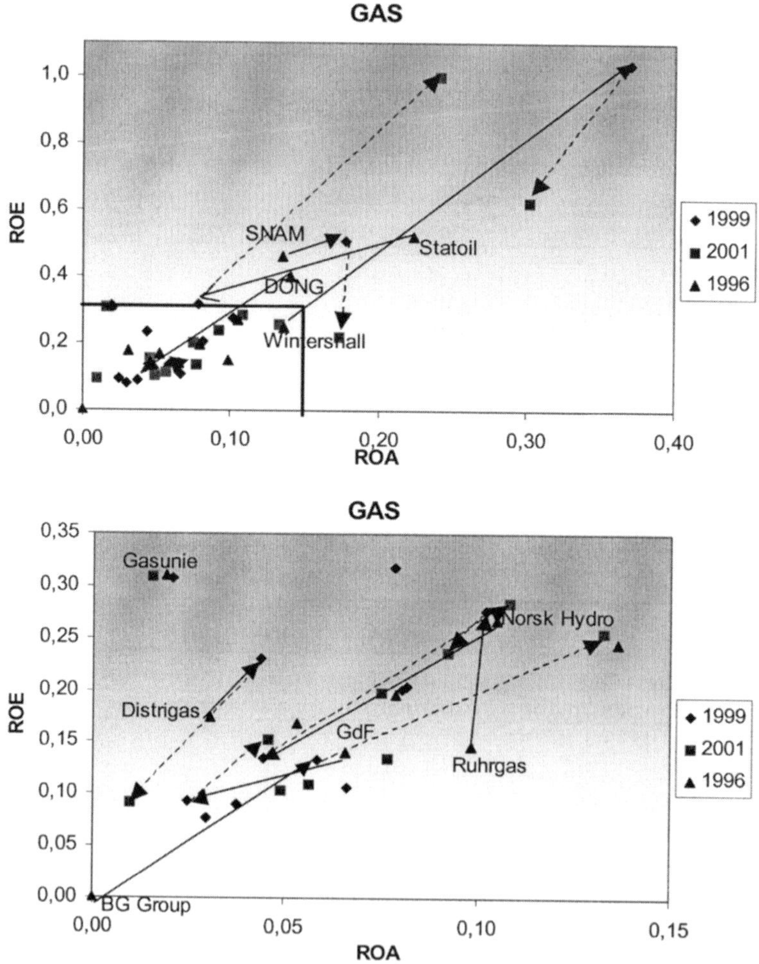

Fig. 10.9. A. ROA and ROE in gas industry, high performance companies. B. ROA and ROE in gas industry, low performance companies.

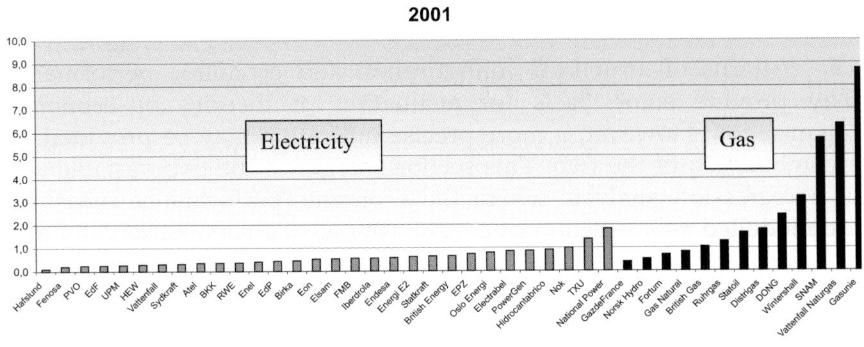

Fig. 10.10. Yield per employee 2001.

Variation in yield per employee, as indicated in Fig. 10.10, is even greater than the ROA and ROE variation. Labour productivity is at its peak in electricity and gas companies like Gasunie, National Power, Vattenfall, Wintershall, Dong, whereas it is far lower in companies like Hafslund, EDF, GDF, Norsk Hydro and Fortum.

10.7. Price Earnings Variation

Considerable spread in the price-earnings ratio also characterises electricity and gas companies listed on the stock exchange. At the top end several UK companies are represented, together with Enron (now out of business), Suez Lyonnaise and Hafslund. Further more five of the top seven has seen an increase from the previous periods. In the Iberian companies one can observe a downward trend from the top year of 1999 which can probably be related to the economic crisis they where facing in Latin America (Fig. 10.11).

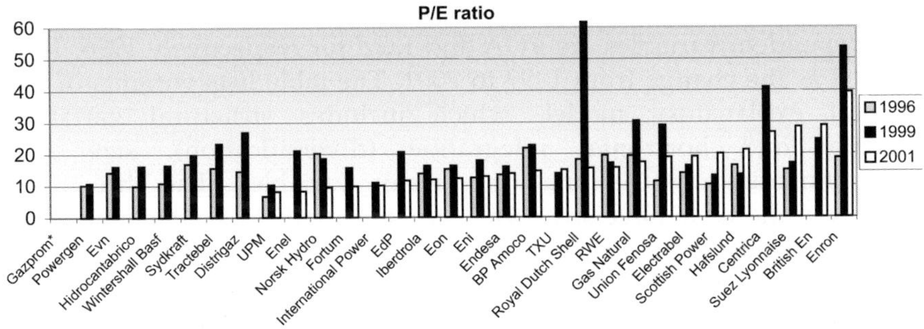

Fig. 10.11. Price-earnings ratio. Shell has a P/E-ratio at almost 330. in 1999. This is not shown because of the scale of the figure.

10.8. Configuration and Performance

The patterns of structural configuration and economic performance may provide some basis for evaluation of theories of economic organisation. However, a more precise indication may be provided by a combination of the two. This section integrates the two variable sets in a regression analysis where the economic performance measures are modelled as a function of organisational structure, nationality and industrial sector variables. The panel data allows us to run the model for three periods in time, and to develop indicators of change as input in the analysis.

From a theoretical point of view, production cost and transaction cost theory would lead us to expect that optimal combinations with higher efficiency would emerge through competitive selection. The national style argument, on the other hand, would lead us to expect variation in business configuration and profitability along national dimension, either because of competitive advantage of peculiar national factor-combinations/institutional factors, which imply different forms of competitive advantage in each national context, or because of implicit or explicit national protectionism. The industry argument, on the other hand would entail specific effects of gas/petroleum and electricity sector on profitability. The capabilities explanation implies more firm-specific factors play a central role. Since we do not have relevant capability indicators in our analysis, this may be seen as a residual explanation, in our case.

10.8.1. Emergent effects and shifting explanations

As an over all observation, the empirical analysis presents us with a picture of emergent effects and a shift from national towards more functional configurations. Table 10.2 shows the results of regression analyses of national, structural and sector factors against profitability. The dependent variables are ROA and ROE for respectively 1996, 1999 as well as the change from 1999 to 2001. The table lists variants of the general explanatory model, which includes structural variables (vertical and horizontal integration (diversification), size and internationalisation), national and sector variables. The three first models in the table have structural variables with 1996, 1999 and 2001 values, respectively. In the three last models, the independent variables are changes in the structural variables (vertical and horizontal integration (diversification), size and internationalisation) from the periods 1996 to 1999, 1999 to 2001 and 1996 to 2001. For further details see appendix.

Table 10.2. The regression models with R^2 and *p*-values.

	ROA			ROE		
Model runs with:	1996	1999	1999–2001	1996	1999	1999–2001
1996 structural variables						
1999 structural variables					0.338 **0.038**	
1901 structural variables						
1996–1999 strucutral change		0.311 **0.070**			0.475 **0.010**	
1999–1901 strucutral change			0.437 **0.007**			0.467 **0.006**
1996–1901 strucutral change						

The table gives the explanatory power (R^2) and *p*-values for all models that provide significant results. As we can see, the explanatory power of the models arises only towards the latter part of the period (1999 and 1999 to 2001). Apart from a significant correlation with 1999 return on equity in a model run with 1999 structural data, the model runs with structural changes are those that carry most explanatory power, both with respect to static profit (ROA/ROE) in 1999 and dynamic development of profit (ROA/ROE) in the 2001–1999 period.

A plausible interpretation of these results may be that profitability becomes systematically targeted in industrial organisation only as the deregulation process develops, whereas the profitability focus may have been far less important under previous planned economy conditions.

Table 10.3 shows how each factor in the explanatory model, contributes to profit or change in profit. Variables 1–13 are national dummies, variable 14 is an industrial variable, variable 15 represents type of ownership and variables 16–19 are structural variables. The table gives *b*-values and *p*-values for ROA and ROE values for 1999, and change in values between 1999 and 2001. As in Table 10.2 it has different combinations of dependent and independent variables for different time periods.

We can observe a time development where the explanatory strength of different factors shifts from the first to the second period. In the first period, the national style factor seems to play a dominant role. Both in the 1999 structure against ROE and 1999–2001 structure against 1999 ROA and ROE, the configurational variables such as vertical integration, size and internationalisation carry little explanatory power. This indicates that national style/institutional specificities in the national context at this stage did have a stronger role in shaping profitability

Table 10.3. Specific results of regression analysis.

	ROA values 1999		ROA values 1999–2001		ROE values 1999		ROE values 1999		ROE values 1999–2001	
	Model runs with 1996–1999 structural change		Model runs with 1999–2001 structural change		Model runs with 1999 values		Model runs with 1996–1999 structural change		Model runs with 1999–2001 structural change	
	Beta	Sign.	Beta	Sign.	Beta	Sign.	Beta	Sign.	Beta	Sign.
1. Norway	.614	.098	−.438	.029	.318	.151	.613	.061	−.406	.041
2. Sweden	.423	.181	−.229	.183	.201	.301	.377	.173	−.212	.213
3. Finland	.470	.126	−.162	.298	.134	.458	.412	.125	−.183	.237
4. Germany	.585	.081	−.295	.101	.934	.000	.674	.025	−276	.120
5. The Netherlands	.182	.511	−.235	.148	.214	.221	.346	.159	−.160	.317
6. Belgium	.399	.149	−.007	.967	.267	.138	.451	.066	−.019	.912
7. Switzerland	.293	.317	−.226	.154	.063	.739	.315	.220	−.246	.121
8. France	.408	.122	−.108	.480	.249	.191	.307	.179	−.103	.499
9. Spain	.628	.112	−.219	.225	.453	.047	.716	.042	−.239	.184
10. Portugal	.157	.434	−.073	.592	.070	.647	.155	.380	.032	.814
11. Italy	.549	.035	−.140	.326	.483	.012	.544	.019	−.140	.322
12. UK	.633	.076	−.141	.432	.431	.040	.639	.042	−.125	.481
13. USA	.637	.130	−.283	.131	.174	.339	.524	.152	−.235	.125
14. Electricity	−.113	.500	.037	.782	−.096	.529	−.111	.449	.050	.704
15. Stateown	−.175	.458	−.022	.906	−.120	.577	−.179	.387	.011	.953
16. Size	−.125	.702	.438	.007	−.448	.040	−.048	.867	.424	.007
17. Vert	.015	.923	−.380	.018	−.008	.958	.135	.326	−.358	.023
18. Div	−.698	.003	.273	.104	.061	.796	−.659	.001	.292	.082
19. Inter	.140	.443	−.449	.009	.337	.087	.050	.754	−.435	.010

*1–13: National variables; 14: industrial variables; 15: ownership variables; 16–19: structural variables.

than functional configuration. The exception is diversification, which already at this point seems to have a clear explanatory effect.

In the next period, however, the configurational factors seem to have played a stronger role, and we can observe factors like size, vertical integration and internationalisation having a major explanatory effect in the models with 1999–2001 variables against the 1999–2001 development in profitability (ROA and ROE). A plausible interpretation may be that this development reflects a gradual transition towards configurational optimisation in a more commercial environment coming out of emergent competition in newly deregulated markets.

As far as the national factors are concerned, the 1999–1996 models show that German and Italian contexts in particular correlate favourably with profitability

As far as configurational factors are concerned, we have already noted that change in diversification in the first period (1996–1999) had a major explanatory effect on 1999 profitability. The effect was clearly negative, both for ROA and ROE, indicating that diversification strategies in this period clearly did not have a short run payoff.

The explanatory power of the configurational variables in the later period (1999–2001) on profitability development in the same period is particularly related to size, vertical integration and internationalisation. While increase in size (measured as revenue) had a positive effect on development of profitability in the same period, the effect of internationalisation and vertical integration was negative, both for ROA and ROE. Diversification, in this period, had only moderate explanatory effect. What we are seeing is, in other words, an emergent advantage of growth in size and emergent disadvantages of vertical integration and internationalisation.

The lack of strong effects in the early period and the changes in explanatory factors should caution against drawing strong conclusions. The broader picture is one of configurational pluralism, and of evolution of effects of configurational as well as national factors on profitability. Industrial organisation in the energy sectors, during a period of deregulation apparently allows a pluralism of forms to co-exist. National institutional contexts clearly play a significant role in the first period after deregulation, probably reflecting the limited internationalisation of the European grid-bound energy markets in the first period after deregulation. The next period displays stronger effects of industrial configuration on profitability change. Interestingly, neither public ownership nor industrial sector seems to carry explanatory power as far as profitability is concerned.

10.9. Concluding Comments

Recapitulating our theoretical point of departure, we have juxtaposed five perspectives on economic organisation: the production/transaction cost, the strategic positioning, the national style, the industrial and the capability perspectives. At a more operative level, we have assumed that if production-and /transaction cost were major driving forces, this would create a functional pressure towards crystallisation of best practice, best structure and standardised performance. On the other hand, we have assumed that the strategic positioning hypothesis would be supported if we observe scaling up of potential market power without obvious scale-, scope- or performance advantages. The national style and industry arguments simply imply that we should find systematic variation in configuration and performance along national and industrial dimensions. The most difficult theoretical argument to evaluate empirically is the capabilities argument, because it raises arguments that are difficult to measure without further micro-data. We have therefore left this argument to be discussed as a residual factor.

A main finding regarding structural configuration of European Energy industry is the pluralism of forms and engagements, which seems to contradict a strongly formulated universal production and transaction cost hypothesis. We find a variety of structural configurations of energy companies side by side along such core structural dimensions as vertical integration, diversification, size and internationalisation. It is, for example, possible to be profitable both as a specialist in one function, but as a vertically integrated company it is also possible to be diversified, but also sectorally focused and still retain profitability. Furthermore, it is possible to be international, but also nationally confined and small as well as large. Similarly, we find extensive variety in performance on several performance indicators sustained across our European energy company sample.

This pluralism in configuration and performance can be interpreted in several ways. First, it may be taken as evidence for strong path dependency of organisational capabilities and national contexts developed under the pre-deregulation national and regional monopoly regimes. Second, it could also be taken to reflect diversity of resource bases that motivate build-up of different systems with different organisational implications. Given the diversity of resources and competencies, one could argue that configurational variety might be sustainable both under competitive as well as non-competitive conditions. Under competitive conditions sustainable diversity would be based on the specific competitive advantage of special resources and competencies. Under non-competitive conditions, configurational variation would allow efficient and non-efficient forms to exist depending on the size of the trade barriers and on the policy focus within the segmented national planning systems. The high diversity not only in structural configuration, but also firm performance supports the interpretation that a considerable part of the variation may be non-competitively driven.

More specifically, our analysis of structural configuration has uncovered not only structural diversity, but also structural moves in different directions. This suggests that functional convergence is not only lacking in existing organisation, but also characterises current structural adaptation under deregulation. Whether this is due to competitive sheltering or to variation in capability and resource bases is, of course, difficult to tell.

When it comes to size and internationalisation, we have found that the main trend is to grow in both directions. Given the dominant position of some of the up-scaling companies like EON and EDF, it is likely that market power may be a dominant concern. The finding strengthens the assumption that size is negatively correlated with scaling up and particularly with differentiation.

Our sample also provides clear evidence of national differentiation, and demonstrates that German and Italian backgrounds are important determinants of company size. This obviously gives considerable support to the national style hypothesis. However, surprisingly, public or private ownership does not carry explanatory power, neither in terms of structure nor performance.

From our analysis we can conclude that the industrial hypothesis is also supported. We have documented that the gas industry outperforms electricity industry by 8.7% versus 5.4% ROA and by 26.5% versus 16.9% for ROE. The higher ROA and ROE figures may reflect less regulatory pressure on parts of the gas industry, particularly the interface with the upstream side. The most striking differences, however, come in the field of labour yield per employee, where the gas-sector outperforms electricity by a factor close to 3 to 1. In any case, these differences reflect extensive differentiation of market structures and capital flows across the two energy sectors.

We have also seen that ROA, in our sample, varies from almost 0% to more than 17%; variation in ROE from 0% to more than 50%. The spread is more or less the same in the electricity and gas sectors. The extensive differences in economic performance between companies indicate substantive market segmentation and/or strong capability-based differentiation.

However, we also find traces of convergence. We have found that our model acquires more explanatory power throughout the period. This might be interpreted as reflecting a development from national planned economy where economic results were strongly shaped by national and regional policies towards rivalry under international market-economy where firms are gradually more competitively exposed.

A plausible explanation for the stronger explanatory power of the model based on structural change rather than static structure may be that the change indicates commercially related structural moves whereas the structure as such is largely inherited from the previous monopoly or planned economy. Nevertheless, the signs of functionally driven convergence under common market pressure are relatively weak when compared to the extensive diversity of company-performance and organisational configuration. This lends support to the claim that nation and industry-specific factors continue to play a major role supplemented by strategic moves towards oligopoly positions as well as probably a considerable portion of capability-driven economic organisation.

Appendix:

List of companies included in the statistical analysis

Atel
Birka
BKK
British Energy
British Gas
Distrigas
DONG
EdF
EdP
Electrabel
Elsam
Endesa
Enel
Energi E2
Enron
EON
EPZ
Fenosa
FMB
Fortum
Fynsværket
Gas Natural
Gasunie
Gaz de France
Gullspång
Hafslund
HEW
Hidrocantabrico
Iberdrola
IVO
Midtkraft
National Power
Neste
Nok
Norsk Hydro
Oslo Energi
PowerGen
PVO
Ruhrgas
RWE
Sjællandske
SNAM
Statkraft
Statoil
Stockholm Energi
Sydkraft
TXU
UPM
Vattenfall
Vattenfall Naturgas
VEBA
VIAG
Wintershall

References

Annual Reports for 50 European Energy companies (1996–2001).

Bain, Joe, S. (1972). *Essays on Price Theory and Industrial Organization*. Little, Brown & Co., Boston.

Eklund, Trond and Knutsen, Knut (1997). *Regnskapsanalyse*. Universitetsforlaget, Oslo.

Masson, Robert, T., Qualls, P., David and Bain, Joe, S. (1976) *Essays on Price Theory and Industrial Organization in Honor of Joe S. Bain*. Mass. Ballinger, Cambridge.

Killion, Mark (2003). *Outlook for Corporate Profits, Which are the Leading Sectors?* GlobalInsight.

Chapter 11

Strategic Configuration: A Casuistic Approach

DOMINIQUE FINON, ATLE MIDTTUN AND TERJE OMLAND, in collaboration with LUTZ MEZ, AUGUSTO RUPEREZ-MICOLA, ROSARIA DI NUCCI, ISABEL SOARES AND STEVE THOMAS

11.1. Introduction

The statistical analysis in the previous chapter has provided interesting insights, but has also raised a number of unanswered questions. In an attempt to plunge a little deeper into organisational and institutional specificities this chapter explores the strategic configuration of a selection of 20 companies in further detail. Given our interest in institutional and cultural embeddedness of business configuration, we have grouped the companies into five regions: the Nordic, the UK, the Continental, the Iberian and the Italian regions in order to analyse the role of contextual factors in industrial configuration. This grouping in regions also reflects the logic of the physical integration of electricity and gas markets because of the existence of permanent bottlenecks on the interconnections. The selected companies are the biggest in their region and include both electricity and gas industries. Oil and gas companies are included here when they are strongly home-based, without a marked transnational nature such as Statoil, DONG, REPSOL and ENI, contrarily to Shell, BP-Amoco or Total. Table 11.1 gives an overview.

This chapter is organised in four sections. The first section explores the strategy-relevant factors in each regional regulatory context. The second section discusses horizontal and vertical positioning, hereafter labelled as functional forms of strategic configuration. The third section discusses the strategic configuration along the geographic dimension. The fourth section examines the financial aspects of configuration.

Table 11.1. Large energy companies grouped by region.

Nordic	UK	Continental	Iberian	Italian
Vattenfall	Scottish Power	E.ON	Endesa	ENI/Snam
Statkraft	Powergen	RWE	Iberdrola	ENEL
Fortum	Centrica	EDF	EdP	Edison
Elsam	National Power	Suez	Gas Natural	
Statoil		Ruhrgas	REPSOL	
Norsk Hydro		GDF		
DONG				

11.2. Strategic impacts of the regulatory context

We have, in previous sections of this book, dealt with regulation more broadly. However, specific aspects of regulation are so crucial to business strategies that they deserve additional elaboration at this point.

The national regulatory context and its evolution interacts strongly with company strategies both with respect to vertical and horizontal integration, diversification into other energy value chains and internationalisation. The underlying culture and approach to competition policy plays a determining role in this respect, either by sustaining effective competition by multiplying players or conversely by accepting contestability of a few dominant firms. Regulation affects strategic configuration in different ways: through the impact of the depth of the reforms in terms of business restrictions and profitability limitations, through direct mandatory changes in horizontal and vertical integration, through national champion policies, through remaining in public ownership and through the financial benefits allowed by regulatory and financial provisions (stranded costs, nuclear decommissioning funds, etc.).

The rapidity and depth of the reforms

The slowness of the reforms in certain countries (France, Belgium, Germany, etc.) favours the home-based companies by protecting their market with a large captive customer base and the security of their annual cash flows for diversifying or internationalising their activities. The company's size and the concentration of market power is greatest in those countries. It constitutes an effective barrier of entry to these markets for the UK and Scandinavian companies which are too 'competitive' and too small to be allowed in. The withdrawal of Fortum, the Finnish multienergy company, from the continental markets was determined by the lack of opportunities to effectively compete.

Conversely the deepness of the reforms in other countries (UK, the Nordic countries, The Netherlands, Italy) decreases the prospect of profitability of the market players. But institutional complexity of some deepened reforms like the Italian (in which the creation of the electricity bourse has been delayed for three years) could serve as an advantage for the national incumbents (ENEL, and SNAM, the former autonomous ENI's subsidiary) both in electricity and gas, during the initial years and confine entries of new players to marginal roles.

Competitive pressures on national players are de facto created by the degree of transparency of the TPA provision, the existence of commercial infrastructures such as organised markets (pools, bourses) and mandated deintegration of the system. The incumbent companies could be incited to exit some activities of the value chain (wire jobs, retail supply) and to engage internationally in countries with profitability, as did the British specialised producers (National Power, later Innogy and International Power, PowerGen), the gas supplier (British Gas), or the transporters (National Grid, Transco) in USA or elsewhere. Conversely some entrants' hit and run could also result from the toughness of the regulation, as was the case with the American investors in UK have decided between 1995 and 2002.

The mandated structural changes

The conception of the competition rules by the regulatory authorities (sectoral and antitrust) determined the mandatory changes in horizontal and vertical integration of the incumbents at the moment of the reforms and thereafter. In some countries regulatory authorities have imposed divestitures in electricity production or gas release programmes (Italy, UK), unbundling of the production and the network businesses (UK, Italy, Spain) and, in certain cases, strong restrictions on integration between production and supply-distribution businesses (UK, Norway). The national competition authorities and the European antitrust division have had in some cases intervened in mergers and acquisitions that have consequences for several national markets. In some cases they have asked for important remedies with implications for industrial organisation for clearing them in countries not affected by initial divestiture. EDF's virtual divestiture of 6 GW after its partial takeover of the German EnBW in 2001 is a case in point. So are the controls of a number of German and Spanish mergers.

Legal unbundling in the gas industries between the wholesale suppliers and the transmission grid has been decided in UK, Italy and Spain (with a clear ownership separation helped by the floating of the grid company). Austria and The Netherlands have engaged in a process of dissociation of the supply business handled by a number of entrants beside the weakened incumbent. The network business has developed its own logic within the framework of a strong regulation.

The depth of the unbundling between the supply and network business in distribution could create opportunities for multienergy strategies in the different market segments for the preceding mono-energy companies, the former gas retailer going towards the electricity business and conversely the electricity distributors going into gas.

Conversely, in some other countries, the concept of fair competition does not preclude the preservation of integrated industrial structures and large-sized companies, as long as the TPA rules are formally transparent and non-discriminatory (Belgium, France, Germany). In these cases, the intensiveness of competition is shaped by the number of accessible producers and large entrant suppliers.

The national champion policies

In some countries (France, Germany, Belgium, Spain, Portugal, Finland, Sweden), tacit or explicit industrial policies have favoured the promotion of one or two national companies protecting them from divestiture provisions and favouring their strategies of mergers and acquisitions in the country. Such favours have in particular been demonstrated through a lax interpretation of the antitrust law. The Spanish Ministry of industry thus actively encouraged the mergers in an initially fragmented electricity sector in the nineties; in the same spirit the German Government bypassed the opposition of the antitrust authorities in E.ON's takeover of Ruhrgas in 2002. In these cases the electricity and gas businesses were seen by governments as a still promising industrial field for multienergy and multiutility globalisation of national enterprises which previously were confined to national or regional territories. The ambitions seem to have been to support the strategies of national champions in the European multi-energy market. Spanish authorities did, however, subsequently adopt a stronger pro-competition policy when the two electricity majors Endesa and Iberdrola were denied a merger in 2000 to create a company which would have dominated 80% of the generation and supply in Spain. They confirmed this position in 2003 with the restrictions put on Gas Natural in its tentative takeover of the second-largest incumbent electricity producer Iberdrola.

Conversely, in some countries (UK, The Netherlands, Norway, Austria), governments have not attached importance to the preservation or the creation of national champions with the size and the financial means to compete with other companies in the so-called global power industry or the European 'single market'. Priority was put on other sectors considered to be more dynamic and innovative for the national economic growth. Similarly, the entries of numerous US and European outsiders which accompanied the disaggregation of the power industry and the value chain recomposition were politically well accepted by governments in UK and The Netherlands.

The degree of stability of public ownership

Although the introduction of the competition is frequently accompanied by privatisation, some countries have maintained public ownership of their main companies (France with a centralised public ownership; Finland, Sweden, Norway, Denmark with a protected multilevel public ownership). Other countries have developed privatisation programmes (ENEL in Italy since 1998, Endesa in Spain between 1995 and 1998; CEGB, Scottish Power and the RECs in UK between 1990 and 1994; supraregional utilities EnBW, Bayernwerk, HEW, BEWAG in Germany in 2000–2003). The public companies are favoured by the preservation of their status because of the deterrence to entries via capital markets. This could be balanced by opportunities created to entries by the privatisation of other companies if the electricity and gas sectors are partially fragmented with regional utilities and municipalities co-existing with the dominant producer–supplier (Finland, Norway, Sweden). Public ownership could also be an obstacle to competition through its influence on access to capital through the implicit financial backing from the State that follows public ownership. Such backing has been suspected in the case of EDF and GDF before their change of legal statute in 2004.

However, public companies also suffer handicaps, as their strategies are also constrained by their public status in different respects. Firstly, their internationalisation could be restricted by some reciprocity conditions imposed by foreign governments, as was the case for EDF in Italy and Spain, and by the impossibility of proceeding to partial acquisition by stockshare exchanges with other companies. Secondly, if their public status is strongly linked to a public service model, some regulatory constraints could be imposed on their internationalisation as is the case for Statkraft (Norway) and on their diversification strategies outside their mono-energy core business towards other energy value chains or networks businesses (water, telecom) as is the case for EDF and GDF in France along the so-called *principe de spécialité* in the French administrative law. Such constraints on diversification have clearly encouraged priority put on aggressive internationalisation in the French companies' industrial strategies.

The regime of provisions (stranded cost, nuclear decommissioning cost)
Partly linked to the influence of the electricity and gas incumbents on rule making, the rules established for compensation of the *stranded costs* in some countries (Spain notably) contribute significantly to the financial resources of companies with strong internationalisation. In a similar logic the regulatory rules for provisioning in view of the future expenses in nuclear decommissioning and waste management are very favourable to the large companies operating a fleet of reactors since the provisions are not externalised in public funds. German electricity companies and the French utility have thus accumulated large nuclear provisions (respectively €35 billion and €28 billion in 2000) that they are free to use for acquisitions abroad. Externalisation of the nuclear funds does not allow Spanish and Swedish electricity producers the same luxury.

11.2.1. Concluding comment

The factors discussed together contribute to highly differentiated versions of the European energy markets. They represent specific anchors for co-evolution of industrial strategies to reproduce European variety. Common competitive exposure, inherent in the EU energy market regulation, constitutes an important underlying driver. However, the institutional variation on critical points remains sufficient to generate extensive strategic differentiation.

National regulatory frameworks provide de facto market buffering effects through such mechanisms as ownership protection and weak competition policy. They also provide specific stimulus effects, through special provisions for capital accumulation, freely available decommissioning funds, stranded assets compensation, etc. The national regulatory frameworks may also entail de facto handicapping of domestic industry through structural reforms, and institutional limitations on strategic configuration.

11.3. Functional Configuration

The broad picture of functional configuration (defined as horizontal and vertical positioning) of the European energy industry lends support to both the specialist and the broader multiindustry arguments. Starting out, in many cases from mono-sectoral engagements either in gas/petroleum or in electricity, a number of European companies moved into wider multienergy and multiutility positions. Simply said, electricity companies are moving into gas and gas companies are entering the electricity business. Conversely there have been very few entries from other sectors, in particular from water services (such as the French Vivendi), compared to what was initially expected at the beginning of the nineties.

In the Iberian market, companies like Iberdrola, Endesa, Union Fenosa and Natural Gas positioned themselves as 'global service operators' or 'customer focused multiutility companies' in the first phase of deregulation only to retreat to more consolidated multi-energy positions following the economic downturn in Latin America in the early 2000s.

The two major German electricity players, as well as the Belgo-French Suez group and its Tractebel subsidiary covering Electrabel and Distrigaz, conglomerates with extensive engagements across a wide range of industries, on the other hand, were already at the outset of deregulation. Commercial adaptation to the market economy therefore came to signify an opposite process towards stronger focus, in E.ON's case towards a multienergy position, and in RWE's case towards a mutliutility position. Ruhrgas, which was focused on the consolidation of a monoenergy strategy has been integrated into E.ON's multienergy portfolio. Suez has continued to manage complex multiutility engagements, with a refocusing on energy, water and environmental services.

The restrictive regulation, which forces French public energy companies to abide by the 'principle of spécialité', precludes the French publicly owned energy companies from broader configuration. Both EDF and GDF thereby have remained locked into sectoral specialist positions in the domestic market up to 2004, but they have compensated that restriction by broad geographic diversification.

The UK industry, following the politically orchestrated deverticalisation at the outset of the electricity market reform, has engaged in vertical reintegration as soon as restrictions were taken away. From positions of specialists in production, distribution or wire businesses, the UK companies have also sought broader multienergy engagements in particular in dual fuel competition.

The large Nordic companies have followed both broader and specialised functional strategies. The merger of petroleum and electricity in Fortum in 1998 clearly signals a strong commitment to a multienergy strategy from the largest Finnish market actor. However, the two Norwegian state companies, Statoil and Statkraft have, with only small exceptions, upheld 'French style' monosectoral energy strategies.

Although clearly characterised by diversity, the European configuration is also characterised by general patterns. From diverse points of departure, and with varying institutional restrictions, European companies have nevertheless followed a wider multifunctional orientation in the late 1990s followed by orientation towards more focused configuration in the early 2000s, presumably strongly affected by responses to the downturn of the hyper-economy and in particular the internet and telecom industry.

We shall briefly run through some of the major European energy companies in the five focus-regions and then consolidate our conclusions.

11.3.1. Functional configuration on the 'continental plate'

The continental scene is dominated by the Big Four (E.ON, RWE, EDF and Suez) with some space left to the gas majors until 2003. Configurations span from vertical integration in a monoenergy filière (EDF in electricity, GDF and Ruhrgas in gas) to conglomerates which have developed multienergy multiservice activities with vertical configuration (RWE, Suez and E.ON) as the main part of their revenues. On this scene there is only limited space for specialists with a narrow focus on a limited set of the value chain-transmission, multienergy distribution-supply, trading or power generation. Examples are merged former municipal distributors such as the Dortmund-based MVV Energy in Germany or some small generator–distributor as the Dutch Essent that succeeds to enter the German market.

Vertical configurations inside conglomerate structures

As far as the three conglomerates are concerned, each of them have a specific history but have followed a parallel focusing on energy and environment activities since 1999 shaped by the conglomerate and multiutility model. The two German groups developed a strategy of consolidation by vertical integration in electricity, a diversification towards gas and water business and an aggressive internationalisation in these activities.

At the end of the nineties RWE was the most active in the electricity chain with a share of around 35% for the electricity business in the turnover in 1996–1999. The company had important activities including lignite production, oil products and chemicals, environmental services, industrial systems, civil engineering and real estate. In 1998 the new Suez-Lyonnaise group with the Electrabel–Tractebel–Distrigas' activities realised 39% of its turnover in the electricity and gas chains, besides other activities in water and environment, financial services, civil works and communications. As far as E.ON is concerned, its activities in the electricity business (from PreussenElektra and Bayernwek) covered only 19% of the 1999 turnover. In addition comes ex-VEBA's important activities in oil and chemical (€26.6 billion in particular with Degussa's chemical business) and in distribution and logistics. Thereto ex-VIAG's business in aluminium and packaging (€11 billion), chemical (€7 billion) and distribution. Every group had some ambitions in telecommunications in the mid-nineties with shareholding in new telecom companies.

However, the consolidation of their conglomerate and multiutility platforms did not exclude some focus elements. E.ON and Suez divested important assets in different activities (distribution, civil engineering, financial services). The funds (€21 billion for E.ON in 2000) were invested in expansions in the energy core business and, in the case of Suez, in the consolidation of the holding control of the Belgian Tractebel (extension of the Suez's shareholding from 50.5% to 98%) and the financing of internationalisation of Tractebel in electricity and gas project development.

In the electricity value chain, following the subsequent mergers of RWE-VEW and VEBA-VIAG in 1999–2000 which allowed horizontal concentration and the two German groups consolidated their business in three ways. Firstly, they merged their subsidiaries' supply business. Secondly, as they were only present in generation, transmission and industrial retailing, they increased their stake in regional utilities by vertical integration in electricity and gas distribution in order to secure access to end-customers. RWE thereby took over five regional utilities in 1999. E.ON followed the same strategy. Thirdly, they searched for synergies with the gas business in Germany (and for RWE in gas distribution in Central Europe in its internationalisation strategy). So RWE took over Thyssengas, the third gas company, in 2000, gathered all its gas stakes in the regional utilities and focused on gas business as a major element in its international expansion strategy. E.ON had already successfully started its gas business positioning before the clearance of its spectacular take over of Ruhrgas in 2002–2003: it inherited Avacon from VEBA and Bayerngas from VIAG and took control on Thüga which holds interest in 130 municipal utilities after its merger with Contigas in 2001 and covers 25% of the gas retailing. The clearance of the merger with Ruhrgas (65% of the gas wholesale) has imposed some important divestitures to E.ON in municipalities but not in Thüga. Furthermore, the merger created an effective vertical configuration of the gas value chain inside E.ON and reinforced the financial capacity of the group. Before remedies imposed for the clearance of the merger, the share of gas business, which included Ruhrgas' sales in the new E.ON-Energy turnover reaches around 30% (around €15 billion on €50 billion in 2002).

When the Suez group was established in 1998, gas and electricity activities were already present in a vertical configuration inside Electrabel and Distrigaz in its European electricity and gas business. Tractebel was Suez's energy holding. The Belgium market, their homebase, was characterised by quasi-integration with Electrabel holding majority stakes in 85% of the municipal electricity and gas distributors and with long-term contracts and skillful sub-contracting supplementing the ownership position. Suez Industrial Services and Elyo complemented the multienergy offers by supplying energy services. Outside Belgium, Electrabel (which became the Tractebel's European business unit) has replicated in different countries, a business model where the development of supply activities are backed by physical assets as in France, Germany, Italy (with its shareholding in the third GenCo InterGen acquired in 2002) and in Spain where it builds up 1600 MW generation capacity. This model is, however, not replicated overseas. In USA Tractebel focuses on independent generation, CHP on site, energy service supply and innovative business in the LNG trade; in Latin America and South East Asia (Thailand), it focuses on the development of energy projects (IPPs, gas distribution development).

While the three conglomerates have shared many features of broad horizontal engagements, they also differ. The difference between the three industrial groups in their positioning outside electricity and gas does not seem to have major impact on their competitive advantages in the energy business, with the exception of some potential with water distribution and environmental services in the supply. RWE and E.ON's strong positions in refinery and oil products distribution do not create synergy with gas and electricity business.

Suez and RWE both stand out with extensive diversification into water service. Suez could base this engagement on its strong traditional position in the water distribution in France, Europe and overseas (20% of its sales) with the former Lyonnaise des Eaux (now Ondeo). This position gives a capital of skills for its investment in energy business outside its home base and consolidates its offers of multiservice to large industrial and commercial customers. For RWE the water distribution is a more recent field of engagement and constitutes part of a diversification in its strategy of refocusing on the service business beyond energy. RWE entered water first with a small investment in urban water distribution in Germany and in Hungary in 1997 and then made a major move with its acquisition of Thames Water in 2001 which increased the profit by one fifth.

Monosectoral configuration

Interestingly, the *monosectoral* strategy which is advocated by much of the neoclassical literature on industrial organisation as the most efficient remains a dominant feature of our sample of European energy companies' strategies primarily because of political and institutional restrictions.

The regulatory constraints of the French public companies have thus dramatically restricted their diversification, in particular into other energy businesses – gas for EDF, which nevertheless has been looking for entries in this business and vice versa for GDF with the exception of cogeneration on site and multienergy supply to eligible consumers and in water services and telecommunications. For the two public vertically integrated companies the strategic focus has been put on internationalisation in electricity business (or respectively gas) and in energy services in order to grow and compensate the loss of market share in the French market of eligible consumers after the first reforms (around 17% of the limited eligible segment (30%) in 2002).

EDF's diversification which corresponds only to 2% of its total sales in 2002, could develop indirectly in the future, through its increasing control of Edison which is the second largest gas supplier in Italy and through EnBW which has taken over in 2002 with ENI the control of GVS, the fourth German gas distributor. In the energy services the presence of EDF has been developed by the transformation of the former Vivendi's energy service subsidiary Dalkia in a common venture (65%/35%) with a turnover of €4 billion with activities in 30 countries including takeovers in Italy and Switzerland.[1] GDF has also sought diversification into energy services, especially in the field of cogeneration in France where it owns a capacity of 550 MW. However, GDF has also achieved broader energy diversification through its takeover of the main energy service companies in Italy (Agip Service, etc.) and in UK (Heat Service). It has prepared entries in generation in 2003 by buying a CCGT unit in UK or developing units in Spain or

[1]This venture has been consolidated by the purchase of 4% of the shares of Vivendi Environment (now Veolia) in 2002.

north of France in partnership and by buying shares in the French IPP SNET.

Finally, as opposed to the other gas companies, GDF has developed an original strategy of vertical integration upstream by acquisition of stakes in gas exploration and production in British, Norwegian and Dutch North Sea offshore fields and in German on-shore exploration and now in Algeria, despite its initial lack of skills in this business. Its goal is to reach a production covering 15% of its marketed gas in 2005. Its incentive is to limit the risk of potential price squeeze between purchase price indexed on oil price and wholesale price. Its stakes in some LNG upstream projects (Snohvit in Norway, Damietta in Egypt), in some transmission companies in Central Europe (SPP in Slovakia) and in some interconnectors responds to the same objective of risk limitation.

11.3.2. Functional configuration in UK

As for the continental companies, the UK market features both horizontally integrated, multifunctional and specialised strategies. The evolution of the UK energy industry is special, however, as the centralised electricity incumbent – CEGB – was split up both horizontally and vertically by political design. For a long time the political authorities limited its reconfiguration through strong policy against vertical integration. The same policy was pursued with the gas incumbent of British Gas including successive splitting (gas release programme, complete unbundling). The combination of structural split and open capital markets lay the UK open to extensive takeovers from foreign companies. A first wave of the US and a second wave of the Continental European takeovers, therefore have been the major shapers of the UK configuration.

Broad multifunctional configuration

Centrica and Scottish Power are two of the prominent and largest UK companies with broad multifunctional strategies.

Since its creation out of British Gas in 1997, *Centrica* has been consistent in its ambition to diversify into services other than gas in the UK market and today represents a broad integration of services in the customer interface of the value chain. The company first entered the electricity supply market thereby moving into a multienergy configuration, but continued to expand into other infrastructure services, and into even wider services thereby taking on a multiservice character. Centrica built up engagements in security, insurance, and in 1996, in financial services with a credit card (Goldfish), in 1997 in automobile service by taking over the Automobile Association (AA), the largest roadside breakdown recovery service and in 2000 in fixed mobile and internet telecom services. In its annual report for 1997, it stated that its aim was 'to be the first choice supplier of energy and services to homes and businesses' and this remains its objective even today. Perhaps its strongest asset is its brand name (British Gas in UK). Centrica's core business remained its retail gas business, particularly the residential sector, where it retained a market share of more than 80%. Its share in the segment of medium and large consumer was much lower, about 24%. Centrica retained some upstream engagements in Morecambe Bay fields, which accounted for 20% of its gas requirements.

Scottish Power represents another broad integration approach, but with an electricity sector point of departure, and with a much stronger upstream anchoring than Centrica. Starting out as a vertically integrated electricity company as opposed to its competitors in electricity generation in England, Scottish Power had, by 1997, diversified into water through the acquisition of Southern Water (for £1.7 billion) and the regional electricity company, Manweb, (for £1.8 billion). It was launched in 1994 as Scottish Telecom in the telecom business, which was, however, soon developed along a fairly independent track. Its subsidiary (7% of the turnover) was renamed and thus, had 49% of its shares floated in 1998.

Scottish Power's UK business was again split into two parts: between generation, trading and supply business and the monopoly transmission and distribution business. Scottish Power had 0.9 million gas consumers compared to 2.5 million electricity consumers. However, like most other companies in Britain that tried to develop as multiutilities, it was not successful and it has now divested its water and telecom interests to concentrate on energy.

Specialists and moderate horizontal integration

Compared to the 'Continental plate', the UK case does not feature the French institutional anchoring for monosectoral specialist configuration. Nevertheless, we may also observe, on the UK scene, some examples of narrower functional configuration.

Powergen's functional engagement has been rather more modest. After the split up of the Central Electricity Generating Board in 1990, the company started, like National Power and British Energy, as a specialised electricity generation unit. After the first regulatory veto in 1996, the ambition to integrate vertically in UK was finally implemented in 1998, by the competition authority's allowance of the acquisition of the East Midlands REC. Powergen also tried to diversify by buying into the US market by attempting unsuccessfully to merge with Houston Industries and then successfully taking over Louisville G&E.

Like Powergen, *National Power* also started out as a specialised generating company. Its first move beyond this position was also to integrate vertically. National Power bought Midlands Electricity's 2.2 million customer retail supply business in 1998. Attempting to broaden out into wider energy engagements, National Power also acquired a retail company with more than 500,000 gas customers. This was a move towards a vertically integrated position, at least in UK. Internationally they were more fragmented, with interests in both vertically integrated companies and in simple generators, particularly in USA. In 2000 National Power split the company into a UK operation, Innogy and an international operation, International Power. Innogy continued to integrate its electricity business, but was obliged by the regulator once more to reduce its generation by selling power plants in 1999–2000. However the company increased its number of customers by buying two more REC retail supply businesses in 2000–2001.

Foreign takeovers

The UK business configuration is special in so far as the major part of the English and Welsh electricity market now is taken over by foreign companies. The extensive foreign ownership relates back to the disintegrated structure designed by deregulation based on static textbook models without commercial sustainability. When the government's Golden Shares in the RECs expired in March 1995, it was clear that the electricity companies would soon be takeover targets. At the same time US utilities were newly freed to invest outside USA and saw acquisition as a way to make easy profits and perhaps as a springboard to Europe. When the easy profits had been won, trading conditions in UK more difficult (pressure of the regulation, requirement of a much clearer separation of distribution and retail supply) and it became clear that UK was not an entry point to the other European markets, most of the US companies pulled out. After 1997, the US-owned RECs were soon all for sale and were picked up by generators and by EDF (London Electric). In some cases, the generators and the most aggressive RECs just bought retail businesses, which were growing in value.

Impulses for restructuration also came from within the UK market. As already mentioned, the UK generators saw acquisition of a REC as a way to bypass a potentially risky wholesale market. Up to 1998 the two large generators were protected from takeover by golden shares (used by the government in 1996 to block a takeover of National Power by the US Southern Co). Both were trying to build international IPP companies and had the objective of growing their international business, so it was comparable in size to their UK business. The international IPP market was smaller and less profitable than expected and by 1998, when they were allowed to buy a REC, neither company had made significant progress in the international market.

Both Innogy and Powergen were however taken over by major European competitors. Innogy was taken over by RWE and Powergen by E.ON in 2002 which also bought TXU-Europe's distribution and supply activities in the same year.

11.3.3. Nordic functional configuration

The Nordic scene covers a large range of functional configurations, spanning from specialists with narrow focus on one sector and a limited set of value chain activities to vertically integrated configurations in mono- or multi-energy systems. One company, Norsk Hydro, stands out as a conglomerate with broad industrial engagements beyond energy.

Coming from the position of a dominant electricity generator in Sweden, *Vattenfall* has gradually moved into complementary positions in energy supply and distribution, through a series of mergers and acquisitions, starting more than a decade before the electricity deregulation. This development towards a *vertically integrated electricity company* was continued after deregulation and the full unbundling of the network. In its branding strategy the company explicitly positioned itself with a three-legged strategy of 'energy market, electricity production and grid management'.[2] With its German acquisitions in 2001–2002, Vattenfall continued its development into a vertically integrated electricity company, but also acquired some of the broader *multiservice* character that characterises the newly acquired German City works.

With the merger of *Immatran Voima (IVO)* and *Neste Petroleum*, the Finnish state orchestrated a broader energy–industrial configuration moving its two state companies from two monosectoral companies to a *broad multienergy company* called *Fortum*. In its mission statements, the company signalled its status as an international energy group, capitalising on the opportunities created by deregulation. On the gas

---

[2]From mission statement in the 1998 yearly report.

side, the mandatory sales of the Finnish gas distribution Gasum, for clearing the merger decreased the vertical integration. On the electricity side, however, Fortum engaged primarily in complementing its dominant generation base with supply acquisitions. Its Norwegian Birka subsidiary is thus characterised by a wide Nordic presence on the supply side, backed up by the mother company's generation capacity and generation and supply capacities acquired through the acquisition of vertically integrated companies.

The Norwegian state company *Statkraft* represents one of the *narrower specialised* Nordic *configurations*. The company moved into the deregulated market economy as a pure generator after having been deprived of the main grid, which was transferred to a separate company without compensation. After a first period of consolidation and trade-oriented strategies, Statkraft has sought to expand vertically through a series of acquisitions and alliances with regional and municipal supply companies in Norway until a refusal in 2002 by the regulatory authority.

The Danish generator *Elsam* resembles Statkraft's initial position, with a fairly *specialised monosectoral configuration*. Following the Danish electricity market deregulation, the West Danish company Elsam has consolidated the six West Danish electricity generators into a generator monopoly in the West Danish market. A similar move has been undertaken with E2 in the East Danish region. Both companies have refrained from moving into retail supply and remain committed to generation and wholesale trading strategies.

With its *conglomerate character, Norsk Hydro* remains an outlier in the Nordic region throughout the 1990s and early 2000s, but moves towards a stronger focus throughout the period. The company profiles itself as an industrial company, anchored in processing of natural resources with a focus on fertilisers, energy and materials. Following the falling petroleum prices in the late 1990s the company worked actively to focus on its strategy. This has resulted in selling out of the petrochemicals division and engagements in seafood. The remaining focus areas are energy, fertilisers and light metals. In the energy field, Norsk Hydro's electricity generation has traditionally served to supply electricity to its own smelting industry, but the deregulated market has given opportunity for more ambitious trading strategies where Norsk Hydro draws on both its generating and consumption functions as well as the company's position as a North Sea-based petroleum producer and gas supplier to continental Europe. However, the still reluctant opening of the downstream gas market has precluded Hydro from taking strong downstream positions. The company as a whole, in fact has a conglomerate character with its engagement in three sectors, but is mostly engaged in the upstream side.

Compared to Norsk Hydro, *Statoil* has remained more *narrowly focused* on petroleum, including oil and gas as well as petrochemicals. However, the company also has considerable downstream engagements, with more than 2000 service stations in nine countries. From the mid 1990s Statoil also signalled an interest in broadening out its engagement to the electricity sector and engaged in electricity retailing based on its retailing competency in petroleum and its trading competency from the gas market. This engagement however, failed, due to difficulties in handling transactions in the electricity market and Statoil has pulled out of this engagement, consolidating its position as a petroleum company with a heavy upstream bias, but also with some branching out into petrochemicals.

The smaller of the Nordic petroleum companies, DONG has remained consistently monosectorally focused on petroleum and gas throughout the 1990s and early 2000s with a public property regime. However, through acquisitions of the Danish natural gas supply company Naturgas Syd, it has taken up a stronger position in the downstream market and controls more than 40% of the natural gas sold in Denmark.

11.3.4. Functional configuration on the Iberian peninsula

In functional terms, the development of leading Iberian companies has gone from monosectoral positions towards broader multisectoral engagements. Companies vary, however, in the scope of their engagement. Some have primarily concentrated on a multienergy focus, while others have taken broader multiutility positions. More, perhaps, than the other regional markets, the Iberian market has responded strongly by signalling consolidation and focus in the early 2000s. The economic problems of Latin America, where Iberian firms have had their most extensive investments may have contributed to this development.

The development towards broader multisectoral configuration is clearly illustrated by the private energy group *Iberdrola* which coming out of a power company position in 1999, proclaimed that Iberdrola now considered itself a 'global services operator in electricity, gas, water and telecommunications'. Iberdrola invested in the gas industry (direct import contracts, stakes in Gas Natural, interests in a regasification terminal, industrial supply). The company invested in Portugal and strongly in emerging markets in Latin America and continued its diversification strategy by acquiring a number of stakes in non-core assets, mainly telecommunications, fiber cable network, water service and a multimedia company. However, these new engagements only summed up to 11%. The downturn of the stock market in 2001 and the Latin American recessions led to a marked shift in Iberdrola's diversification strategy, quite parallel to the change in its extensive internationalisation. The company's non-energy related business was clearly of minor importance, despite the fact that the sub-units in the water industry and the property and real estate business increased their profits substantially.

The former public electricity company *Endesa*'s functional strategy and development are quite similar to that of Iberdrola. From 1995 to 2001 the company grew continuously, becoming a global player in terms of generating capacity as well as its customer base. Endesa and its affiliates operate in twelve countries across three continents. Over the years it expanded its operations into telecommunications as well as other non-core business branches, forming strategic alliances and investing heavily into the acquisition of new companies. In telecommunications, Endesa holds almost 30% of the Auna, a company that combines a number of companies in the telecommunications sector and controls Retevision, the second largest Spanish telephone company. This multiutility strategy would have been very much strengthened if the proposed merger between Endesa and Iberdrola had not been prevented by highly unfavourable remedies imposed by the Spanish competition authorities which led the Boards of Directors of the two companies to cancel the project. In the aftermath of this refusal Endesa sold one of its power subsidiary Viesgo (2365 MW) to the Italian ENEL. In the wake of the global economic recession and the particularly difficult business environment of Latin America, the expansionist vision of Endesa shifted to a more modest focus on profitability and the electricity business and customer service.

Like their electricity counterparts, the Spanish petroleum and gas companies also took fairly expansive strategic moves in the late 1990s. However, the petroleum-dominated REPSOL and the gas-dominated Gas Natural have chosen different expansion paths, knowing that REPSOL has an important ownership position in Gas Natural. It was initially 45% and could indicate a verticalisation strategy similar to ENI's in gas supply. But for financial reasons it had to reduce its ownership share from 45% to 24% to compensate losses in its costly internationalisation strategy. REPSOL has expanded further into petrochemicals, following the petroleum value chain, whereas Gas Natural has sought to develop a multiutility strategy by broadening its customer interface to other sectors.

The case of the two minor Spanish electricity producers, *Union Fenosa* and *Hidrocantabrico* with respective market shares of 9% and 4%, also illustrates the opportunities for strategic moves in the opening market. They have mainly remained focused on their home market, being partly under the control of foreign operators (National Power controlled 25% of Fenosa until 2000, and EDP and EnBW each control 40% of the shares). But they have organised a very aggressive entry on the gas market. Union Fenosa signed imported gas contracts in 2000, and participated in Egyptian projects before creating a joint venture (Fenosa Gas) with ENI; it had the ambition to be the second largest gas supplier in Spain. Hidrocantabrico took over Naturcorp the third Spanish gas distributor in 2003 and has a market share of 10%, being the second largest supplier in Spain.

Gas Natural which had a quasi-monopoly in supply through different regional subsidiaries worked to expand the gas business in Spain, 'improving' their vertical integration, engaging in CCGT installation and the development of a stronger international presence, taking advantage of opportunities in Latin America and Morocco. Gas Natural expanded in its traditional market creating distribution and supply companies when it reaches an area (20 firms in 2001). These firms got the Gas Natural brand closer to the consumer. In addition, Gas Natural has flagged ambitions to move further into what it terms a 'customer-focused multiutility'. For that purpose it established new business divisions to take care of the new ventures in electricity generation and supply, telecom and integral home services. Taking advantage of the electricity market deregulation, Gas Natural has become the largest new entrant and holds 3% of the Spanish electricity market in 2002. In this logic it aggressively attempted a hostile takeover of Iberdrola in 2003, despite the opposition of its strategic shareholder REPSOL, in order to become the first energy group in Spain, but it failed.

REPSOL increased its degree of vertical integration with the acquisition of upstream assets, expanded in Latin America to position the firm as an integrated oil and gas provider and developed power generation capabilities based on vertical integration between gas and electricity in this region. It took position in natural gas production and LNG exports in Trinidad with BP with an export contract signed with Gas Natural. REPSOL's upstream integration in petroleum improved significantly, with the merger in 1998–1999 with YPF the Argentinean national oil company, which was primarily an oil and gas producer. The merger obviously boosted REPSOL's internationalisation, and also the diversification into refining and petrochemicals.

Finally progress was also made in the integration of electricity and gas.

In Portugal the newly privatised electricity incumbent EDP has favoured a strategy of internationalisation (presence in Spain with 3% stake in Iberdrola and a 40% stake in Hidrocantabrico in 2001 as well as an engagement in a regional distributor in Brazil), rather than the multiservice–multienergy strategy. More recently, in 2003, with the agreement of the Portuguese government, there has been a move towards a merger of EDP and GALP, the Portuguese gas company which is partly owned by ENI.

11.3.5. Italian functional configuration

The functional configuration of the Italian companies are characterised by strong electricity and gas convergence and a move towards verticalisation in their respective non-core energy business. The three leading companies have, however, moved from very different positions. ENEL and ENI, coming out respectively of electricity and gas/petroleum, have moved to complement their energy portfolios, in part financed by resources freed in mandatory downscaling of activity in their core sectors. Edison, however, has moved from a chemical-agro-energy conglomerate position into stronger energy focus in response to opportunities created by the mandated divestitures of the incumbents. While a point of departure for both ENI and ENEL, mono-sectoral specialisation has not proven to be a sustainable configuration for Italian energy industry. However, neither has conglomeration. The multienergy/multiutility position thereby seems to be a point of convergence.

As already mentioned, *ENEL* started out as an integrated national electricity company but was mandated, through the deregulation reform, to reduce its vertical integration in electricity transmission with the floatation of its subsidiary TERNA, and to divest one-third of its generation capacity and some distribution networks. So in 1998, the company commenced diversifying horizontally and started enlarging its business scope by setting up a telecommunications company, Wind Telecomunicazioni, of which ENEL owned 51% and by taking interest in InfoStrada, as well as by putting new efforts in the gas and water sectors. In the gas value chain, ENEL took control over the second gas distributor Camuzzi (1.7 million clients, 10% market share) in 2001 and merged it with 40 local gas distributors in ENEL Diztributione Gas. Furthermore, ENEL developed upstream engagements by the acquisition of the main international coal marketer Masefield Coal and by an alliance with British Gas to develop a LNG terminal and two gas production and transport projects producing countries. The model followed was the 'multiutilities/network' company.[3]

As many other European energy companies, ENEL was also pushed into stronger focus from the early 2000s. The path followed until 2001 of a multiutility orientation (gas, electricity, heat, water, telecommunications and waste) was revised in 2002, as the Board changed. The divestment of former engagement in water marked a return to more traditional core business. ENEL, however, has maintained its presence in telecommunications. As opposed to E.ON, RWE, Endesa and Iberdrola that have all sold interests in telecom in 2000–2003. ENEL maintains its telecom position with a majority shareholding in Wind, the third largest operator in mobile telephone (with France Telecom till 2003 and Deutsche Telekom till 2001) and Infostrada, the second largest operator in fixed-line telephony. But, even being atypical as one of the remaining electric utility in this field, it is about to reduce its position in this activity in 2004 in order to follow the main stream of refocusing on its energy business. Waste management in which it is the second largest operator in Italy (Elettroambiante) had been another field of steady diversification. ENEL has, however, not maintained its water activity where it has sold out. Electricity generation and supply, however, remained the core business, covering about 75% of the revenue in 2001, together with gas distribution and sales activities where ENEL controls about 11% of the domestic gas market. So ENEL still maintained a considerable breadth in its activity portfolio, which included: undertakings in the energy, telecommunications, media, information technology and utilities networks (electricity, gas, district-heating and telecommunications).

The Italian gas/petroleum company *ENI* is a company with a scale and scope far beyond the Italian market. Today, ENI is one of the world's largest integrated energy groups (world's sixth largest oil company), operating in and around 70 countries. The company which is still 30% controlled by the Italian State is present in the oil and natural gas business, the electricity generation sector, the petrochemicals and in the service and engineering sectors, where it has achieved

[3]The broadening of functional engagements also had organisational implications. ENEL SpA became a financial holding company, while the service divisions became independent companies, streamlining their organisation and focusing operations on specific business areas. The reorganisation into a holding structure allowed the individual units to pursue more specialised strategies and responsibility for own operational decisions, leaving management of economic optimisation and operational synergy optimisation within the business areas.

leading positions at the international level. In 2001, its sales of natural gas in Italy totalled 59 billion cubic metres, 3 billion cubic metres in Europe destined to Italy and 4 billion cubic metres in secondary distribution abroad.

The strategy followed in the gas segment has been to develop natural gas sales in Europe by integrating all phases of the natural gas chain, maximising volumes sold in Italy, while keeping within the ceiling set by laws (50% of retail sales). For this purpose ENI merged its wholesale gas subsidiary SNAM[4] in 2002 and also increased its interest in the holding Italgas to 54.3% stock shares (40% of market share with 6.8 million clients).

A part of the internationalisation reflects a vertical integration strategy to control the risk of the upstream gas value chain in production and transport. ENI could use its experience in the development of large oil and gas projects and its particular positioning in the Mediterranean countries (Algeria, Lybia, Egypt, Turkey) to be active in gas production and transportation. ENI has interests in a LNG chain in Nigeria (10.4% in the Bony project), in the development of a gas field and a LNG chain in Egypt (33% in the Damietta project of 4.5 bcm in operation in 2004). This interest is reinforced in 2003 by the acquisition of 50% Union Fenosa Gas shares and the transfer of Edison's shares (35%) in the other Egyptian project Iduke.

ENI's development strategy aimed at exploiting the entire value chain of natural gas also includes the objective of expanding its presence in the electricity generation sector and to make a dual fuel offer to Italian industrial consumers. Enipower is the third largest Italian operator (in 2003) with an overall installed capacity of over 1000 MW, 3800 MW under construction and 2000 MW to be approved. This will give ENI a considerable position in a broader multifunctional approach to the Italian market.

The Milan-based agro-energy conglomerate Monte Edison's primary industrial presence lies in the chemical, mineral, electromechanical, textile and glass sectors with some activities in electricity generation with the control of two IPPs (Sondel, ISE). The liberalisation of the energy market enabled it to return to its historical energy business focus from the pre-1962 nationalisation. Edison's functional configuration comes out of a complex pattern of mergers, acquisitions and

[4]The high pressure gas pipelines belonging to ENI (96% of the national network) were transferred to SNAM-Rete which was separated from the group after an initial floating of 40.2% of its capital in December 2001 as part of a regulation-imposed restriction on market share. The rest is planned to be put up for sale later.

internal structural adjustments in 2000–2003. The present Edison was set up after Italenergia, led by the industrial group Fiat and Electricité de France, took over Edison's former parent Montedison in 2001. The outcome of the subsequent merger between Montedison, Edison, Sondel and Fiat Energia gave life to a new company that took the *Edison* name.

The new functional configuration has implied active deconglomeration towards more focused multienergy and multiutility positions. In the 1990s and early 2000s Edison developed strong upstream positions, particularly in the LNG chain (with an Italian terminal in development), in exploration and production in the gas business on the basis of its skills in hydrocarbons production (it purchased the Deutsche Shell's Italian gas reserves and assets in the eighties; it is active in E&P project in six countries) and in power generation in Italy.

Edison is multienergy-oriented in its strategy with a focus on the wholesale and large consumer markets. It developed its generation assets from an IPP basis (ISE, Sondel) and the acquisition of the second former ENEL's Genco Eurogen by Edipower (with Edison's control of 40% shares). Even if Edison did not develop a complete vertical integration in the two value chains, its strategy is based on a vigorous concentration on the energy businesses: from 25% of its revenue in 2001 to 83% in 2002 while multiutility engagements in telecommunication and water remained more marginal. This implied considerable divestitures in diverse activities (agrobusiness, waste technologies, engineering, chemical) for €7 billion in 2002, which decreases by half its turnover from 2001 to 2003.

11.3.6. Concluding comment

Starting from different points of departure, there seems to be some convergence towards a middle ground of multiutility and/or multienergy configuration for several European energy companies. However, both specialised monosectoral energy companies as well as single function companies continue to develop alongside.

In addition to commercial market forces, which seem to have pushed conglomerates into a stronger focus, institutional constraints as well as public ownership constraints have definitely also contributed to shape strategic configuration. The Italian mandatory demands on sell-out of incumbent positions in both the gas and electricity value chains has clearly stimulated multienergy configuration. Likewise, the French service public model has stimulated continuation of the service specialist model. However, specialised business models have also emerged commercially as the supplier model exemplified by Centrica in UK and the trader model until recently exemplified by Enron which specialised in wholesale trading without any backing on physical assets. The difficulties met by this model in 2001–2002 has made attractiveness to vertical configuration production/supply and gas/electricity convergence schemes much more attractive.

11.4. Geographic Configuration

The impact of institutional elements are quite discernable also when it comes to geographic aspects of strategic configuration. European firms have tended to take very specific geographic foci when they internationalise, indicating that cultural and/or institutional capabilities play an important role in industrial configuration:

- The UK electricity and gas industry has, over the 1990s and early 2000s, had a strong propensity to place its foreign investments in the US or Canada.
- The Iberian industry has had an equally strong propensity to invest heavily in Latin America.
- The Italian companies have only recently invested outside the peninsula, being focused on the local adaptation to a complex liberalisation process. The exception has been the national oil and gas company, which has traditionally invested in different regions.
- Nordic industry has, with the exception of the Swedish company Vattenfall, tended to invest dominantly within the Nordic countries and the Baltic Rim and to refocus on the regional market when larger internationalisation has been tested. The Norwegian gas/petroleum industry, with its broader global engagement, is here in line with the other oil multinationals kept out of the analysis.
- The continental majors are less homogeneous, the German ones having in a first stage a modest strategy focused on Central Europe and Baltic Rim before enlarging their scope to the major UK market, meanwhile the public French groups and the private Belgium ones have adopted a global strategy, before refocusing partly on Western Europe in the last stage.

Nevertheless, there are also similarities. First, internationalisation has tended to proceed in two waves: a widely dispersed internationalisation, followed by more focused foreign investment strategies.

- The first stage (1996–1999) was characterised by seizing every opportunity in Europe and focusing on a number of regions in the world: buying stakes in privatised companies (distribution, integrated companies) at a level which gives some management power, purchasing some production assets, and in some limited cases developing greenfield production projects, as in USA, Asian countries and some Latin American countries (Argentina, Mexico, etc.). The opportunities were created by the local privatisations and divestitures. The companies did not search for weak alliances and small stocksharings in foreign companies, but for significant stakes for exerting managerial control and orienting strategies. The limited opportunities in Europe have been an incentive to widen the geographic scope, as the emerging markets were perceived as more attractive than the saturated European markets because of their potential of growth. The European companies were influenced by the American independents' or utilities' strategies which were the most aggressive in the development of the global power industry during this period.

- The second stage (2000–2002) was marked by the definition of the more focused industrial strategies marked by the refocusing on the adjacent national markets which allowed physical exchange with domestic capacities; as well as other highly dynamic and accessible European markets. Only then did they develop positions in other regions. This refocusing was encouraged by the crisis of the 'Enron business model' based on a large international setting, and the virtual trade without important ownership of physical assets, and the financial crisis of the American utilities and independents (AES, CMS, Entergy, Mission, TXU, etc.). The latter divested their main international assets in Europe in 2002 and left room for the European buyers. The macroeconomic and financial crisis in Latin America served to reinforce focus European consolidation.

The analysis of regional configuration for each of the five regions is given in the following five sections.

11.4.1. Continental companies (France, Germany, Benelux)

In the continental market the Big Four (EDF, E.ON, RWE and Suez) and the two gas majors (Ruhrgas, GDF) have defined internationalisation as one of the primary goals in their energy activities in 1998, the year of the amplification of these strategies under the effects of the reforms.

During the first stage, EDF and the Suez-Tractebel group which has a peculiar strategy of asset development in numerous regions were among the more aggressive competitors on the worldwide electricity market and gas assets, or company with stakes with the American players (AES, Entergy, etc.). The German companies joined this group more recently, but with a stronger focus on the European Union. Differences could be observed in the tempo and scope of the internationalisation strategies of the Big Four and the two gas majors. Firstly, the French public utilities were more aggressive and more diversified geographically in their internationalisation strategies than the German electric and gas companies between 1996 and 2001.[5] The latter oriented their foreign investments in different activities – gas transport and distribution, water distribution – with minor international investment in their core electric business.

Second, the German companies were geographically more focused on the Central and Eastern Europe than the European Union up to 2001. In 2001–2002 after the mergers RWE-VEW and Veba-VIAG they developed ambitions towards the most open European markets. Thirdly, as opposed to Tractebel-Suez and the two French utilities, the German electricity companies have fewer involvements in the 'global power' business of IPP development (only a number of projects in South-East Asia and Latin America). They did not take the risk of acquisition of companies in these areas. Table 11.2 presents an overview of the geographic configuration of the Continental majors.

The German energy companies

Taking the Continental pattern of internationalisation into somewhat more detail, the German electricity companies, RWE and E.ON (before Bayernwerk and PreussenElektra) have lately developed equity shares holding in the adjacent countries in electricity, gas and water. It is remarkable to note that no German giant made offer when the electricity sector was privatised in UK or when opportunities of acquisition were created by the American utilities' strategic moves from Europe, UK in particular, before 2002.

RWE began with shareholding in the Hungarian electricity and gas distributors (ELMU, ELMASZ, TIGASZ, Matra) and the Czech and Polish gas companies (STE, Praska) in 1996–1997 and consolidated these positions by new acquisitions (ZEK and KB-Gaz Stettin in Poland, DDGAZ in Hungary) between 1999 and 2001. It took control

---

[5]The conglomerates E.ON's and RWE's higher internationalisation ratio (respectively 47% and 34% in 2001) than EDF's and GDF's ratio (respect. 23.3% and 15%) are explainable by the high rate of internationalisation of their non-energy activities (chemical, electromechanical, mechanical equipment, civil works, etc.) by 2001.

Table 11.2. Geographical configuration of the continental companies in energy and services.

	E.ON	RWE	Rurhgas	EDF	GDF	Suez-Tractebel
Western Europe	UK	UK	Sweden	UK	Germany	Germany
	NL	NL	Finland	Germany	Italy	Italy
Sweden	Swiss		Italy	UK	NL	
Swiss	Luxembourg		Spain	UK	France	
Finland	UK		Sweden	Portugal	Spain	
Spain	Spain		Austria		Swiss	
			Swiss		Spain	
Central Europe	Hungary	Czech	Hungary	Hungary	Hungary	Hungary
		Hungary	Czech	Poland	Slovakia	Poland
		Poland	Poland	Slovakia		
			Romania			
			Slovakia			
			Estonia			
			Latvia			
USA and Canada	USA	USA			Canada	USA
Latin America and Asia				Argentina	Egypt	Brazil
				Brazil	India	Thailand
				China	Mexico	Morocco
				Egypt	Uruguay	Argentina
				Mexico		Chile
				South		Peru
				East Asia		Thailand
						Singapore
						Korea

of two Luxembourg electricity companies and a Dutch distributor (Nederland N.V.) in 1999 and, jointly with EDF, acquired shares of Motor Colombus which controls ATEL in Switzerland (sold out in 2004). But the effective scaling up of RWE's international expansion occurred in 2001 firstly in the water sector with the acquisition of Thames Water in UK and USA and secondly in the energy business in 2002 with the acquisition of Innogy (€4.8 billion) in UK and the Czech national gas company Transgas (€4 billion), which gave it a strategic position in the European gas trade.

E.ON-Energy (before BayernWerk and PreussenElektra) followed a similar internationalisation track. The most significant investments abroad were PreussenElektra's share holding in Sydkraft (which progressively increased from 20.6% in 1996 to 60% in 2001) and the acquisition of one of the four Dutch producers EZH. Bayernwek (Viag) has acquired positions in Austria, Switzerland (Watt, FMB), Italy and the Czech Republic, E.ON was more aggressive just after the merger in 2001. With a large financial surface it expanded its control of Sydkraft and took a foothold in Finland (Espoon Säkhlo) in 2001. Above all in 2002 it acquired Powergen (UK) for €8 billion and TXU Europe mainly located in UK for €2.2 billion. It gave E.ON-Energy a good positioning on the UK market with assets in production, network and supply, a total of 8.8 million of customers and entries on the US market (with the LGE company).

Despite its central position in Europe, *Ruhrgas* also followed a moderate strategy of international development before its acquisition by E.ON. This is reflected in its quite low ratio of internationalisation (11%). Ruhrgas' focus was on central Europe and the Baltic region, partly because of the limitation of opportunities to acquire assets or companies in the other European Union countries. It took position in Hungary (90% of DDGàz in 1997), Poland (50% of IRB Spolzka in 1999), Romania (29% of SC Congaz in 2001), Slovakia (17% of SPP). In Scandinavia and the Baltic region it took position in Sweden (17% of Vattenfall Natural Gas), Finland (20% of Gasum Oy), Latvia and Estonia. It did not search partial vertical integration by taking positions in the international gas production or in the LNG trade to manage risks, as GDF does. However it developed financial participation in Gazprom stock (5% in 2001) for that reason.

The Suez group was largely internationalised. In 1997, the year it was created by merger, it consolidated this positioning by developing appropriate skills in its three core businesses, water and energy service businesses backed by an original finance culture to select opportunities. So in parallel with a presence in sixty countries in water distribution (70% of its subsidiary Ondeo's turnover), the group is active in Europe, via Electrabel, on the basis of project developments in generation (several CCGTs in Italy, Germany, Spain in particular), and with total or partial ownership in IPPs as in The Netherlands (EPON), France (CNR, SHEM), Italy (Intergen with Acea-Roma), Poland and Hungary; it develops supply activities in the countries where it has physical assets or where it controls some distributors as in Germany with the control of two large statdwerke. It is also active in USA in independent production, CHP and trading, as well as in South East Asia (mainly Thailand) with several IPP projects and in Latin America where it owns majority shareholding in a large hydro generation company Gerasul in Brazil. In the international gas business Suez has a position in the LNG trade in USA and in the development of distribution networks in different developing countries (Argentina, Chile, Peru, Thailand, Singapore, South Korea). It is developing a position in Spain in a strategy of electricity and gas convergence, via its water service subsidiary Hisusa which acquired 5% of Gas Natural in 2002.[6]

The French public mono-energy companies

Along with the 'principle of specialité' which constrains their diversification strategies, the French public monosectoral companies are incited to expand internationally. After a first period of inconsistent acquisition strategies, EDF after 1999 tended to focus its acquisitions on Europe and to acquire total or majority control. It became a major integrated player on the British market through the acquisitions of London Electricity and then three supply businesses and distribution networks. It has minority shares positions, but is sufficiently large to exert influence, in Germany with EnBW (34.5%), in Italy with ItalenergiaBis (18%) which controls Edison, in Sweden with Graninge (36% sold out in 2003), in Switzerland with ATEL (20%) and in Austria with ESTAG. The initial positions taken in Central Europe (distribution in Hungary and Slovakia, generation in Poland) were much less strategic. EDF's 'global power company' adventure is paved by developments of important generation projects in China, Egypt, Mexico, Vietnam and Laos and important acquisitions in regional distribution in Brazil (Light) and Argentina (Edenor) where it has largely been affected in 2001 by country-risk.

GDF has followed a parallel path, but with less opportunities in Europe. It made some acquisitions in regional distribution in Germany (Gasag, Preussag Energy, EMB), in Austria, in Portugal, and in the gas supply in UK. It has also acquired positions in Italy where in 2003 it has taken control of two regional distributors (33% of Arialgaz and 40% of Italcogim) and part of a Lybian contract (1.5 bcm/y) in the framework of the gas release programme. It has also important positions in energy services in Italy and UK. It has stakes in Central Europe (Slovakia, Hungary in distribution). Elsewhere it is active in distribution in Latin America (Mexico, Argentina, Uruguay) and in Canada. It is also active in gas production with some stakes in the British, Norwegian and Dutch off-shore, German on-shore and in Algeria, and in LNG liquefaction terminals in Egypt and in Norway.

[6]Suez-Tractebel holds also 2% of shares of Repsol, which itself holds 24% of the shares of Gas Natural. It has directly contracted in 2002 a 1 bcm/y import contract with Algeria for supplying its future Spanish clients.

11.4.2. The British companies

The geographic configuration of the largest UK companies has been coloured by their specialist orientation, their medium size and their market-oriented and financing culture. As exemplified by Powergen, Scottish Power and National Power, after an initial period of broad experimentation with engagements in several European locations and over several continents, the large UK firms have tended to consolidate their US positions at the expense of other foreign engagements. Centrica, which only emerged out of restructuring of British Gas in 1995–1996 chose to reverse the order and entered into its US engagement prior to some recent European developments.

In the case of *Powergen*, the company started its international engagements by making a number of foreign investments, mostly in independent power projects in Germany, Hungary, Australia, India, Indonesia and Portugal. Powergen tried to diversify by buying into the US market first by attempting unsuccessfully to merge with Houston Industries and then successfully taking over Louisville G&E in 1999. It saw the US as a much more lucrative market than the rest of the world. Partly reflecting this change of focus and partly to fund the purchase, the assets in Asia-Pacific, Portugal, Germany and Hungary were also put up for sale.

Until 1999 *National Power* had an aim to become a leading global power company. They had until then invested both in UK and abroad, with the purpose of reaching their strategic goal. In UK, they repositioned their operations into an integrated energy busines with operations in Portugal, the Czech Republic, Pakistan, Australia and America among others. As seen before, the international businesses of National Power were split from its national engagement, and regrouped in International Power. The national and international engagements were then floated independently as two companies.

By 1998/1999 *Scottish Power* for the first time broadened its geographic focus with the takeover bid for Pacificorp in USA on which its geographic configuration has been fairly focused. The company has never made any attempt to penetrate the European mainland markets and it quickly divested Pacificorp's non-US interests, notably its Australian holdings in the power sector, despite the fact that with this American activity the UK operations represented only 45% of turnover.

Centrica began internationalisation three years after its split out from British Gas in 1997. The company, in 2000, made its first foreign acquisitions, buying two energy retailing companies: Direct Energy Marketing Limited in Canada (860,000 clients), and Energy America

Table 11.3. Geographical configuration of the British electricity and gas companies.

	National Power (Innogy)	Powergen	Scottish Power	Centrica
Western and Central Europe	Portugal (IPP)	Germany (IPP)*		Belgium
	Czech Rep. (IPP)	Hungary (IPP)* Portugal (IPP)*		
USA and Canada	USA(IPPs)	Louisville G&E	Pacific Corp	USA (supply) Canada
Others	Australia (IPPs) Pakistan (IPPs)	India (IPP)* Indonesia (IPP)*		

*Powergen interests in IPPs sold in 1999.

(400,000 customers) in USA. Since this date, Centrica stepped up its acquisitions in USA (Texas, 1 million customers) and in Canada, and, for the first time, in Europe by acquiring 50% interest in a Belgian energy supply company (Luminus).

The geographic configuration of the UK companies clearly reflects the importance of cultural links. With some variations the Anglo-American axis seems to have been of greater importance than the geographical and institutional linkages to continental Europe. We observe the same links for National Grid which take the control of two transmission companies in USA. This cultural affinity, went both ways in the 1990s, when the US firms were also major investors in UK energy industry. However, the second wave of foreign investments in UK have been European, and in a longer time perspective this may also come to affect the geographic configuration of former British firms abroad (Table 11.3).

11.4.3. The Nordic energy companies

The major Nordic electricity companies have, after some experimentation with broader geographic engagements, generally consolidated a regional focus, the exception being Vattenfall which has taken a major position in the German market. Larger oil and gas actors like Statoil and particularly the energy-metals-agricultural conglomerate, Norsk Hydro, have increasingly expanded into European and global engagements.

Electricity-dominated companies

Fortum has positioned itself as a leading Baltic Rim petroleum company, and an electricity company with a broad Nordic focus. In the oil businesses, besides the Baltic engagements in oil products supplies, Fortum has stakes in oil and gas fields in Norway (sold to ENI in 2003) and Oman. In the electricity sector, it pursued a broad Nordic and also broader North European and Global approach in the 1990s with an engagement in the Swedish regional distributor Birka Energi together with Stockholm Energy and through a market agreement with Stora Enso, mainly in Sweden. While Fortum (IVO) started out as a Finnish generation company, the focus, on the other Nordic countries has been, above all on the supply side. But it developed positions in generation and in supply in UK and Germany, as well as the presence in markets of Central Europe (stakes in Hungary) and South East Asia in IPP, as part of a selective global strategy. In the early 2000s however, Fortum has concentrated on consolidating its Nordic strategy, and has sold out power plants in Hungary and UK as well as its North-German holdings because of the limitations of the competition on the continental market.

Among the larger Nordic electricity companies, *Elsam* (DK) and *Statkraft* (Norway) are the most geographically focused, with a more or less exclusive home market engagement. Statkraft's geographic focus has moved from a domestic/Norwegian through a North European faze and back to a Nordic/Norwegian approach. Statkraft's European positioning has predominantly been trade-focused. In addition to this, it has also taken a major financial position in the second largest Swedish generator/supplier Sydkraft. However, the position of E.ON as the other foreign partner has precluded Statkraft from utilising its Sydkraft-position for an industrial strategy purpose.

Of the major Nordic electricity companies, *Vattenfall* has most successfully established a North European base. Vattenfall first penetrated into the Finnish market by buying the third largest Finnish distribution company HSÖ. It also developed a North European strategy with German positioning through Vasa Energi and with positions also in Poland, Lithuania and the Czech Republic. In 2001 Vattenfall made a major move into a North European position by integrating the four German companies HEW, VEAG, LAUBAG and BEWAG into a single subsidiary. This move doubled the company's turnover and Vattenfall became the third largest player on the German market.

Petroleum-dominated companies

Looking more specifically at the company developments in gas/petroleum industry, we can observe that small petroleum companies like the Danish *DONG* may consolidate and broaden a North Sea position, but are not prepared to move broadly into European or global strategies.

The energy–metals–fertiliser conglomerate *Norsk Hydro*, on the other hand, is one of the Nordic energy companies that have clearly aimed beyond the regional approach and expanded its global positioning in selected fields (aluminium in USA and in Germany, fertiliser in Brazil). Although its oil and gas production is mostly located in Norway, it develops important gas trading and supply activities on the European market.

Statoil has numerous global energy engagements, with business operations in 25 countries, a station services business in 9 North European countries and a focus restricted to petroleum and to a much less extent to gas. Based on its North Sea fields, Statoil is a major supplier of natural gas to Europe, delivering 14 billion cubic meters of its production in 2000. However, it is only weakly present downstream with some stakes in a number of regional transmission companies in Continental and Central Europe.

Figure 11.2 presents an overview of the major geographical extensions of Nordic configuration (Table 11.4).

11.4.4. Iberian electricity and gas companies

The characteristic of the Iberian internationalisation is both its relative volume and its particular Latin American focus. Following international deregulation the largest Spanish and Portuguese companies have invested extensively in their old South American colonies, whereas their European engagements have been far more modest. For many companies the international expansion was part of very expansive strategies for global multifunctional positioning in the 1990s and early 2000s. However, due to the economic setback in the Argentinean and Brazilian economies as well as to a pressure on focused consolidation from bubble-weary capital markets, we can also observe major retreats from the Iberian South American engagements, leading over to much more focused strategies aimed at geographical and functional consolidation. The oil company *REPSOL* (€5.8 billion with YPF in 2001) is illustrative of the move towards Latin American countries with its takeover of the Argentinean oil company YPF in 1998–1999 which doubled its turnover but also created some financial vulnerability under macroeconomic instability.

In the electricity sector, from 1996 to 2001 *Endesa* also proceeded into expensive South-American acquisitions which have given Endesa

Table 11.4. Geographical configuration of the Nordic electricity and gas companies.

	Fortum	Vatten fall	DONG	Elsam	Statkraft	Statoil	Norsk Hydro
Western Europe	Belgium Denmark Germany* Norway Sweden UK* Irland*	Germany Finland Denmark The Netherlands Norway*	UK Norway Sweden		Sweden Germany The Netherlands Austria Denmark	Norway Denmark Germany Belgium France Ireland Sweden UK	Global activities in more than 60 countries
Central and Eastern Europe	Hungary* Poland Estonia Latvia Lithuania	Poland The Czech Republic Latvia Lithuania Estonia	Lithuania			Poland	
USA and Canada	USA Canada					USA	
Latin America, Oceania and Asia	Malaysia Middle East Russia Thailand	Bolivia* Thailand*			Laos Peru Nepal	Azerbaijan Russia Angola Brazil China Mexico Nigeria Venezuela	

* Sold out.

a leading position in electricity generation, transmission and distribution mainly in Argentina, Chile, Colombia, Peru, before turning its attention to the European countries. It took stakes in Edenor, the North Buenos Aires area distribution company in Argentina, in a generation company and the Lima distributor in Peru. Endesa also took control of Enersis, Chile's largest private power group and therby also Edesur, Edenor's direct rival in Argentina. As for REPSOL, Argentina's devaluation of its currency brought significant losses to the company and also led to a strategy revision with the expansionist ambition being downgraded. The focus has thereafter turned domestic and to a much lesser extent European. Endesa expanded its operations in Italy with the acquisition of the first ENEL's divested Genco Elettrogen and in France with the purchase of stakes in SNET (33%, then 55%), the first French IPP.

Iberdrola has also followed the pattern of strong Latin American engagement. In 1995, the company started its international expansion as a majority shareholder with the acquisition of two Bolivian distribution companies, then acquired minority interests in two electricity companies in Argentina and in electricity generators and distributors in Chile, Colombia and Guatemala. Iberdrola invested strongly in Brazil where it led consortia which owned two major distribution companies and purchased a hydropower utility. Roughly 45% of Iberdrola's customers were situated outside Spain and Europe. It also invested in stakes in some companies in gas distribution in Brazil and Colombia. In Europe Iberdrola has only taken one position in Portugal, with a crossed shareholdings (4%) with the Portugal's major electricity provider EDP, with the long-term perspective of a common Iberian power market. Following the downturn of the business cycle, the geographical focus also for Iberdrola tended switch back to Spain, and to a limited extent Northern Africa.

Gas Natural has also followed the general Iberian pattern of extensive investments in Latin America. Gas Natural has engaged in a steady internationalisation strategy with focus on the large Latin American countries. It has acquired stakes in several distribution firms and/or concessions to develop network and to provide gas in the major Latin American cities, Rio de Janeiro, Mexico City, Sao Paulo and Buenos Aires. (In 1997, Latin American countries already accounted for 43% of Gas Natural's customers). Some of these firms are fully owned subsidiaries and others involve partial arrangements with Spanish partners Iberdrola or Repsol-YPF which also owns 24% of Gas Natural's capital. In addition, Gas Natural has been engaged strongly in the development of the Magreb–Europe gas pipeline which is a means of securing contractual Algerian gas supply. But it is not

Table 11.5. Geographic configuration of the Spanish and Italian energy companies.

	Endesa	Iberdrola	Gas Natural	ENEL (Electricity)	ENI* (Gas downstream)
Western Europe	Italy France	Portugal		Spain	Germany Spain Portugal Greece
Central Europe					Slovenia Hungary Croatia
Latin America and others	Argentina Chile Colombia Peru	Argentina Brazil Chile Colombia Guatemala Mexico	Brazil Argentina Mexico	USA (renewables)	Argentina Brazil

*Shareholding in gas and electricity.

clearly involved in the development of new LNG chains abroad, like the third electricity company Union Fenosa in Egypt and Repsol in Trinidad, even if Gas Natural is very active in the development of new regasification terminals in Spain (Table 11.5).

11.4.5. The Italian electricity and gas companies

Apart from ENI which is largely internationalised and operates in seventy countries for oil and gas activities, internationalisation has not been the electricity utility ENEL's first objective before it reached a strategic consolidation on its home market. The multienergy entrant Edison which is the second player in electricity and gas has been completely focused on the restructuring of its energy business and was not active outside the Italian market. A part of the internationalisation of the Italian companies respond to the objective of controlling the upstream risk in the gas value chain in production and transport.

ENI has used its experience in the development of large oil and gas projects and its particular positioning in the Mediterranean countries (Algeria, Lybia, Egypt, Turkey) to be active in gas production and transportation. Compared to Total which is also an important producer in Europe but without important assets downstream on its home-based, ENI which merged in 2001 with its former gas subsidiary SNAM, could use this advantage for bargaining with foreign government and industrial partners and to establish itself in a number of LNG export projects.[7] Interestingly ENI's international strategy in the gas value chain downstream (distribution-supply) is parallel to GDF and Ruhrgas's up to 2002. Even if it came late, it sought first to seize opportunities in Central Europe (Slovenia in 1994, Hungary in 2000) and then in the European Union in the more attractive markets. It entered the Iberian market with the acquisition of the majority shares of the Portuguese Galpenergia for penetrating the Spanish market in 2000–2002 and then created a common joint venture (50–50) in gas business with the third Spanish utilities Union Fenosa Gas. It also entered the German market with the takeover with EnBW of the fourth regional gas distributor GVS.

ENEL is a latecomer in internationalisation. The single effective acquisition in its electricity business was Viesgo in 2001, the fifth largest Spanish producer-distributor (2.4 GW) affiliated with Endesa.[8] Further acquisitions are envisaged in France or Russia. Another acquisition ChiEnergy, Energy Global International in USA, reflects the specialisation of ENEL in the renewables generation, which places ENEL in the forefront of the major green producers. Finally apart from its direct gas import contracts with Nigeria, Russia and Algeria, ENEL established an alliance with British Gas for exploring two partnerships in two gas export projects associated to the creation of a LNG regasification terminal in Italy.

Edison on its side has foreign activities mainly corresponding to its skills in gas exploration and production (shares in gas fields in Algeria, North Sea, Vietnam, Iran) and responding to the necessity of risk control in the gas bulk supply (in Egypt 50% of share in gas field with British Gas and in 35.5% in the Damietta LNG project of 4.7 bcm). Some shares in generation projects in Brazil and Egypt can also be mentioned as secondary assets.

11.4.6. Concluding comments on geographic configuration

Having clearly documented the two waves development as a predominant pattern of internationalisation for European energy

---

[7]As mentioned above, ENI has interest in a LNG chain in Nigeria (10.4% in the Bony project), in the development of a gas field and two LNG chains in Egypt (33% in the Damietta project of 4.7 bcm and 85% in the second Egyptian LNG project Iduke). In Northern Europe the purchase of Fortum Petroleum and Gas in 2003 gives ENI an interesting position concerning the North Sea gas, given its import projects from Norway. Finally the participation with Gazprom in the Black Sea pipe-line Blue Stream Line towards the Turkish market gives it the future control on 8 bcm of transport capacity towards Near East.

[8]It corresponds to the Endesa's symmetrical purchase of Elettrogen, in Italy.

industry, the question naturally arises how it could be explained. One interpretation is to see it as the result of a learning process, where companies, after a phase of early experimentation, later have discovered a more sustainable internationalisation strategy. Given the uncertainty of learning processes, it is also conceivable that there might be elements of herding around popular fashions.

The first wave of European internationalisation with only limited experience to build on could, in this perspective, be seen as a trial and error process leading to failures for certain types of strategies such as trading without physical assets in the framework of multienergy/utility strategy or to develop merchant plants without initial long term contracts (the so-called Enron model). The painful learning would subsequently lead over to a more conservative positioning.

A second explanation of the shifting positions from the 'wide scope' first wave to the 'more focused' second wave of internationalisation is to see it as having to do with growth and diversification strategies under respectively the *hype-market* of the 'new' economy by 2000 and a depressed market in the early 2000s. On the one hand, taking positions in promising growth markets might seem attractive under expansive ambitions and under assumption that ICT technology would allow coordination of complex engagements. On the other hand, the general economic setback and the falling prestige of the 'new economy' might dictate a more focused internationalisation strategy to gain operative credibility.

Even if the general move in and out of expansive international positions can be interpreted within the two above perspectives, the specific location of these investments, nevertheless, indicate that cultural and institutional capabilities play an important role in industrial configuration. Similar findings are well documented in the international foreign investment literature. The Iberian energy companies, with their extensive investments in Latin America give us the strongest examples, but experiences in other regions such as the investment of UK companies in the US and vice versa, also point in the same direction. This effect may be due to the general cultural familiarity, and also to specific institutional features such as the weight of the 'market paradigm' and similar enforcement of the market rules that characterises both UK and USA. National styles and cultural contexts may, in other words paradoxically survive as core premises for industrial development in an economy through culturally specific internationalisation strategies where national borders are more and more transcended.

However, large companies like Suez and partly EDF have, in line with the large petroleum companies developed multinational strategies that transcend cultural contexts. Possibly, therefore, the strong cultural ties in internationalisation may gradually erode. Moreover, the recent takeover of major parts of UK industry by continental European companies (E.ON, EDF, RWE) may change the Anglo-Saxon ties and forge a broader European orientation. Also, the Iberian companies' losses in Latin America have provoked a retreat to their home-market with a new look at the European opportunities for Endesa.

From the point of view of the market players, therefore, the vision of a unified European market for energy still seems too optimistic in terms of capital markets (when we speak of electricity and gas companies) which would be the pendant of a supposed integrated markets of electricity and gas. One might observe instead a certain move to consolidate strategic positions in regional market domains.

11.5. Financial Configuration

The strategic positioning described in the previous sections affects the capital structure and return of energy firms and therefore deserves a financial comment. The capital structure is obviously affected as companies take on debt to manage restructuring and acquisitions. The medium-sized British companies together with some of the continental majors like EDF, Suez and RWE and some Spanish companies have made large acquisitions and structural changes and this shows up in their financial configuration. Most of the Nordic and Italian companies have been less expansive in their strategic positioning in the European energy market, and have therefore retained a more well-balanced financial structure.

Strategic moves may also obviously influence returns. The increased engagement of Iberian companies in Latin America is a case in point as is EDF's and the Suez group's positioning in the same region. As the economic crisis in Argentina and Brazil activated the downside of these investments' currency exposure for revenues in their Latin subsidiaries, the main financial outcome was reduced.

Strategic conservatism of some companies have also had positive effects on their financial structures and performances. Those which remain in the gas business (Ruhrgas before 2003, GDF, SNAM before 2002, Gasunie) diversified moderately after the reforms, the exception being British Centrica which resulted from mandatory ownership unbundling of the former incumbent. As they had higher profitability in the unbundled gas transmission and supply business than the electric utilities before the reforms, their financial performances remain higher than the electricity incumbents and the other

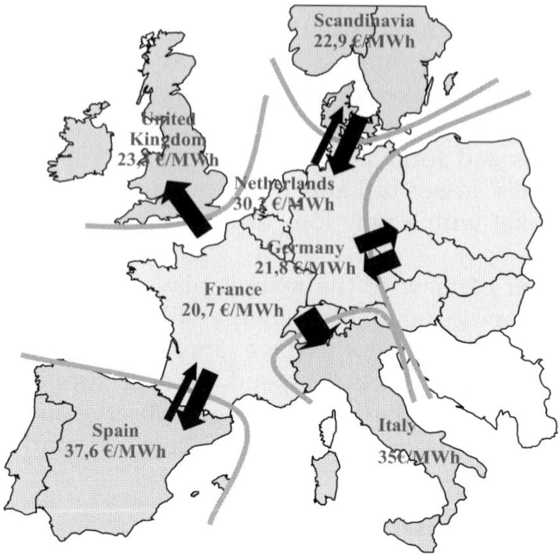

Fig. 11.1. Electricity price differences between regional markets (Averaged price on the period February–November 2002) *Source*: Nordpool and Platts estimates.

multienergy entrants, because of their contractual oil price indexation, milder competitive exposure and slower opening up of the national gas market. Given the new mechanisms of price formation and extensive regional price variations, the price development must be brought in as a major parameter of profitability. Energy prices are obviously also a major factor in determining profitability in energy industry.

As indicated in Fig. 11.1, the Spanish, Italian and Dutch prices have been extensively higher than the continental European, Nordic and UK prices. The Nordic prices used to be lower, as indicated in Fig. 11.1, but increased occasionally to slightly over the continental level due to, scarcer generation capacity after 10 years without capacity investment and climatic and hydrological variations. The UK prices have fallen dramatically after NETA and the continental prices have also seen a fall in France, Germany and Belgium after the liberalisation as a result of overcapacity and a first period of oligopolistic competition.

In electricity and gas industry competition, which has been instilled by reforms, has tended to push market prices downwards and thereby to restrict rentability in competitive activities in a situation of overcapacity. Strict regulation of network businesses has limited profit margins also in this part of the value chain. But the situation varies widely between countries in relation to the competitive

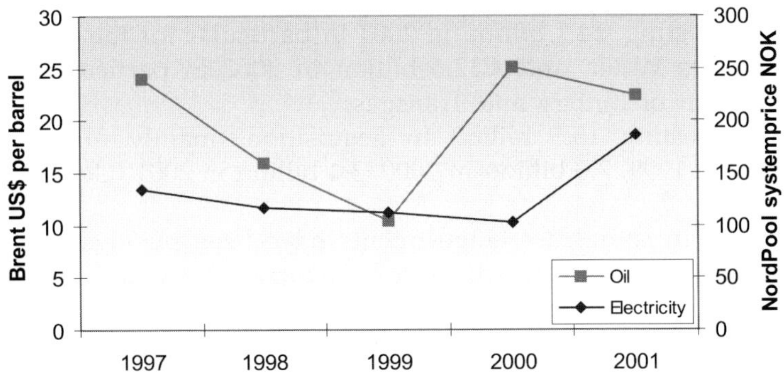

Fig. 11.2. Evolution of oil prices on 1997–2001.

exposure mandated by the market reforms, the intensiveness of the competitive forces and the supply and demand balance (in particular situation of overcapacity).

Moreover for oil and gas companies, yearly performances depend upon the period of the oil price cycle, given that the wholesale gas price is still determined for a long time by the contractual price indexation (Fig. 11.2). We shall briefly review some of the major patterns of financial development in leading firms in each of our five focus-regions and will use the material from the statistical Chapter 10 as a benchmark.

11.5.1. The continental companies

Both the capital-structure and the profitability of the continental majors varies extensively throughout the late 1990s and early 2000s as result of major differences in financial strategies and acquisition patterns.

Indeed the international expansion of the companies has been costly, with some criticized overvaluation and has included such moves as:

- EDF investing €20 billion between 1998 and 2001 for its acquisitions abroad;[9]
- E.ON.Energie investing €20.1 billion in 2002 three quarters of which for the acquisition of PowerGen and Ruhrgas (€10 billion for the latter);

[9]EDF's investment in UK in 2001–2002 (€5 billion) has been made by the LE group partly by financing by indebtedness. It had proceeded in the same way in Brazil and Argentina in 1999–2001 when it developed interest in other companies.

- RWE investing €4.1 billion in 2001 in particular for the acquisition of Thames Water, and €12.6 billion in 2002 in particular for the acquisition of Innogy and Transgas;
- Suez investing 15.5 billion in acquisition (mainly for Tractebel stakes) in 1999, 7.6 billion in 2000, 3.4 billion in 2001 (Gerasul, etc.).

However, in some cases investments have been matched by major divestitures. Suez thus divested financial activities and civil engineering in 1999 and 2001 for around 3 billion per year; E.ON divested assets for €20 billion in 2000 in particular in telecommunications activities. Moreover some of these giant companies could use the cumulated provisions for nuclear decommissioning and waste management (35 billion for the four German companies in 2000, 28 billion for EDF) in order to finance their ambitious strategy, as these funds were not externalised from the utilities' books, as they were in Spain and Sweden. Nevertheless a substantive part of the acquisitions were also debt-financed. The critical financial situation which resulted from the large past investment in costly and risky acquisitions obliged them to follow up their acquisition policy with divestitures.

Partly because of different investment/divestment patterns, but also due to institutional factors, the equity to asset ratio of the six major continental companies varies between the low level of 0.10 for EDF to 0.30–0.40 for E.ON and Ruhrgas in 2001.

The low equity/asset ratio of EDF was in part politically determined by the government that was opposed to financing EDF's and GDF's expansion by increasing the State's equity.[10] With the exception of E.ON which divested substantially, the asset basis of the major companies has increased between 1998 and 2001 as a result of important acquisitions abroad or, for the German companies, downstream integration in gas and electricity distribution (Fig. 11.3).

The debt/equity ratio increased significantly between 1999 and 2001 for RWE and EDF, generally as a consequence of major acquisitions.

[10]Notice that the EDF's equity reaches in 2003 only €19 billion for EDF for an asset value of €147 billion. The GDF's equity reaches €3.8 billion for an asset value of €9.2 billion for an asset value of €15.2 billion.

The French government refused to back up the financing of EDF and GDF by increasing the equity of the two companies and put a limit on financial investment. The financing of expansion was strained by the regulatory cap imposed upon the financial resources for acquisition by the 'contrats d'entreprise' which formalises the relation between them and the public administration. For EDF it limits investment in acquisition to a level of 8 billion between 1997 and 2000 and 19 billion between 2001 and 2004. GDF has also faced a similar constraint with a limit of €1.5 billion per year.

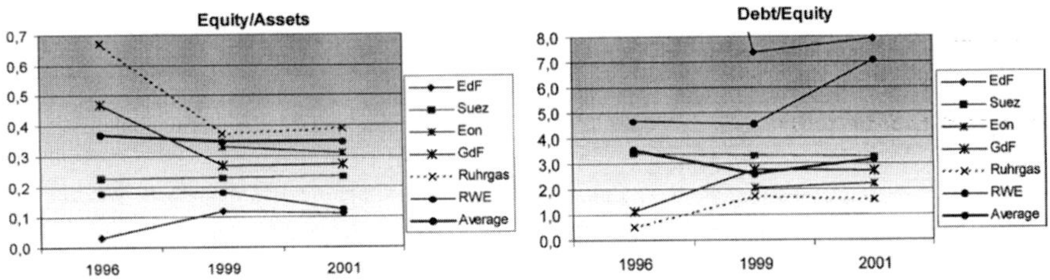

Fig. 11.3. Equity to assets and debt to equity ratios for the continental region. *EDF had in 1996 a debt/equity ratio of 30.

Both of these companies following E.ON and Suez decided to stop their investment in acquisition and mergers in 2003. The gas companies have remained more financially stable with a ratio set around 2.6 for GDF and 1.6 for Ruhrgas.[11]

The profitability of the continental majors also varies extensively. Generally, the European majors perform under average, with Suez and EDF in the lower end of the spectrum. The return on equity varies within a wide range of 0.10–0.30 in 2001, the lowest level being registered for EDF in 2001. Return on assets shows the same variation. Suez figures in the middle range between 1999–2001. However in 2002, because of the necessity of provision for the loss of share value of the Latin American and Asian assets in water distribution (and at a lower level in electricity), Suez's implement accounts present a deficit of €900 million. Suez proceeded in 2003 to an important programme of divestitures of €12 billion, because its debt/equity ratio of 3.0 was considered as too high by the shareholders (it is reduced to half in 2004). The gas companies and in particular Ruhrgas are clearly better performers.

A general factor behind this development may be the price decrease of 20–40% for the industrial segment on the continental markets since the opening of the electricity market to mid-2003, after which price increases to €30/MWh. The exception is the Dutch market which has featured exuberant prices under a restrictively managed international trade regime. Furthermore, the extensive and costly acquisitions that were part of these companies' strategic positioning in the long run may not have paid off in efficiency terms in the short run.

[11]The companies have different ratings by Standard and Poors. Despite important indebtedness the public companies keep a quite interesting rate without the explicit warrant of the government because of the implicit back-up of the government in case of financial distress.

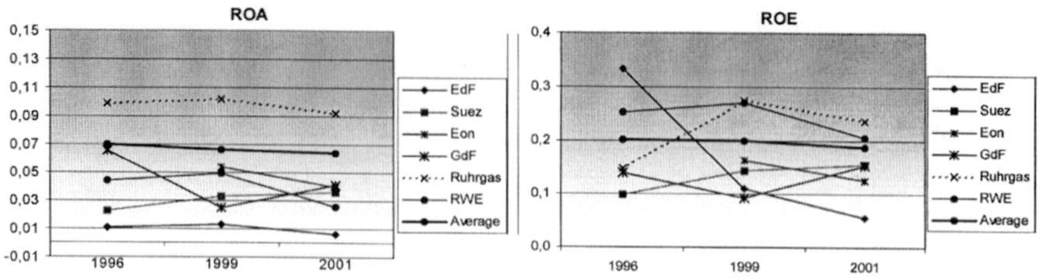

Fig. 11.4. Return on assets and equity for the continental region.

In Europe the acquisitions may have been too costly, as in the case of the British, Spanish, German and Italian companies which are not in good financial situations. For instance the acquisition of 45.5% of EnBW by EDF has cost around €3 billion and its losses in 2003 has reduced EDF's result by €0.6 billion. The exposure to the Latin America crisis (Argentina, Brazil) in 2000 and 2001, which particularly hit EDF and Suez may also be part of the explanation. These difficulties have obliged EDF and Suez to stop their acquisition strategy since 2002 and to reduce their exposition by some divestiture in the risky countries.[12]

11.5.2. The British companies

UK energy industry has been characterised by a substantial increase in debt and falling profitability as measured by ROA. Compared to other European energy companies, the UK-based companies score lower than the European average. In terms of debt development the most dramatic move has been National Power from 1999 to 2001, following its demerger into Innogy and International Power and measured against Innogy in 2001. However, also the nuclear producer, British Energy, has figured consistent weak solidity over the whole period (Fig. 11.5).

Powergen's acquisition of East Midland in 1998 also had a dramatic effect on the balance sheets, as did Scottish Power's acquisition of Southern Water and Manweb, increasing the debt/equity ratio by 70% between 1995 and 1997 and doubling it again after the acquisition of Pacificorp in 1999 (Fig. 11.6).

From positions well over the European average, the return of the UK companies, as measured by ROA, has fallen extensively in 2001. Their relative weakness has been amplified by the consequences of the passage to the NETA in April 2002, in particular the sharp decrease in

[12]Suez divested massively in water business in different risk-exposed countries.

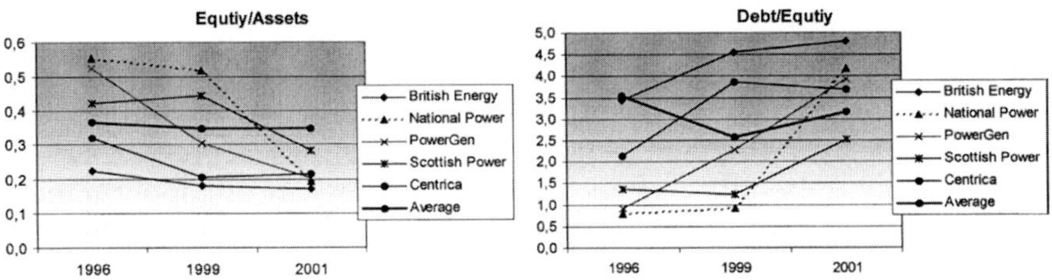

Fig. 11.5. Equity to assets and debt to equity ratios for the UK. *In 2001, National Power is represented by Innogy.

the wholesales prices from an annual averaged price of €30/MWh to €20/MWh. The producers, which have developed a vertical integration, have been less affected than British Energy, the nuclear assets owner, which was not integrated in the supply and had to be backed by loans from the government to avoid bankruptcy in 2002.

National Power has kept relatively high financial returns, compared to other UK energy companies and also to the European average. The demerger that was undertaken will of course influence the development from 1999. The 'new' Innogy kept most of the assets from National Power, while the equity and profit were reduced towards 2001. This explains the downward trend in ROA, while the ROE has increased.

The financial condition of Centrica at the time of its launch was poor. While the underlying business was profitable, exceptional charges left heavy losses overall. The exceptional profits were only possible because of Centrica's ownership of its Morecambe gas fields, which made a profit of £566 m.

The development of high debt and the falling profitability as measured against the asset base raises questions as to whether a high level of mergers and acquisitions, in particular foreign takeovers in UK have geared company values and debt too high to be adequately served under tough regulatory conditions. Downturn in electricity

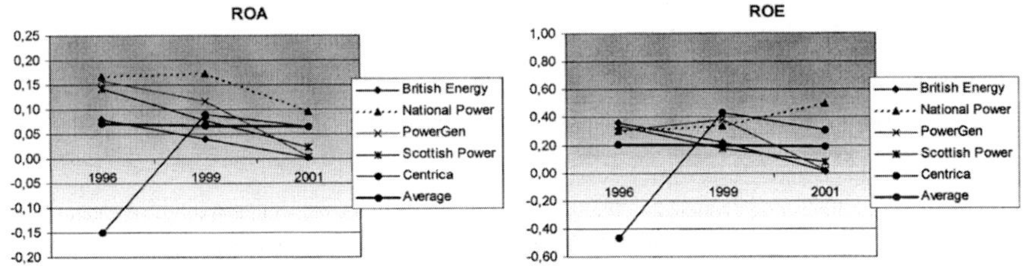

Fig. 11.6. Return on assets and equity for the UK. *In 2001, National Power is represented by Innogy.

market prices in response to new electricity trade arrangements may also have played an important role.

Possibly the recent UK acquisitions by continental majors should also be seen in a strategic perspective, where they serve as part of commercial empire-building in the future European oligopoly games just as much as commercial investments for short-term profitability.

11.5.3. The Nordic companies

As a general pattern, Nordic energy industry is characterised by high solidity and average profitability, indicating, perhaps, that the high level of public ownership has played a limiting role on expansive financial gearing. However, variation in financial structure is rapidly increasing, reflecting marked shifts in company strategies.

Notably the Swedish state owned company Vattenfall has expanded dramatically into the North European market, followed by distinct changes in its financial position. As a consequence, Vattenfall has left the typical Nordic profile with a priority on consolidation, gradual organisational changes, and a relatively low risk profile. The equity-to-assets ratio for 2001 which dropped to 0.226 (down from 0.416 in 1999)[13] can be related to the acquisitions in Germany and Poland, which where mostly loan-financed (Fig. 11.7).

The Finnish state owned company Fortum also initially embarked on an expansive European strategy, but withdrew from this aspiration and shifted to a Nordic focus with more moderate effects on the equity-to-assets ratio.

The Norwegian state company, Statkraft has largely financed its latest investments in the Norwegian market by public transfer without weakening its solidity. The state sponsored acquisition strategy has led to a relatively stable and high financial equity-to-asset ratio for Statkraft in 2001.[14] State ownership has also given Statkraft strong credit ratings.[15] The Danish company Elsam, together with the oil

---

[13]The lower the equity-to-assets ratio, the higher the risk (of financial distress) carried by the shareholders. Since risk is priced in the capital markets, increased risk implies a higher cost of capital. A higher capital cost deteriorates prospects for the future investment.

[14]The company is characterised by its tight bonds to its owner, the Norwegian State that has provided Statkraft with NOK 16 billion earmarked for acquisitions.

[15]This is mainly due to provisions in the laws that regulate state-owned enterprises, which says that this type of company cannot file for bankruptcy. In the event of bankruptcy, the Norwegian state is responsible for Statkraft's creditors receiving full cover.

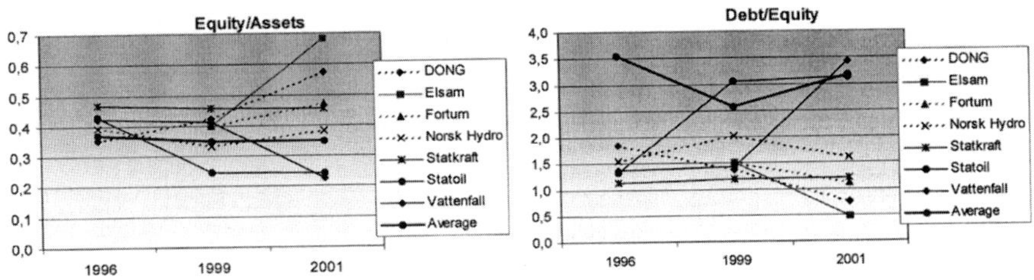

Fig. 11.7. Equity to assets and debt to equity ratios for the Nordic region.

company DONG mark the extreme solidity-end of the Nordic spectrum with an equity-to-assets ratio as high as 0.6–0.7.

On the low solidity end of the Nordic spectrum, Statoil, the Norwegian petroleum company, figures together with Vattenfall with an equity to asset ratio of 0.25. Compared to Statoil, Norsk Hydro had a considerably higher equity-to-assets ratio, scoring just under 0.40 in 2001. The lower score in 1999 was due to the acquisition of part of the Norwegian petroleum company Saga that was mainly financed through increased debt. Compared to the equity to asset figures, the variation in return-on-equity ratio is larger among the Nordic energy firms (Fig. 11.5).

While the Nordic solidity figures are generally higher than average in Europe, the Nordic companies score around the average in terms of profitability as measured by ROA and ROE. For Statoil, the return on equity ratio follows the pattern of the other petroleum companies until 2001, when Statoil increased their ratio from 0.31 to 0.99, This development was the result of both an increased debt and a beneficial tax treatment (in June Statoil realised a non-taxable gain of approximately NOK 1.4 billion).[16] Norsk Hydro and DONG show some of the same patterns as Statoil, but did not manage to take the same advantage of the rise in oil price from 1999 to 2001. In the case of the conglomerate Norsk Hydro, low prices in other sectors in its conglomerate portfolio served to bring the ratio down in 1999 (Fig. 11.8).

The Finnish integrated energy company Fortum saw an increase in return on equity in 1999, even though the oil prices were lower than in 1998. However, this can be related to cash flow from the sales of

[16]Statoil's acquisition of part of the Norwegian state's direct financial interests on the Norwegian shelf has led to a lower equity base in 2001, which then again has increased the debt/equity ratio. The acquisition was financed through a public offering of shares for NOK 12.9 billion, issuance of new debt of NOK 9 billion and the remainder from existing cash and short-term borrowings.

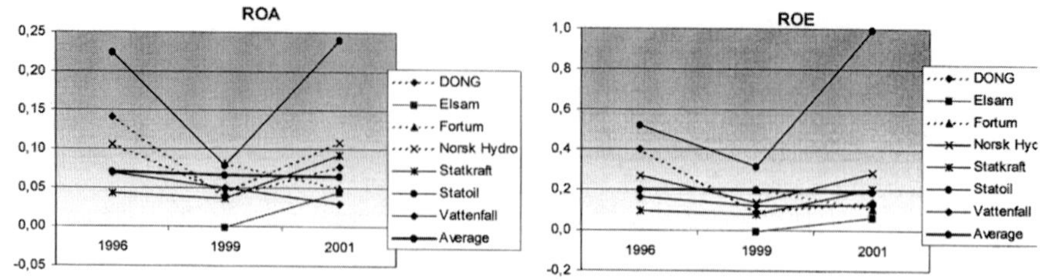

Fig. 11.8. Return on assets and equity for the Nordic region.

their shares in Gasum,[17] their domestic gas subsidiary, in response to the demands from the EU Commission following the IVO–Neste merger.

The electricity companies have had a more stable development of return on assets throughout the period, probably due to a more stable electricity price. Vattenfall saw a decrease in return on assets from 1997 to 1999, but then managed to show a rising trend. The payoff from the company's extensive German acquisitions in the later part of the period has, however entailed lower increase in return on assets than its Nordic competitors.

Statkraft, which performed below Vattenfall in the first part of the period (between 1996 and 2001) took full advantage of the rising 2001 prices as well as a compensation due to an agreement with E.ON, which terminated the power exchange agreement related to the Viking Cable.

Elsam was clearly an underperformer in the Nordic electricity market. It started out with a negative result in 1999, but then managed to create a surplus in the following period, as the rationalisation from the merger of the regional generation companies into the Elsam group seemed to start paying off.

11.5.4. The Iberian companies

The Iberian energy companies show much of the same pattern as the Nordic energy industry as far as the capital structure is concerned. At the low end of the equity/assets ratio Endesa and EDP scores below

[17]The European Commission required Fortum to divest in Gasum because of the merger between IVO and Neste. The Finnish natural gas company as part of the EU Commission's request, following the merger of IVO and Neste. As a result of the divestment, Fortum realised €1.1 billion in cash and recorded €646 million in capital gain.

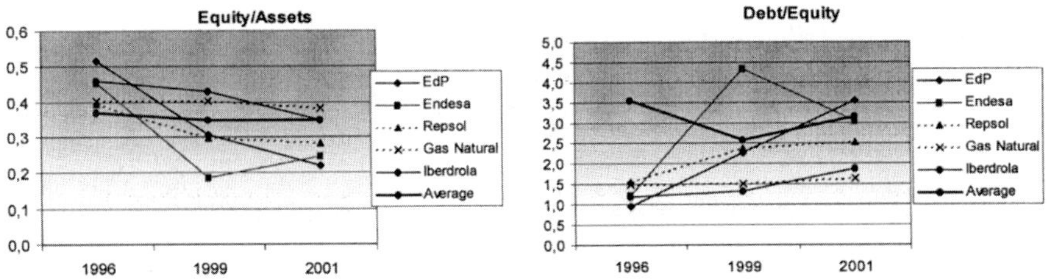

Fig. 11.9. Equity to assets and debt to equity ratios for the Iberian region.

0.25 and at the upper side Gas Natural manages 0.35 or higher. Iberdrola, Gas Natural and REPSOL have remained stable above average, while Endesa and EDP have dropped between 1996 and 2001 from 0.46 to 0.24 respectively 0.51 to 0.22. For the debt/equity ratio, that pattern is much the same as seen in equity/assets.

Apparently, many of the Iberian companies have managed to develop their expansive overseas strategies without rigging debt too high. The most expansive debt financing is to be found with Endesa with its acquisitions in Latin America, EDP which acquired a company in Brazil and 40% stakes of Hidrocantrabrico at high prices in 2000, and REPSOL which acquired YPF in Argentina. The levels of equity/assets and debt/equity are quite similar to EDF, RWE and Suez (around 2.5–3.0 for the debt/equity) (Fig. 11.9).

When it comes to profitability, the Iberian firms are under-performers. Particularly EDP, REPSOL and Endesa have performed at low levels both as far as ROA and ROE is concerned. The explanation is not the price of electricity which is fairly high by European standards, but the low profitability of the Latin America engagements and the importance of provisions for the so-called stranded costs. Gas Natural remains the exception with a close to the European average, which is explainable by the protection of the gas market up to 2001 and the high gas price in Spain, which allowed relatively large investments in infrastructures development (Fig. 11.10).

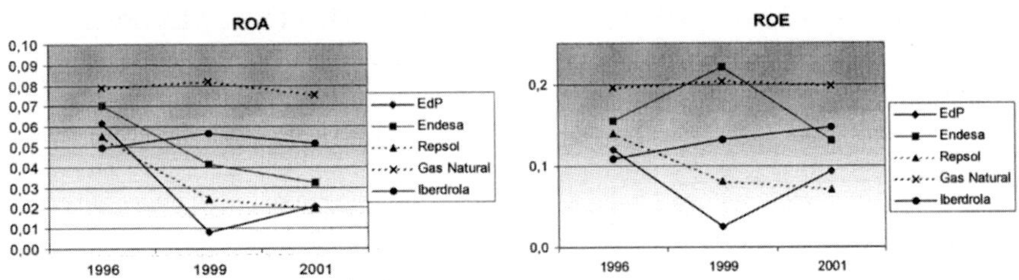

Fig. 11.10. Return on assets and equity for the Iberian region.

11.5.5. The Italian region

The Italian region is characterised both by high solidity and high profit resulting from the protection of the national electricity and gas markets.

ENEL figures in the lower end of the solidity spectrum with an initial debt of €66 billion in 1996 which decreased to €43 billion in 2001 with a diminishing debt/equity ratio decreasing from 2.1 to 1.66 in 1999 and increasing again to 2.03. Montedison (Edison since 2002) was also heavily endebted in 2001 before beginning its strategic focusing on the multienergy business. In the higher performance end the oil/gas companies ENI and SNAM (the gas subsidiary merged with it in 2002), have equity/asset rates rising to 0.9 and 0.8 and with debt-equity ratios falling to close to 0.

The extensive divestitures following regulatory imposition of limited market shares in their incumbent sector has freed extensive capital for financing strategic expansion into new sectors or internationally (Fig. 11.11).

In terms of profitability, ENEL figures in the lower end of the profit spectrum with a ROA of around 0.05 and ROE of 0.13–0.21. The ROE improved slightly, with the decreasing asset-base since 1999, and its telecom business reaching profitability by 2001. Edison featured profit at the same level as ENEL in the first year of its creation.

Like the global oil and gas companies, ENI had a comfortable ROA of 0.35 and ROE of 0.40 in 2001. SNAM also presented good results up to 2001. Given ENI's size (revenue of €52.6 billion), financial costs of its international development in gas projects as well as its acquisition of gas downstream in the Iberian peninsula in 2000 and in Germany in 2002 have not caused even a temporary decrease in profitability (Fig. 11.12).

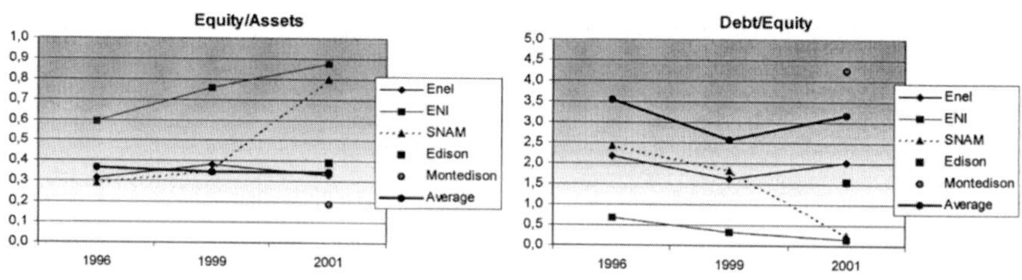

Fig. 11.11. Equity to assets and debt to equity ratios for the Italian region.

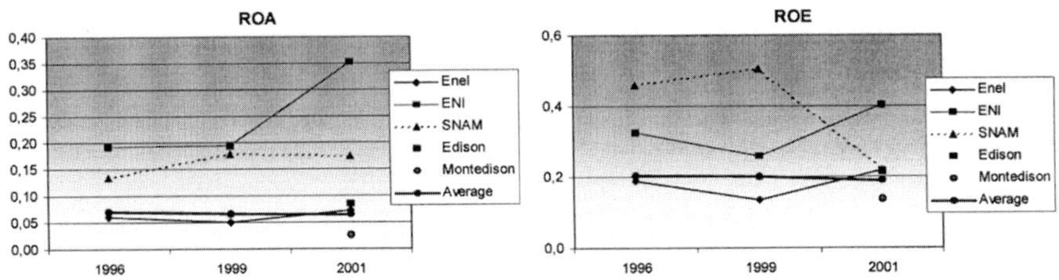

Fig. 11.12. Return on assets and equity for the Italian region.

11.6. Conclusion

Some general conclusions from the statistical analysis in the previous chapter have been that the pluralism of forms and engagements, seems to contradict a strongly formulated universal production and transaction cost hypothesis, and may be taken as evidence either for strong path dependency or for large diversity of the resource bases.

The statistical analysis has also provided clear evidence of both national and sectoral differentiation, which obviously lends support to the national style and industrial organisation hypotheses. However, we have also found a gradual increase in the explanatory power of functional configuration on economic performance over time, indicating that increasing commercial pressure, as markets mature, may exert a homogenising influence on enterprises. The pressure from the competitive market continues to be far from compelling, however, allowing firms to integrate both vertically and horizontally on a large scale at the same time as there are clear economic returns to specialisation.

The more casuistic approach of this chapter supports many of the conclusions from the statistical analysis, but adds more texture. Summing up the findings in this chapter, what we see emerging in the European energy market is a co-evolution of regulatory framework and company strategies modulating the general competitive forces to national style. Regulatory path-dependency contributes to highly differentiated versions of the European energy markets. Various blends of institutional structuration of markets create differentiated contexts for industrial configuration, where companies adapt commercially to market environments, but using specific institutional elements actively/intentionally or passively/unintentionally as platforms for strategic positioning. They represent specific anchors for co-evolution of industrial strategies to reproduce European variety.

We shall briefly summarise some of the main patterns explored in the previous sections on regulation, functional and geographic configuration and financial configuration and then draw some analytical conclusions.

11.6.1. Regulatory contexts, path-dependency and national styles

Configurational pluralism may be taken as evidence for strong path-dependency of regulatory contexts under the pre-deregulation national and regional monopoly regimes, which still linger on through implicit trade barriers. Varieties of configurational forms reflect the limited competitive pressure in some countries with less explicit but still efficient trade barriers, which allows efficient and non-efficient forms to exist side by side.

Specific institutional influences on strategic configuration of firms has been focused through such factors as: the rapidity and depth of the reforms, direct mandatory changes in horizontal and vertical integration; national champion policies; and the remaining public ownership. The slowness of the reforms in certain countries (France, Belgium, Germany, etc.) has favoured the home-based companies by protecting their market with a large captive customer base and the security of their annual cash flows for diversifying or internationalising their activities. Tacit or explicit industrial policies and weak competition policy have favoured the promotion of one or two national companies favouring their strategies of mergers and acquisitions in the country.

Conversely, the deepness of the reforms and strong competition policy in other countries (UK, Nordic countries, The Netherlands, Italy) decreases the prospect of profitability of the market players: in some of these countries regulatory authorities have imposed divestitures in electricity production or gas release programmes (Italy, UK), unbundling of the production/import and the network businesses (UK, Italy, etc.), and, in certain cases, strong restrictions on integration between production/generation and supply-distribution businesses (UK, Norway).

Although the introduction of competition is frequently accompanied by privatisation, some countries have maintained public ownership of their main companies. However, public companies also suffer handicaps, as their strategies are also constrained by their public status in different respects: restriction in diversification in France, reciprocity barriers for international expansion, etc.

11.6.2. Functional configuration and learning

Faced with the opportunities created by market liberalisation and disintegration of value chains and sales of public companies, former electricity and gas companies tested different strategies, including the global power business model, the multiutility/multiservice model, the trading model without physical assets, etc. Empirical evidence has already brought us to conclude, in the preceding chapter, that the production and transaction costs were not the major driving forces for a crystallisation of best practices, optimal structure and optimal size. The incumbents' and entrants' learning have brought them to select new strategies after confronting the fragilities of the specialist model in trading and in merchant production, the risk of diversification in telecom in terms of capabilities and business risk, the country-risk in certain regions, and the barriers to entries in different European regions such as the continental markets (Germany, France, etc.). The latecomers in internationalisation could draw lessons from the failures of earlier business models.

Moreover, beyond the market learning experience, our analysis of horizontal and vertical positioning lends some support to the general competition argument in the sense that there seems to be convergence towards a common middle ground. After a first period of diversity of functional configuration, there is convergence towards the diversified and vertically integrated configuration. Starting out, in many cases from monosectoral engagements either in gas/petroleum or in electricity, a number of European companies moved into wider multienergy and multiutility positions. We observe movements in Belgium, Germany, Italy, Spain, Portugal and Finland, which are more explainable by strategic positioning than by productive efficiency. Interestingly, the *sectoral specialist* strategy, which is advocated by much of the neoclassical literature on industrial organisation as the most efficient, remains on the European energy strategy agenda primarily because of political and institutional restrictions. French firms would obviously have been multisectoral without regulatory restrictions on diversification. In UK the production and transaction cost arguments could explain the vertical integration movement after the change of the market design in 2001, given the new market risks to manage and the consecutive relaxation of the competition policy.

11.6.3. Geographic configuration and learning

When it comes to size and internationalisation, we have found that the main trend has been is to grow in both directions. But the paths have been quite diverse. Firstly the internationalisation of some companies was influenced by cultural/and institutional capabilities. The UK electricity and gas industry has, over the 1990s and early 2000s had a strong propensity to place its foreign investments in the US or Canada with a similar market culture. The Iberian industry has had an equally strong propensity to invest heavily in Latin America both in big-sized and medium-sized companies. Secondly the empirical analysis has clearly documented two stages of internationalisation of energy industry and differences between big-sized and medium-sized companies.

The first stage (1996–1999) was characterised by seizing every opportunity in Europe and focusing on a number of regions in the world. These strategies were pursued even by some medium sized companies like Fortum. The second stage (2000–2002) was marked by more focused industrial strategies with a refocusing on the adjacent national markets which allowed physical exchange with domestic capacities as well as other highly dynamic and accessible European markets. The medium-sized companies have tended to focus on their national and regional markets.

An economic explanation of the shifting positions of the big-sized companies from the 'wide scope' first wave to the 'more focused' second wave of internationalisation could clearly be built on the different conditions for growth and diversification strategies under respectively the *hype-market* of the 'new' economy by 2000 and a depressed market in the early 2000s. On the one hand, taking positions in promising growth markets might seem attractive under expansive ambitions and under assumption that ICT technology would allow coordination of complex engagements. On the other hand, the general economic setback and the falling prestige of the new economy might dictate a more focused internationalisation strategy to gain operative credibility.

11.6.4. Financial configuration

As we have seen from the previous section, the strategic positioning described in previous sections affects the capital structure and return of energy firms. We have seen that the UK medium-sized

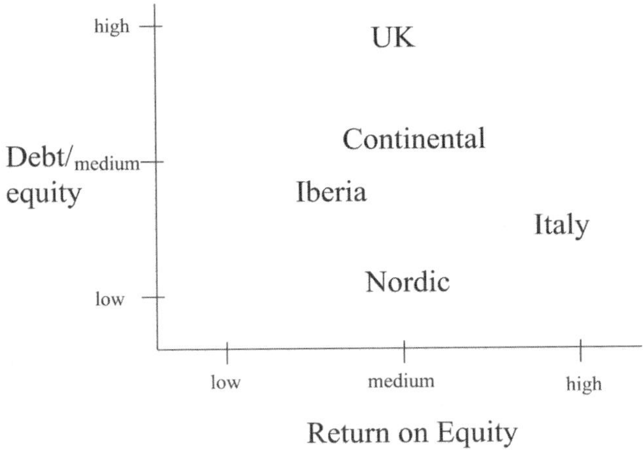

Fig. 11.13. Summary of financial configuration, regional averages.

companies together with some of the continental majors like EdF, Suez and RWE and some Spanish companies have made larger acquisitions and structural changes and this shows up in their financial configuration. Most of the Nordic and Italian companies have been less expansive in their strategic positioning in the European energy market, and have therefore retained a more well-balanced financial structure.

We have also seen that strategic positioning also obviously has influenced returns. The increased positioning of Iberian companies in Latin America is a case in point as are EDF's and the Suez group's positioning in the same region. As the economic crisis in Argentina and Brazil activated the downside of these investments' currency exposure in their Latin subsidiaries, the main financial outcome was reduced. Strategic conservatism of some companies have also had positive effects on their financial structures and performances.

The regional variation of capital structure and profitability indicates that institutional context matters. As indicated in Fig. 11.13, in terms of solidity the UK and the Nordic regions stand out as the counter-poles. The former with the highest debt/equity rates and the latter with the lowest debt/equity rates in our study. In terms of returns, the counterpoles are the Iberaian and the Italian regions with respectively medium to low and high returns. The continental region takes a middle position along both axes.

Interestingly, both the Iberian and Italian regions are high price regions for electricity, and the regional differences cannot therefore be explained with price differentials in the underlying power market. In both cases the gas/petroleum companies outperform electricity companies, and the differentials may have to do with differences embedded in the gas market.

In terms of financial solidity, we have pointed to the consecutive merger and acquisition wave that has characterised the UK industry, and that seems to be one of the major factors behind the debt burden. The expansive US investment strategies undertaken by the UK companies is another important explanatory factor.

The Nordic energy industry, which stands out as the counterpole to the UK has, with a dominant public ownership, on average been less open to mergers and acquisitions, and less expansive in their positioning strategies than their UK counterparts.

However, the regional patterns cover extensive company variation, indicating that explanatory factors do not lie at the institutional level alone. In the Continental region, the divergence is large between RWE and EDF, with a debt equity rate between 7 and 8, on the one hand, and the rest of this on the average.

11.6.5. Concluding note

Taken together, the evidence from both the chapters on strategies is that, following deregulation, the large European energy companies have been through extensive commercial experimentation, with sequential attempts to globalise and regionalise, to spread out into multiutility positions but also backing down to more limited multienergy configurations. Yet, there is obviously sufficient diversity in resource bases and institutional conditions to allow competitive survival of a multitude of business concepts.

With the move from global to regional focus, the continuous scaling up in energy industry increases its potential market control. Positioning for strategic market power therefore, is obviously on the agenda of the largest actors, motivating scale and scope beyond simple efficiency constraints.

In order to understand this complexity we obviously need a fairly rich conceptual framework, maybe also beyond the introductory perspectives presented in this section. However, the institutional and strategic embeddedness of business configuration also necessitates integrated analyses of regulatory models, market structures and business strategies. Drawing on all the three sections of this book, we will attempt some broader reflections of this type in the concluding chapter.

Section IV

Chapter 12
Reshaping European Energy Industry: Patterns and Challenges

DOMINIQUE FINON AND ATLE MIDTTUN

12.1. Introduction

Since the launch of the European energy market, deregulation policy in the late 1990s has been a dynamic interplay among regulatory initiatives, business strategies and an evolving market structure. In a situation with limited European consensus on strong deregulation policy and extensive national subsidiarity in their implementation, the policy impact may be highly dependent on the European Commission's ability to shape markets and trigger business strategies. Reverting to the core issues raised throughout this book – Europeanisation, liberalisation and industrial integration – this chapter proceeds to an integrated analysis by summing up the insights across all the three sections and discusses the major challenges facing the European electricity and gas industry today.

12.2. Europeanisation, Liberalisation and Integration: an Integrated Analysis

The first analytical focus is on Europeanisation, contrasted against the continuation of national governance. The discussion here links the Europeanisation issue up to the national style versus globalisation debate, arguing for more nuanced and relevant interpretation of the two counterpoles based on the evolutionary theory.

The second focus is on liberalisation. The focus here is on how market-based European integration may come about both in a competitive and various non-competitive modes, notably through oligopolisation. Based on the joint evidence from the analyses of regulation, markets and business strategies, it is argued that non-competitive integration represents a possible outcome of the European energy deregulation exercise. This theme is pursued further in the final part of the chapter on the challenges of oligopolisation.

The third focus on industrial integration obviously relates to the aforementioned themes. Integration is discussed at three levels: at the trade level, referring to intercountry electricity and gas trade, at the capital market level, referring to mergers and acquisitions across national borders and at the business strategy level, referring to the overall positioning goals of the companies. As a summing up of the previous foci, the current European energy industry is presented as an interplay between business strategies, market structures and institutional contexts that co-evolve under different political economies. The dynamic interplay among regulation, markets and business strategies also indicates that the sequencing of politics, markets and business responses is central to the outcomes. Both as far as Europeanisation and liberalisation are concerned, the reform results have therefore to be judged over considerable time.

12.2.1. Europeanisation or national style

Given the still fairly brittle EU construction, and the deep cultural and institutional embeddedness of European nation states, a central question is obviously if and how the European energy systems can be integrated under a common commercial order. Analytically this question can be (systematically) related to the general debate over internationalisation versus national style where the literature conveys two quite different pictures. On the one hand, international integrationists like Ohmae (1985, 1995) or Doz and Pralahad (1993) argue that the modern economy, in our case the internal EU market, is moving towards seamless integration and convergence of firm strategy and behaviour. On the other hand, the national styles argument (Whitley, 1992, 1999; Hollingsworth and Boyer, 1997; Hall and Soskice, 2001) points to national institutions, markets and business practices as still exerting a dominant influence.

Implicitly, and sometimes also explicitly, the national styles literature draws on a path dependency argument, asserting that industrial systems cannot develop independently of previous events and that local positive loops propagate traditional patterns into future strategic decisions (David, 1993). This interpretation implies that small events may play an important role in determining the future course of long-term development. Thus, the path dependency and national styles literatures foresee that institutional, social and organisational factors will continue to produce national differences in strategic orientations, even under international competitive conditions (Whitley, 1999; Maurice and Sorge, 2000; Zeitlin and Herrigel, 2001).

Given the close interplay between economic, political and institutional dimensions (Polanyi, 1944; Djelic and Quack, 2003) this debate has a bearing on all the three levels of our analysis – regulation, markets and business strategies. At the economic level the focus is on how international trade, foreign investments, global/European capital markets and firm internationalisation impact on and interact with national markets. At the political level the focus is on regulation and institutions and how they impact on the national economic behaviour, via the coercion of rule-making at the European level.

We are concerned, however, that the national style versus globalisation debate may too crudely dichotomise the issue and fail to capture the richer texture of our empirical analysis. To move beyond this stark dichotomy, we have added a supplementary evolutionary perspective which allows us to reformulate the two positions in a more nuanced way.

Drawing on the evolutionary theory (Aldrich, 1999) and its distinction among the three functions – variation, selection and retention – allows us to reassess and specify further the globalisation versus national styles controversy.[1]

This more nuanced conceptual framework obviously gives room for a more complex interpretation as it is conceivable that one or more functions are governed by national style, while other functions are driven by processes at the global or European level. Reformulating the positions in terms of the evolutionary framework:

- The strong Europeanisation position would see international convergence on all the three counts. Assuming a European-wide regulated market, this position implies that selection is universal throughout the whole market system and that retention takes place at the international level and/or through national bodies adhering strictly to international regulatory blueprints. Firms' strategies and practices converge to a globally defined most efficient optimum. Assuming full factor mobility, an ultra strong convergence

[1]In a market-based system, variation is created by strategic entrepreneurship and its ability to generate new products and business models. Selection, or the ability to pick out the most successful among that variety refers most directly to the market level. Retention, or the ability to institutionalise new practices/products in stable and efficient production and governance systems refers first and foremost to the regulation level.

hypothesis will also eliminate national diversity in generating new products and business practices except for diversity clearly linked to natural resource endowments.
- The strong national styles position points instead to the differentiation on all the three dimensions. It assumes extensive national control over market selection, and institutional/regulatory retention mechanisms that function at the national level with full autonomy from international decision-making. National variation of business strategies and business practices may be argued not only as responses to national selection and cushioning industry against international competition, but also on the basis of a conception of efficiency which includes path dependency and the competitive advantage of specific national resources, competencies and practices.
- In between these extremes it is, as already mentioned, possible to construct several intermediary positions. We can, for instance, imagine Europeanisation of selection mechanisms, in particular by coercion and isomorphism (in the sense of Di Maggio and Powell, 1991) in the context of considerable functional exposure to international competition (the market pressure hypothesis). At the same time variation of business strategies might be based on national style, recognising the national and even firm diversity in variation or generation of new products and business practices and/or peculiar national institutional forms of retention. However, the literature on institutional isomorphism also points out that there may be other forces of standardisation, such as imitation, spread of cognitive models through professional networks etc. that constitute a competing avenue to European uniformity (Di Maggio and Powel, 1991).

Summing up the empirical analysis in the three core sections of the book does indeed, provide a fairly mixed picture of European energy systems in terms of the globalisation/Europeanisation versus national styles debate:

(i) At the level of *business strategies*, we have observed a pluralism of business models and strategic engagements which seemingly contradicts at least a strongly formulated convergence position. The signs of functionally driven convergence under common market pressure are relatively weak when compared to the extensive diversity of company performance and organisational configuration. This lends support to the claim that nation- and industry-specific factors continue to play a major role in continuing to reproduce a variety of business models and industrial configurations. While this may be explained both by institutional and resource differences, it could also be explained by differences in capabilities at the firm level.

Interestingly, the sectoral specialist strategy, which is advocated by much of the neoclassical literature on industrial organisation as the most efficient, remains on the European energy strategy agenda primarily because of the political and institutional restrictions, notably in France.

Nevertheless, the time series analysis in Chapter 10 indicates some tendencies towards convergence of business strategies over time. After a first period of diversity, we have observed some convergence towards a common middle ground in terms of horizontal and vertical configuration, in particular a multienergy profile, indicating isomorphic pressures on business over time.

The informing pressure of market selection in the European arena is also highly debatable. The regional variation of capital structure and profitability indicate widely different operative conditions. We have, for instance, documented that the UK and the Nordic regions stand out with widely different capital structure. The former with the highest debt/equity rates and the latter with the lowest debt/equity rates in our analysis. In terms of returns, the counterpoles are the Iberian and the Italian regions with medium to low and high returns respectively. The co-existence of such diversity within the energy market indicates highly differentiated competitive pressure across the European market space on core economic variables.

However, national identity does not seem to be the only basis for strategic configuration. We have found that the oil and gas companies systematically outperform electricity companies, probably due to the differences embedded in the electricity and gas markets such as industrial structures, gas companies' non-ownership of the capital intensive upstream, oil price indexation in the price formation mechanisms, etc.

(ii) As far as *market development* is concerned, functional convergence of the economic (first order) regulation framework under the momentum of the Europeanisation of the electricity and gas markets should in the outset provide considerable European integration, based on functionally equivalent regulation in different arenas (European Union, national parliaments and regulators). Ideally, this should create a uniform selection mechanism for business projects throughout the European area.

However, while international price formation has taken place in certain regions (notably the Nordic region as well as across the French and German borders for electricity), electricity and gas markets still remain mainly multidomestic. In spite of a considerable commercial challenge, absence of a European-wide marketplace also clearly impedes new entrants. Quasi-vertical integration by pre-reform long-term contracts in gas industries creates the same barriers.

The regulatory context or institutional retention itself also contains ambiguity. On the one hand, measures such as adoption of regulated TPA, creation of autonomous or independent regulatory authorities (except in Germany), have commonly accelerated openness of the final markets. Even market-oriented regulatory frameworks for gas, that were put in place long after the electricity market deregulation, have been deepened between 2000 and 2002 beyond the minimalist level of the initial reforms.

Nevertheless, political and institutional variables, in particular the attachment to public service and the willingness to maintain national champions, represent important factors that serve to maintain national as well as sectoral differentiation. While economic (first order) regulation has seemingly come quickly in place, political (second order) regulation converges much more slowly.

Moreover, national differences in gas resources endowment and dependency on remote external sources have also contributed to the resistance to liberalisation reforms in the gas sector.

National variation has also surfaced clearly in the diversity of competition regulation between countries and impeded a common conception of industrial organisation.

(iii) However, the institutional and market analysis also clearly highlights the *contextual dimension*. In order to have the desired European integrative effect of a uniform selection mechanism, reforms must be consistent. It is not sufficient to adopt one radical regulatory measure such as a total market opening if the set of other measures (type of TPA, unbundling, balancing regimes, etc.) and the industrial organisation are not consistent. Market integration is thereby constrained by the peculiar combination of technical and institutional parameters, which gives a national style bias.

The *regulatory context* thus contains considerable duality, as seen from a globalisation versus national style point of view. While the political second-order regulation retains predominantly national structuration, the first-order economic regulation institutionalises a stronger pan-European retention. Furthermore, in spite of regulatory similarities, differences in industrial structures between countries interplay with their regulatory contexts in specific ways and thereby contribute to the national variation.

The empirical analysis in the previous sections, thus, definitively supports the intermediary mixed position, where there is considerable institutional convergence at some points, notably regulated TPA (except Germany till 2004), and considerable formal market opening (except France), which presumably embodies some uniforming pressure by market selection. These converging trends are strengthened by the transcription of the 2003 Directive, which imposes such market rules in the national laws. Some convergence in business configuration also supports the drive towards Europeanisation.

However, there is still considerable functional differentiation due to various incomplete integration of complementary market-enhancing elements which strongly modifies the unified selection pressure of the European markets. So extensive and deep structure differences embedded in commercial traditions and general institutional set-up retain differentiated market structures and market practices. These differences also trickle down into technical institutional details that contribute to integration.

12.2.2. Liberalisation and integration

Our analysis has shown the Europeanisation of energy markets to be strongly challenged by national institutional peculiarities due to national styles of business configuration. Similar challenges also confront the second element of EU's new energy policy, namely competition. Even though the European deregulation is phrased in competitive market terms, there is a strong heritage of large-scale oligopolistic and monopolisitic industry which represents a fundamental competitive challenge to the new energy market design. Integration could be the answer. Liberalisation is the fundament of the market integration between countries, foreign trade being one major element of the competitive forces on each national market. It is also the fundament of the cross-sectoral integration by legally allowing new business strategies of multienergy and multiservice diversification to the former mono-energy monopolies. However, integration by capital markets and business strategy is also a factor that reduces competition at the national and European levels.

Variety of competition models

The combination of strategic positioning from major European companies and national industrial policy ambitions may prove a difficult match for a EU with relatively weak structural policy tools. The tension between Europeanisation and liberalisation developments in the European electricity market may be schematically represented in a two-dimensional matrix as in Fig. 12.1, with degree of effective

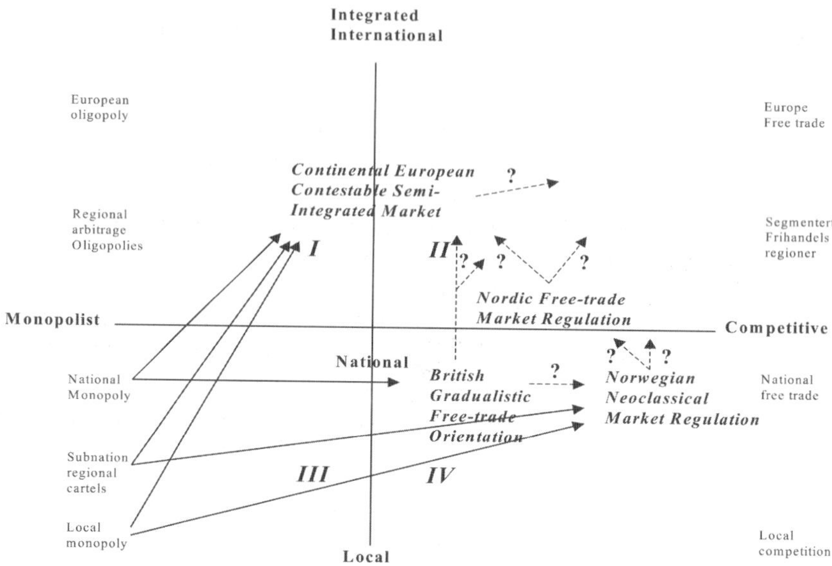

Fig. 12.1. Market opening and competition models.

competition along the horizontal axis and geographic expansion of the market along the vertical axis.

To sum up, developments in dominant Continental European markets such as France and Germany, point to a trajectory away from national and monopolistic planned economy (quadrant III) towards a European semi-competitive and semi-integrated market system (between quadrants I and III), characterised in line with the contestable market theory rather than the liberal free-trade model.

Other countries, such as England and Norway, have made unilateral moves into deregulated positions (in quadrant IV). In the Nordic case, a further expansion into quadrant II has taken place, but this position is challenged by the scaling up of companies as well as weak international grid connections.

Whether the European liberalisation project may in the longer run lead to fuller integration, open trade and competitive markets (quadrant II) still remains a fairly open question. The new 2003 gas and electricity directives and the so-called Florence and Madrid processes indicate ambitions towards a future stronger competitive market policy in Europe. However, at the same time major European companies have kept growing at a spectacular rate.

As pointed out in the section on regulation, the European countries strike very different balances between internationalisation and competition. As already mentioned, the European scene features at least three different regulatory models:

- The market-oriented model which is characterised by the promotion of competition and market efficiency as the primary objectives of the reforms. When this model is implemented on an international basis, like in the Nordic countries, it converges towards quadrant II in Fig. 12.1.
- The public/national interest-oriented model at the other extreme of the spectrum (quadrant I/III), which emphasizes these objectives to restrict the opening up of the market to the minimum mandated by the Directives and to define specific protection for pursuing distributional objectives and the possibility to develop energy and environmental policies. This model typically remains national and non-competitive in scope.
- The mixed model or national industry-oriented model, which is characterised by the explicit combination between liberalised measures and the willingness to implement industrial and redistributive policies. This model typically supports internationally oriented business strategies, but into the oligopolistic (quadrant I/II) positions.

As pointed out in the section on market, limitations in competition at the European level is given, particularly by the lack of homogenisation of industrial structures between countries. This is an issue on which the EU has only limited capacity to intervene beyond the case of mergers and acquisitions where it may propose incentives remedies. The diversity of competition regulation between countries does not facilitate common conception of industrial organisation of national electricity and gas industries.

The increasing business engagement across gas and electricity industries increases the potential market control of European energy firms. Starting out, in many cases from mono-sectoral engagements either in gas/petroleum or electricity, a number of European companies have moved into wider multienergy and multiutility positions. Across the European space, we have observed such movements in Belgium, Germany, Italy, Spain, Portugal and Finland. Such moves seem more motivated by strategic positioning than by a strive for productive efficiency, and may also have been encouraged by industrial policy, in particular national champion policy, in the respective companies' home markets.

Integration of industries and markets

Deregulation of both gas and electricity industries allows freer commercial integration across energy sectors, if such integration proves commercially attractive and opens the opportunity to cross national borders for the energy companies. Market-based integration could progressively result from changes at three levels – electricity and gas trade, capital markets and business strategies.

As far as trade is concerned, the expanding use of gas-based electricity generation, should introduce a trade-arbitrage based motivation for cross-sectoral integration. However, we have noted strong technical and institutional limitations on European cross-country electricity trade in spite of extensive development of electricity exchanges and commercial capabilities. With the exception of UK, far less open direct market-based trade has taken place in the gas sector. The later implementation of the gas directive may be one explanation. However, whereas electricity trade is seriously limited by technical grid-constraints, gas interconnections are extensive and should not pose technical barriers to trade. Trade in gas should, therefore, at the outset have more favourable conditions. The unequal distribution of gas resources and the ensuing geopolitical bargaining between the supplier and the consumer nations and companies offers a possible explanation.

As far as capital markets and business strategies are concerned, we have seen extensive attempts at creating broader energy groups involving both gas and electricity. We have seen integration of electricity and gas wholesale and transmission in the large E.ON–Ruhrgas merger, as it is already the case in the Suez group (Tractebel–Electrabel–Distrigaz) in Belgium and to a lesser extent with RWE and its control of Thyssengas in Germany. We have also seen extensive cross-energy acquisitions in Italy and the Iberian peninsula.

As yet, integration by capital market and by business strategy appears to be on an easier track than commercial integration through trade. When national and local governments decide to privatise their electricity and gas companies, there are no obstacles to entries for foreign European companies through acquisition, if European and national competition laws are respected, in terms of market concentration.

Nevertheless, integration across energy sectors remains strongly coloured by national style. The institutional variety – in particular in the network and interconnection access regulation, the regulatory institutions, the market institutions (power or gas exchanges), but also the incumbents' home-based positions after the reform – adds to technical constraints to limit the scope of integration of national markets. Besides the market rules, the accessibility of the markets is also conditioned by the incumbents' position.

The institutional variation across European national contexts clearly shapes different conditions for inter-sectoral business configuration. The French principle of 'spécialité' has thus retained mono-sectoral enterprises both in gas and electricity as far as the domestic market is concerned. In addition, national institutions and policies also strongly influence the allocation and control of industrial strategy in Europe. We have seen that some governments' answer to European promotion of market integration has been to create and reinforce national champions. These countries preserve incumbents by choosing a minimalist regulatory framework without reforming their industrial structures and consider foreign trade as the main competitive forces. They bet on market integration but do not lower barriers to trade resulting from the incumbents' market dominance.

Other countries have headed for comprehensive regulation and willingness to establish competition in the home market. Competition by foreign trade here is a complementary element to internal competitive forces. The electricity and gas industries in this group of countries are typically more fragmented and privatised. The national champions of the first group of states have therefore responded to the promotion of a European market by buying companies put on sale in this second group. The result is an industrial concentration at the European level which could affect market efficiency.

As yet, the cross-sectoral integration in European energy markets has not included strong moves to integrate regional European gas/electricity groups with the more globally-oriented petroleum industry. Besides gas production, the global European-based oil companies Shell, BP-Amoco and Total are developing activities in direct gas wholesale supplies and in electricity generation downstream in parallel with Exxon-Mobil. For them Europe is in competition with other regions for such development because the companies choose between the most profitable opportunities worldwide. However, for the European medium-sized oil/gas companies development in gas mid-stream, gas/electricity supply and trading is being actively sought via new ownership positions. This is the case for ENI with its division ENI Gas & Power which merged with the gas supply of the former ENI's gas subsidiary, for REPSOL with its stakes in Gas Natural in Spain, for Fortum and Norsk Hydro with their combined electricity and gas business. If global oil/gas companies and the medium-sized groups consolidate strategically, they may pose extensive commercial and regulatory challenges.

12.2.4. Dynamic interplay of regulation, markets and strategies levels

Since the launch of the European energy market deregulation policy in the late 1990s, there exists a dynamic interplay among regulatory initiatives, business strategies and market structure. In the three core sections of this book we have concluded independently by pointing the following trends:

- Despite some elements of path dependency the convergence in regulatory models has been surprisingly rapid as far as economic regulation is concerned. Conversely, the convergence on public service regulation has been slower because of a stronger path-dependency than for economic regulation.
- The de facto market opening across national borders, however, has been reluctant in certain countries. So after the first wave of reforms, the respective European electricity and gas markets remain mainly multidomestic with some exemplary exceptions, in particular the Nordic electricity markets.
- Business strategy development throughout the first half-decade of deregulation has been highly dynamic. However, the dynamics has operated in several directions with the emergence of a pluralism of business forms and engagements. Energy companies with highly varying degrees of vertical integration, diversification, size and internationalisation exist side by side in the European energy markets.

These levels interact, however, and the overall assessment of the EU energy deregulation must consider not only processes at each of these levels, but also the interplay between them. In a situation with limited European consensus on strong deregulation policy and extensive national subsidiarity in their implementation, the policy impact of the EU deregulation initiative may be highly dependent on its market and business triggering effect.

Experiences from the dynamic process initiated by a 'soft' European deregulation and highly discretionary national applications show that the interplay between the initial negotiated agreements and subsequent market dynamics may, in certain cases have taken the reform beyond the initial requirements. As some companies have seen their interests served by fully taking part in the new commercial economy, they have implicitly come to take the reform further by themselves, thereby pushing others in the same direction. Soft negotiated regulation, which itself is insufficient to further liberalisation and European integration could thus be only a first step in a chain reaction. In the longer run, as we have seen in the market and strategy sections, this has, in some cases, triggered spin-off-effects in the form of market opening far beyond the original minimum requirements.

The reform process may, in this way, trigger a trickling down effect through regulation on business strategies and markets. Under these circumstances even soft policies can have strong results, if they trigger complementary strategies and structural evolution. On the other hand, the lack of dynamic interplay of triggering effects could effectively block reform.

Interpreted in this dynamic framework, the tension between national style versus European integration can be seen as rivalry between competing regulatory regimes with competing triggering effects into business strategies and market structure.

Depending on the relative strength of the 'Europeanisation' vis-a-vis nationalistic impulses, business strategies may drive market evolution towards Europeanisation or national orientation of the energy systems. However, business strategies and/or market development may also shape policy as regulatory regimes are shaped with the interests of the incumbent companies as a major premise. Regulation policy thereby becomes more reactive than pro-active with greater influence from the marketplace.

The launch of the so-called 'acceleration' directives in 2003, which took the deregulation reform several steps further than the original electricity and gas Directives of the 1990s is clearly a product of such an interactive triggering effect. However, the attractiveness for some companies of oligopolistic or semi-oligopolistic positions may bias the process towards liberalisation and European integration without competition. For example, the largest German generation/transmission companies (born from the mergers of the former so-called supra-regionals) have, accordingly, been in a position to accumulate large capital assets employable for long-term strategic positioning converging with political goals, with few political limitations on their strategic planning. This has given them many of the privileges of state companies, without the latter's political constraints.

12.3. Challenges, Dilemmas and Outlook

After a decade or more of deregulation (in pioneering markets) and half a decade for the rest of the EU area, European industry and policy makers are discovering not only the possibilities but also the challenges of the new market order. This challenge has both economic and political aspects.

First, a central challenge comes out of the market system itself. As we have already noted, European energy industry has been exploiting its new freedom of international market configuration to scale up in concentration, and in some cases, to a level where the competitiveness of the market itself could be affected. The question is if European competition policy can follow suit.

Second, deregulation of electricity industry in Europe has tended to start with a grace period of energy surplus inherited from the previously expansive coordinated economies and further amplified by better resource utilisation from extended international trade. The regulatory challenge has therefore primarily been to arrange market rules and industrial structures such as to allocate the existing generation fleets for the electricity supply to the consumers in an efficient way. However, as energy demand increases, due to economic growth, the challenge of providing new capacity surfaces.

Finally, the energy sectors face extensive environmental challenges as ambitious climate and renewables policies are targeting these industries. Major issues here are the management of collective responsibilities across national boundaries, but also the management of the interface between the environmental policy field and the core energy markets. While the core markets are increasingly operated under a commercial/competitive mode, with increasing European standardisation, the environmental policies tend to have stronger national designs. Complex interaction effects between energy and environmental markets and policy instruments in both fields also constitute a challenge to the future management of energy industry.

12.3.1. The challenge of oligopolistic market power

Extensive scaling up in size by the largest energy companies, often backed by reinforcing national champion policies places the challenge of oligopolistic market power high on the European energy agenda.

The EU treaty aims at integrating markets and enhancing competition in the interest of consumers and at eliminating distortions caused by Member States pursuing national interests. Ironically, driven by their specific national styles, the response of some governments to measures designed to promote a European market has been to create or to reinforce national champions, and the response of some of these national champions has sometimes been to scale up to a level that threatens market competition. This has partly been achieved through acquisition of companies put on sale as a result of European liberalisation, privatisation and unbundling.

The main argument for maintaining the national companies intact, or encouraging horizontal integration even after deregulation, has been that they will be competitively exposed in a larger European market. However, the premise for this argument is that there are no significant technical or institutional barriers to trade on a European-wide basis. Furthermore, as a continual scaling up would eventually also create oligopolies at a European level, company size would have to be limited at some stage. The emergence of giant firms in electricity and gas is therefore considered by a number of critics to present a risk of exercise of dominant position both at the national and European levels. As there is reason to believe that large companies also influence government policy, it might be difficult to exercise effective control, once the companies become highly politically dominant.

The competitive dynamics of European energy markets is not, however, only challenged by disparate policy approaches from member countries, but also by disparate theoretical models. The combination makes for a plurality of interest/model combinations that does not exactly strengthen attempts at the integrated European competition policy.

12.3.1.1. Theoretical divergence

Classically the debate over competition doctrines and market concentration is a point of dissension between two general conceptions of competition in line with two views on electricity and gas markets concentration.

The first view is the so-called 'standard neoclassical view' for which horizontal and vertical deintegration is necessary for effective competition. It is based on the structuralist view of the competition developed by the Harvard tradition along which there is a causal link between the market structure and the level of welfare (Clark, 1940; Kayser and Turner, 1959). This tradition only takes into account economic justifications of concentration that afford benefits to consumers and buyers.

From this point of view, the dominant position of large gas and electricity companies, at the national level, is an obstacle to effective competition. Effective competition, in turn, is seen to give positive price effects for the consumers. The argument points out the risk that, even without abusing their dominant position, the mere presence of national champions condition the accessibility of national markets and de facto dissuade entries by their competitive advantages as incumbents.

In electricity and gas markets the general competition policy argument is coupled to industry-specific issues. Extensive market power may easily impede entrants when the lack of liquidity in wholesale and balancing markets impedes new entrants.

The second view criticizes the a priori hypothesis that the anti-competitive effect of vertical integration or horizontal concentration which creates or consolidates a dominant position, necessarily overshadows any contribution to economic progress (Jacquemin, 1991). This is partly rooted in the Chicago school and the contestable markets theory, which considers that there is no absolute and significant relation between concentration and market power. This school contends that competition is not predominantly determined by the industrial structures, but by the credible threat of entries allowed by the absence of regulatory and industrial barriers (in particular non-discriminatory access to the network).

A complementary view advocates a pragmatic approach to competition based on the observation of actual firm behaviour, thereby also avoiding the standard neoclassical position's a priori focus on structure. It refers to dynamic efficiency, in particular the capacity to develop economies of scale and scope, to the capacity to invest in the development of capital-intensive equipment, and the innovative capacity of the firms. Concentration could be justified to further dynamic efficiency, particularly in high-tech industries.

At the industry-specific level, the defenders of market concentration from a dynamic pragmatic position, consider that, as in all network industries, the advantages of competition are preserved as soon as the two conditions are met – a transparent and non-discriminatory access to the grids and sufficient interconnection capacities to allow extensive interaction between national markets. Integration across national markets is therefore the main argument for defending concentration within them. This ensures potential competition, which provides consumers with a real choice either in the country or in the neighbouring market.

12.3.1.2. Policies and strategies

Within the fairly heterogeneous European energy markets, our previous analysis has shown that governments have pursued highly diverse structural policies and that industry has developed strategies with highly different structural effects. On the one hand, certain countries, such as UK have launched strong structural fragmentation and deverticalisation policies, other countries, such as Norway have inherited a decentralised municipalistic structure and adopted the radical third party access and end user market opening in the electricity sector. On the other hand, countries such as France and Germany have allowed their industry to exploit scale and scope economies, to assume the risk of investment in large operations of capital-intensive electric equipment and gas infrastructures.

While the former group of countries have tended to emphasise the neoclassical virtues of decentralised competition, the latter group has emphasised contestability theory and international bargaining power to control price risk in the contractual gas purchase. This was presumably the German government's rationale for accepting the acquisition of Ruhrgas by E.ON. Pro-centralisation policies are also self-reinforcing as they are promoted to allow national champions to compete, in an equitable way, with the other European giants in their respective markets and to rival in the European capital market for company acquisitions.

Furthermore, the influence of big-sized companies creates potential risk of regulatory and governmental capture. Given their formidable resources and their dominant position in the national economy, they can influence rule-making and policy-making processes in the concerned countries, in line with Chandler's hypothesis of the influence of big industrial companies in the shaping of the regulation (Chandler, 1977). Critics refer to the influence of the largest German companies on the antitrust moderation of the German government in 2001–2002 and to the French reluctance to introduce pro-competitive legislation under the EDF's influence, as examples of this risk. Other critics see the large integrated companies challenging rational political responses to climate change risk, thereby blocking the promotion of environmental friendly technologies (in particular the renewables, cogeneration and energy efficiency). The large integrated companies have also been accused of challenging the nuclear phase out policies and intensively lobbying for the recreation of ex ante conditions for nuclear investment by controlling the market.

Differences in market cultures and competition doctrines do not easily leave room for a European consensus on competition policy which would mandate important pro-competitive restructuring of the electricity and gas industries. Moreover, since competition policy players may find analytical support from several schools of competition theory, there is no intellectual agreement that would allow an acceptable basis for consensus (Oppenheimer, 2003). Even the pragmatic test of competition effectiveness by the price alignment on the short-term marginal cost (see for instance, the use of the Lerner index) is not considered acceptable for every situation. This is especially the case when we have a production function with important share of fixed costs, or in the case of hydropower, where pricing is a function of speculation over the future reservoir fillings.

Some observers have pointed out that, under the influence of certain national governments and industrial lobbies, a set of rules reformulating the European antitrust authority's competences have limited its regulatory power. This particularly concerns the clarification of the approach to mergers (EC, 2002) and a decentralised enforcement of competition laws, which gives more rights to the authorities of the Member States (Oppenheimer, 2003).

12.3.1.3. Elements of European convergence

Nevertheless, there is a movement towards a compromise between the two opposite competition theory positions. A European convergence process seems to move towards creating the conditions for real contestability in electricity and gas markets with ex ante regulation of the access to networks, and to promote the ex post control by competition authorities at the national and European levels.

Successive EU directives promoting strict unbundling (legally and functionally) between the competitive activities – power generation, power and gas supply – and natural monopoly activities – the transmission and distribution networks are significant steps in this direction. This legislation suppresses different ways for the incumbents to exert market power. The 2003 directives on electricity and gas markets are the latest and most explicit examples of this development and serve to reinforce workable competition, even in countries with large incumbents.

The directive process and the so-called Florence and Madrid Fora have also pushed for unification of information rules and charges for use of networks (and storage in gas), which must be fair, transparent and of a binding nature, including interconnection capacities. However, the limited technical capacity will maintain constraints on cross-border exchange and integration of markets, particularly in electricity. This justifies financial incentives made by the European institutions (Commission, European Bank of Investment) to the development of new inter-country infrastructures because of the investment risk under volatile markets.

In the more general competition policy field, the development of incentive remedies, such as divestures of assets, have tended to replace structural remedies, giving a set of more practicable policy tools. Most of the requests for clearing of the mergers have been severe and in certain cases have had dissuasive effects on the M&A attempts, the best example for being the failure of the tentative merger between the two Spanish power majors in 2000. In the future the problem of control of M&A and abuse of dominant position will remain an important issue.

Furthermore, two additional elements alleviate the difficulties raised by the emergence of big-sized players. First, even in sectors with the remaining large incumbents, the introduction of an effective competition or at least an effective contestability may weaken the political legitimacy of the former incumbents as promoters of national policies. Second the 'Europeanisation' of energy and environmental policies allows pluralistic policy-making process where the giant companies may have their influence more balanced by competing interests, governments and other acting lobbies than in the national context.

The creation of a sectoral regulator at the European level would ultimately be an even stronger guarantee to enforce the accessibility to networks and the development of trade. However, it is hard to imagine the establishment of such an authority, which would imply change of basic principles in the European Treaty, in particular the principle of subsidiarity in this field.

3.2. Investment deficiencies in competitive markets

The question of investment in generation and transmission in competitive electricity and gas markets is increasingly being recognised as a major challenge to deregulated electricity industry and discussed in many European countries. Many of the continental European electricity markets are still in the grace period, as they were liberalised in a situation of large overcapacity. However, some of them have met short periods of price spikes – the Scandinavian peninsula in the winters of 1996 and 2002–2003, The Netherlands during some periods for reasons of transitory rules, Spain during winter 2001–2002 for hydro reasons, Germany in December 2001 and France and Germany during the heat episode of summer 2003.

The seriousness of this challenge is debated, however, and there are several opinions about possible solutions, ranging from a pure laisser-faire position to reintroduction of elements of planned economy. Given the prominence of the energy sectors as general infrastructure, price formation and availability potentially figure high on the political agenda. Taking a broad political and economic view, where the socio-political acceptability and not only economic efficiency is of concern.

The difficulty posed by the deregulated electricity market, particularly as the market reaches scarcity and new capacity becomes profitable, is that the economic and political equilibria may not always

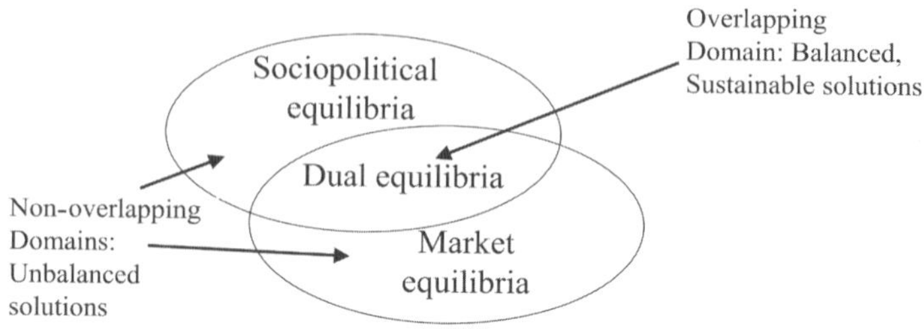

Fig. 12.2. Economic and sociopolitical equilibria.

overlap. The market process may generate prices that politicians find unacceptable, while such prices may be necessary to trigger new investment. There may be, in other words, a set of economic equilibria that fall outside of the sociopolitical domain and a set of sociopolitical equilibria that fall outside of the economic domain (Fig. 12.1). Some pure economic market solutions with a high degree of economic efficiency may imply socially unacceptable distributive effects and not qualify on the social criteria. On the other hand, solutions where the economic realities are neglected would be examples of unilateral sociopolitical equilibria where economic efficiency conditions are not met (Fig. 12.2).

Policy approaches

In principle, there are several approaches to cope with scarcity and investment challenges described above. Approaches may be characterised both along a technological and a governance dimension. Along the technological dimension, some approaches implicitly or explicitly assume the context of the established large-scale technological systems, while others rely on complementary decentralised technological development. Along the governance dimension, some approaches imply the use of coordinated governance, while others imply governance through competitive markets.

Policy of competitive exposure

From a pure market perspective one might conceive adequate competitive solutions, as there will always be a demand–supply balance, given the right price. However, from a sociopolitical point of view, the worry is that pure laisser-faire market governance would impose extreme price volatility and even possible shortages, given inelastic instantaneous electricity demand. Part of the social and political concern would be with the large transfer of wealth from consumers to producers, while capacity is added in response to shortages. This could violate the social equilibrium conditions (Fig. 12.1) and create a real problem of social acceptability, as shown by the social controversies after the recurrent price spikes in some American states in 1999–2001 and above all the Californian crisis, as well as in Norway in winter 2002–2003.

Policy of reversal to investment coordination

The challenges from the perceived negative effects of competitive exposure under large-scale technical lock-in have led to calls for stronger elements of coordinated intervention in the market economy. Even under fairly liberal deregulated market economy, reliability has in part been considered a public good, to be provided by the system operator in real time. In complement to the supply of this public good, there are debates around the supply of 'capacity adequacy' during peak, which could be considered either to be manageable under market logic or to be provided as a public good. In the second case diverse solutions are conceivable:

- a Pigouvian treatment with capacity payments on every transaction
- a Coasian solution with capacity obligations during extreme peak hours combined with a capacity market as a flexibility instrument (as discussed in USA in the definition process of the Federal Energy Regulatory Commission's Standard Market Design proposal)
- or else a governmental coordination for guaranteeing the level of reserve capacity by auctioning for long-term contracts.

Moreover, in some European countries there could be an institutional move to allow public authorities to programme and auction out long-term contracts with guaranteed price for the installation also of base-load equipment. The 2003 Directive gives legitimacy to this move in the name of energy security or environmental protection. These are typical countries where the technological paradigm in electricity remains focused mainly on large-scale technical systems, in particular because of the persistence of the nuclear option. This is the case in France where such measures were included in legislation in 2000 and the Belgian legislation voted thereafter.

With the coordinated approach to investment, this model may avoid the volatility of the previous approach, but possibly at some efficiency expenses. In particular, use of the auctioning instrument for peak equipment and base-load development reduces the scope for endogenous market function in the long term, as it could create overcapacity and therefore depress the market price. Moreover, it could probably also create incentives for free-riding by agents ready to invest on their own but now choosing the unrisky long-term contracts under the bidding process.

Policy of coordinated grid investment

The perceived challenge of underinvestment under competitive exposure has also led to proposals for stronger grid interconnection, especially between regions with complementary resources and/or complementary supply–demand patterns.

The European power markets are currently not very integrated. There are numerous transmission constraints all over Europe, including the Nordic area which has exhibited extensive price differences over the last few years. It can be argued that grid upgrading and new transmission connections may help to stabilise prices during peak periods. Increased grid capacities may as well mitigate market power problems in certain regions. Accompanied with harmonised access rules, tariffs and balancing arrangements, a grid expansion path may have certain benefits that could motivate governmental support.

However, the necessary transmission investments would be large and need heavy subsidies. Because of the complex grid configurations and physical interactions, dimensioning of the necessary investments would also need detailed planning, evaluation and licensing procedures. The experiences in the Nordic area over the last few years have also shown that the presence of transmission capacity may not guarantee the utilisation of the capacity, as transmission capacities between countries run the risk of being underutilised when congestion occurs within the individual national grids.

Endogenous market coordination and negotiated governance as a policy strategy

It could be argued that there is yet a policy option for price-hike control within the market governance model, even under large-scale technology lock-in, namely that of oligopolistic coordination under a negotiated political economy. One could see signs of oligopolistic configuration in several European markets as mergers and acquisitions concentrate the markets in the hands of a few large actors.

The horizontal concentration movement in some countries like Spain, Sweden, Finland and Germany may be interpreted as a classic response of the market to a risk exposure. In France, the defence of a strong vertical and horizontal integration is clearly linked to the will to preserve the capability to invest in highly capital-intensive equipment.

A club of large actors may first seek to stabilise prices at a higher average level than the competitive solution. However, such a club would also have an interest in stabilising the system performance within the social equilibrium space, knowing that the excessive price spikes might call for regulatory intervention. The club could therefore be motivated to supply necessary capacity to secure smooth performance and would not have to take the competitive market effects of some overcapacity due to coordination of a 'gentleman's' agreement type. An implicit or explicit 'gentleman's' agreement with policy makers or society would then be not to push the price too high, in return for which government would not press for stronger regulatory intervention.

Options of decentralised and demand-side solutions

Besides the large centralised large-scale solutions a number of small-scale and demand-side solutions stand out as interesting supplements. The advantage for many of the options of decentralised and demand-side solutions is that they compete against the combined grid and generation costs. Institutional facilitation of such measures, therefore, remains among the most attractive and recommendable policy measures to be undertaken to meet the investment challenge.

Renewables and energy efficiency policies as unintended investment policy

Support for the so-called new renewable energies, which is now taking on quite sizeable proportions on the European agenda, could potentially contribute extensively to European energy supply. In addition, come energy efficiency approaches, which contribute, to lowering demand. While the technologies in focus are generally small-scale, the mode of governance is coordinated, as financing typically comes from public budgets or specialised levies and access rights.

Support schemes for the promotion of renewables and CHP for electricity generation have been extended in a number of countries motivated by environmental protection, i.e., to the preservation of a public good, in particular climate stability. However, such schemes have also been motivated by innovation policy and the need to bring forward a generation of new energy technologies for a sustainable future.

The European Commission has also been very active in these fields, and is trying to coordinate a variety of national support schemes including a range of instruments such as feed-in tariffs, exchangeable quotas and bidding for long-term contracts. Taken together, these governmental support schemes for diffusion of new renewable technologies could have a significant effect on the development of new capacities in European electricity markets. The ambitious goal of increasing European electricity supply for renewables from 12% to 22% by 2012 could contribute extensively to boost the electricity supply capacity. This would clearly not only help to postpone price hikes, but also obviously postpone the period where the market signals would reveal to the ordinary market players the profitability of investing in new equipment or to enter by investing in generation units.

Complex interaction effects

If pursued simultaneously, the policy approaches outlined above are obviously going to have complex interaction effects. Whether the policy process will allow this to be taken into consideration is, however, another matter. The price hikes lead to calls for coordinated investment in generation and grid interconnections. Obviously these initiatives, if implemented, could have cumulative effects that might lead to overinvestment.

As alternative coordinated economy interventions, they would first have to be judged up against each other. The effects of grid investments would, furthermore, have to be evaluated against the background of price and capacity developments in neighbouring markets.

Most importantly, however, both grid investment and generation capacity investment strategies would tend to undermine private, market-based investment, as they are likely to maintain prices under the LTMC threshold. Use of public procurement strategies for price-hedging purposes could therefore easily imply a permanently sustained public investment role. Furthermore, in the context of an integrated market, such investments would have to be coordinated by all the involved countries, unless one country alone should be willing to carry the burden of the total system investment costs.

Less demanding, in terms of international policy coordination, are the options of decentralised and demand-side solutions. A more realistic threat of substitution from decentralised solutions would possibly and implicitly cap central system prices and thereby help maintain the social equilibrium.

12.3.3. Environmental policies and liberalisation reforms

In addition to the competition policy challenges and the problem of price spikes and investment in new capacity, the interface between energy industry and environmental policy raises a number of challenges.

The close interplay between energy and environmental policies implies that environmental policies and markets have important effects on energy development and vice versa. There is also spillover and learning across the two sectors in terms of instruments and policy approaches. The challenge coming out of this is that energy policy and energy industry business strategy must, in a fundamental sense, incorporate and respond to environmental policy.

Policy issues

As a point of departure, the European environmental policy, just as the energy policy is exposed to both Europeanisation and marketisation. The European environmental policy is based on the European Union's general mandate to intervene in environmental issues that have broader international scope. This mandate is complementary with its mandate of integrating markets. But the move towards a coherent and market-oriented European approach desired by the supranational level has been problematic as Member States with nationally designed greening policies have been reluctant to give up control, and the status is therefore one of a precarious balance between national and European order (Midttun and Kofoed, 2001). Original initiatives of energy companies in favour of the promotion of renewables in a number of countries, in particular with the voluntary trade of green certificates, have been also a factor of differentiation.

However, the balance between national and European governments is struck differently in the energy sector and the environmental field. While electricity and gas, even under the previously noted restrictions, are increasingly operated under a commercial/competitive mode with increasing European standardisation, the environmental policies tend to have stronger national designs in spite of supranational institutions' attempts towards institutional standardisation.

Taking green electricity schemes as an example, a number of countries which have adopted feed-in tariffs are not ready to change policy instruments. And, even when there is agreement in principle on market-oriented instruments, such as quotas combined with green certificates, as in Denmark, Sweden, Belgium, The Netherlands, Italy and UK, there are differences in details with a wide array of specific renewable and technology definitions. These differences serve as barriers for exchanges between countries by limiting the scope of standardisation of 'products'.[2]

The general challenge facing energy industry, as it moves towards European integration is that it is moving towards European integration in a competitive mode and still has to face nationally oriented environmental policies and policy instruments. This results from the fact that Member States with strong vested interests in nationally designed greening policies have been reluctant to give up policy control.

Effects of environmental policy on energy industry and markets

The European environmental policy diversity translates into a number of specific effects on energy industry and the function of energy markets at the operative level: emission standards, ecological taxation, 'cap and trade' regulation all have an impact on the competitiveness of gas as a fuel in the electricity generation. They also influence the relative costs of each electricity generator and their position in the spot markets, given the variety in size and technology of their respective fleets of generation units. Moreover, in terms of market efficiency, environmental regulation and plant-siting requirements could deter investment, and thereby adversely affect the market stability and the reliability of supply, as observed during the Californian market crisis.

In addition come the effects of trading instruments on the power market. First, strategic positioning in the permit market could

[2]Attempts at strong European environmental policy integration, such as the European carbon tax directive, have failed. Nevertheless, some milder coordination attempts have been successful, including a Directive, in 2003, on GHG emission ceiling and establishing emission trading in a number of industrial sectors including electricity industry. In parallel, a Directive on renewables with indicative targets was passed in December 2001, and could be complemented in 2004 by a Directive on CHP promotion and one on energy efficiency commitments. A minimalistic energy tax agreement has also set common environmentally oriented taxes on the European agenda.

translate into strategic control in the energy market. Dominant producers could, for instance, refuse to sell emission permits to potential competitors or collude to make the permit price increase. However, the integration of electricity markets and the future European CO_2 permits market might facilitate risk reduction. Furthermore, for countries with trading permits, the price volatility of the emission rights (SO_2, NO_X, CO_2) could co-vary with price spikes in power market and thereby amplify power price volatility.

Last but not the least, mandatory CHP policies, investment in renewables and energy efficiency could affect the electricity markets by reducing the scope for competition among conventional generators, as they contribute to the development of new generation capacities. Measures to achieve increased mandatory ratio of renewables in electricity generation (RES-E) from 14% to 21% in 2010 (the indicative European goal in the 2001 Directive on renewables) include such measures as the doubling of CHP share of electricity generation from 12% to 20%, as proposed in the German CHP Act, to be financed through tax exemptions. As a consequence, the regular wholesale markets could be reduced to 60% of the generation. Moreover the mandatory ratio of energy savings of 1% per year (as imposed upon the British suppliers by the Energy Efficiency Commitment) could imply a reduction in annual growth of the electricity market by 50%.

Furthermore, as just pointed out about the market incentives to invest in generation, mandatory investment in decentralised co-generation and in RES-E will postpone price spikes and contribute to the development of overall generation capacities. They obviously postpone the period where the market signals would reveal to the ordinary market players the profitability of investing in new equipment or of entering by investing in generation units. The large environmental externalities, as well as international coordination already achieved on investment in new renewable, would seem to make this strategy more attractive than a public coordination of conventional investments in generation capacity. At the end of the day, while deregulation and competitive exposure of electricity industry still remains a dominant European policy concern, a balance will have to be struck against the other dominant policy, the environmental protection.

Effects of energy deregulation on environmental policy

Deregulation and competitive exposure of electricity industry may, however, also strongly affect environmental performance both in a negative and positive way. Competition may, for instance, select less capital intensive equipment with short payback time. This could affect the mix of technologies both in favour of less emitting technologies, in the case of transition from coal- to gas-based generation. However, it could also work the other way in the case of transition from hydro- or nuclear- to gas-based electricity generation.

Furthermore, in the absence of new policies to promote the use of renewables, CHP or energy efficiency, competition is likely to lead to higher level of CO_2 from the electricity sector than otherwise would have occurred given the capital intensiveness of these solutions and the absence of scale effects.

Competition may also limit the applicability of previous environmental regulation and the way firms choose to comply with it. Without the possibility to pass through the costs in the regulated tariffs, electricity producers may loose incentives for demand-side management without mandatory policy. The inherent incentives that relied upon the monopolistic utilities' avoided costs in generation and networks, completely disappear with the deintegration of the value chain, and the obligation has to be shifted on to the supplier. With market liberalisation, the burden of these environmental policies must not distort competition by overcharging incumbents or competitors on which mandated RES-E buy-back or DSM obligations would be imposed. Obligations must therefore be equitably distributed between the suppliers (for instance a quota of RES-E or energy efficiency programme as a proportion of the electricity sale) (Wuppertal Institute, 2002). Or, if the obligation is imposed on one particular competitor, the cost of the obligation has to be assessed by an independent agent and its financing has to be externalised by a tax on every traded kWh for feeding a special fund.

Under the new regime, generators would be more concerned with minimising the costs of environmental compliance in order to compete more effectively on the market. Nevertheless, when emissions trade instruments are implemented, competition could enhance the incentives for energy industry to take advantage of emission trading.

Green markets as a new commercial opportunity

Besides the mutual interaction effects between energy and environmental regulation, the environmental arena also provides new business opportunities. This includes firstly the trade opportunities when the suppliers are mandated to invest by quotas of renewables and energy efficiency and could exchange the corresponding green or white certificates. It also includes the growing market for green energy both within industry and households on a voluntary basis, given that commercial actors have taken direct initiatives to serve these new market segments without waiting for government or EU-based regulation in a number of countries. These initiatives are motivated by potential surplus value to be harvested through product differentiation, where green electricity meets a willingness to pay from idealistically oriented consumers as well needs from business and industrial companies to secure environmental certification of their total production process, where energy is an important component.

One part of the green market segment is based on least cost solutions, potentially optimising on the total European green electricity resource base. The focus is here clearly on volume assuming a standardised quality of green electricity. The Swedish 'Bra miljøval' initiative is here a case in point. With a marginal cost addition of less than 5% on the energy cost, it is supplying green energy, mostly from large-scale hydro, to predominantly voluntary industrial end-users.

The other part of the green market segment is more specialised. The question is here how green electricity niches can be carved out in Europe so as to match differentiated multidimensional consumption functions with corresponding production functions? The focus here is clearly on quality, which again implies that the total volume of the green market is spread over a number of quality segments with less concern about the volume and liquidity within each quality segment.[3] Seen from an energy industry business perspective, the green energy markets represent a potential source of surplus income. They do, as well, represent incentives to turn future investment towards renewable sources, although the premium in the industrial mass market is at present very low. Gas, in this perspective, comes in as an 'intermediary' fuel with positive qualities when compared to coal, but fails to qualify as renewable. With CO_2 reinjection into the reservoirs, gas-based electricity generation could qualify in the CO_2 markets and thereby also attain a surplus.

12.4. Concluding Comments

The three sections of this book have illustrated how Europeanisation, liberalisation and cross-sectoral integration interact in a dynamic way through the interplay among the regulation, markets and business

[3]The Swiss 'Solarstrom vom EW' (Solar power from your local utility) is an example of a qualitative nice approach, with prices per kWh 4–7 times higher than the price of ordinary electricity (Wüstenhagen, 2003).

strategies. The complex policy interplay between the European and national levels serves to give this development a highly interactive character, where several questions are left unresolved.

The first pending question relates to the relationship between what we have termed *first-* and *second-order regulation*. How do the new European deregulation initiatives interface with deeper-rooted European industrial and cultural traditions?

The second question is, as we have seen, related to the oligopolisation of markets. How may competition policy resolve the rapid integration of European energy industry?

Third, and related to the previous question, how will Europe resolve the capacity expansion challenge where several routes remain to be explored?

Finally, how will we meet the formidable environmental challenges at the same time as the energy sectors are under extensive regulatory transition?

In the long run, the promise of resolving these questions, while going through a European, as well as an intersectoral integration process, lies in the dynamic interplay between regulatory and commercial forces. Boundary transcending commercial initiatives unleashed by initially cautious regulation may feedback into bolder reforms and trigger developments of far wider implications. However, the long-run sustainability of this development relies on its ability to meet economic and social equilibrium conditions along the path and produce reasonably distributed welfare to all the major stakeholder groups.

References

Aldrich, H. (1999). *Organisations Evolving*. Sage, New York.

Chandler, A.D. (1977). *The Visible Hand. The Managerial Revolution in the American Business*. Harvard University Press, Cambrige (Mass.).

Clark, J.M. (1940). Towards a concept of workable competition. *American Economic Review* (30).

David, P.A. (1993). Path dependence and predictability in dynamic systems with local network externalities: a paradigm for historical economics. In Dominique Foray and Christopher Freeman, Eds., *Technology and the Wealth of Nations*. Pinter Publishers, London.

Djelic, M.L. and Quack, S. (2003). *Globalization and Institutions: Redefining the Rules of the Economic Game, New Horizons in Institutional and Evolutionary Economics*. Edward Elgar, Cheltenham.

Di Maggio, P. and Powell, W.W. (1991). The iron cage revisited: institutional isomorphism and collective rationality in organizational analysis. In Di Maggio, P. and Powell, W.W., Eds., *The New Institutionalism in Organisations*. University of Chicago Press, Chicago.

Doz, B. and Pralahad, J. (1993). Managing DMNCs: a search for a new paradigm. In S. Ghoshal and E. Westney, Eds., *Organisational Theory and the Multinational Corporation.* Macmillan, London.

EC (2002). *Green Paper on Review of the EC Merger Regulation,* January.

Hall, P.A. and Soskice, D. (2001). *Varieties of Capitalism: The Institutional Foundations of Comparative Advantage,* Oxford University Press, Oxford, 2001.

Hollingsworth, J.R. and Boyer, R. (1997). *Contemporary Capitalism: The Embeddedness of Institutions. Cambridge Studies in Contemporary Politics.* Cambridge University Press, Cambridge, 1997.

IEA. *Competition in Electricity Markets.* OECD/IEA, Paris.

Jacquemin, A. (1991). *Mergers and Competition Policy in the European Community.* Basic Blackwell.

Kaysen, C. and Turner, D.F. (1959). *Antitrust Policy: An Economic and Legal Analysis.* Harvard University Press, Cambridge, Mass.

Maurice, M. and Sorge, A. (2000). *Embedding Organizations: Societal Analysis of Actors, Organizations, and Socio-Economic Context, Advances in Organization Studies;* John Benjamins Pub., Amsterdam.

Midttun, A. and Koefoed, A.L. (2001). The effectiveness and negotiability of envirtonmental regulation. *International Journal of Regulation and Governance* 1(1): 79–111.

Ohmae, K. (1985). *Triad Power.* The Free Press/Macmillan, New York.

Ohmae, K. (1995). *The End of the Nation State: The Rise of Regional Economies.* Harper Collins Publishers, London.

Oppenheimer, M. (2003). *EU Competition Law Briefing,* Issue 6, January.

Polanyi, K. (1944). *The Great Transformation.* Farrar Rinehart, New York, xiii, 305 s.

Turvey, R. (2003). Ensuring adequate generation capacity. *Utilities Policy* 11: 95–102.

Whitley, R. (1992). *European Business Systems: Firms and Markets in their National Contexts.* Sage, London.

Whitley, R. (1999). *Divergent Capitalisms: The Social Structuring and Change of Business Systems.*

Wüstenhagen, R., Markand, J., and Tuffer, B. (2003). Diffusion of green power products in Switzerland. *Energy Policy* 37(7): 621.

Zeitlin, J. and Herrigel, G. (2000). *Americanization and its Limits: Reworking US Technology and Management in Post-war Europe and Japan.* Oxford University Press, Oxford.

Index